Results and Problems in Cell Differentiation

A Series of Topical Volumes in Developmental Biology

14

Editors

W. Hennig, Nijmegen and U. Scheer, Würzburg

Results and Problems in Cell Differentiation

Structure and Function of Eukaryotic Chromosomes

Edited by W. Hennig

With 99 Figures

Springer-Verlag
Berlin Heidelberg New York
London Paris Tokyo

Professor Dr. Wolfgang Hennig
University of Nijmegen, Department of Genetics
Faculty of Sciences
Toernooiveld, 6525 ED Nijmegen
The Netherlands

ISBN 3-540-17717-5 Springer-Verlag Berlin Heidelberg New York
ISBN 0-387-17717-5 Springer-Verlag New York Berlin Heidelberg

Library of Congress Cataloging in Publication Data.
Structure and function of eukaryotic chromosomes. (Results and problems in cell differentiation; 14). Includes bibliographies and index. 1. Chromosomes. 2. Eukaryotic cells. I. Hennig, Wolfgang, 1941–. II. Series. QH607.R4 vol. 14 574.87′612 s 87-9447 [QH600] [574.87′322]

Typesetting, printing and binding: Brühlsche Universitätsdruckerei, Giessen
2131/3130-543210

This book is dedicated to
Professor WOLFGANG BEERMANN
in honor of his 65th birthday

Preface

In 1985 I discussed with John Sommerville the state of chromosome research. We both felt that it would be worthwhile organizing a workshop for colleagues active in chromosome research, since the exciting new developments in this field obtained with the new molecular techniques often appeared to be insufficiently integrated into classical knowledge.

When Peter Hausen and I considered that the 65th birthday of Professor Wolfgang Beermann ought to be celebrated with some kind of scientific activity, emphasizing the importance and widespread influence of his scientific work, I remembered our earlier discussion. We felt that the birthday of W. Beermann would be an excellent opportunity to assemble scientists from all over the world and to try and achieve some kind of integration of classic and modern aspects of chromosome research. The opening and closing words spoken by Mel Green and Joe Gall are enclosed, since they emphasize one of the chapters written in chromosome research.

This volume is a document of a symposium organized in 1986 in Tübingen. However, it is not the documentation of a symposium in the usual sense. I have asked the invited speakers of this symposium to prepare a manuscript which meets the original intention and tries to integrate the classic knowledge from cytology and genetics with molecular data. Although some topics of chromosome research have occasionally been treated in other publications, a comprehensive assembly of the present knowledge has not yet been available. I hope that this volume alters the situation by providing an up-to-date account of various fields of chromosome research.

In various contributions it will become evident that often fundamentally new ideas are still necessary to arrive at a real understanding of the phenomena. This is particularly true for all questions on chromosome structure. Understanding gene regulation is closely connected to understanding chromosome structure. This shows that we are still far away from even understanding gene function. This book may help to induce new approaches stimulated by the recognition of the unsatisfactory state of our insight.

For several reasons it has not been possible to achieve a complete coverage of all the divergent aspects of chromosome structure and function. In particular, the important contributions of yeast chromosomes, which have not been appreciated by the classic cytologists as suitable material for research, are not represented. I hope that this volume will nevertheless prove valuable for all those fascinated by chromosomes, by questions on gene function and by cellular differentiation processes.

I am extremely grateful to all contributors for their dedication and co-operation. I feel very excited to read the various chapters on a research line which has been the focus of my own scientific career and which began in the 1960's in W. Beermann's Department in the extremely fascinating and stimulating environment of internationally leading colleagues. Only today can I fully appreciate the primary importance and value of such a scientific environment for post-graduate education.

Nijmegen, April 1987 WOLFGANG HENNIG

Contents

The *white* Locus of *Drosophila melanogaster*

By B. H. Judd (With 2 Figures)

The Genetic and Molecular Organization of the Dense Cluster of Functionally Related, Vital Genes in the DOPA Decarboxylase Region of the *Drosophila melanogaster* Genome

By T. R. F. Wright (With 2 Figures)

Heat Shock: Puffs and Response to Environmental Stress

By M. L. Pardue, W. G. Bendena, and J. C. Garbe (With 5 Figures)

The Y Chromosomal Lampbrush Loops of *Drosophila*

By W. Hennig (With 6 Figures)

Contributions of Electron Microscopic Spreading Preparations ("Miller Spreads") to the Analysis of Chromosome Structure

By U. Scheer (With 9 Figures)

Replication of DNA in Eukaryotic Chromosomes

By J. H. Taylor (With 4 Figures)

Gene Amplification in Dipteran Chromosomes

By A. Spradling (With 5 Figures)

The Significance of Plant Transposable Elements in Biological Processes

By Zs. Schwarz-Sommer (With 4 Figures)

Arrangement of Chromosomes in Interphase Cell Nuclei

By H. Saumweber (With 7 Figures)

Heterochromatin and the Phenomenon of Chromosome Banding

By D. Schweizer, J. Loidl, and B. Hamilton (With 5 Figures)

Multiple Nonhistone Protein-DNA Complexes in Chromatin Regulate the Cell- and Stage-Specific Activity of an Eukaryotic Gene

By A. E. Sippel, U. Borgmeyer, A. W. Püschel, R. A. W. Rupp, A. Stief, U. Strech-Jurk, and M. Theisen (With 4 Figures)

Genetics of Sex Determination in Eukaryotes

By R. Nöthiger and M. Steinmann-Zwicky (With 10 Figures)

Applications of Basic Chromosome Research in Biotechnology and Medicine

By P. M. M. Rae (With 4 Figures)

Introduction

M. M. GREEN [1]

"There is no science without fancy and no art without facts"

V. Nabokov

I am enormously pleased and honored to introduce this book, partly based on a Symposium which was convened in part to belatedly celebrate the 65th birthday of our colleague and friend Wolfgang Beermann, to wish him well and many more birthdays, in part to recognize and to bring up-to-date his genuinely significant research contributions to the subject of "Eukaryote Chromosome Structure and Function", and in part to wish him Godspeed, as he contemplates emeritus status.

I have for some time pondered the question: Am I qualified to make this introduction? For reasons which I shall spell out promptly, I conclude I am qualified. Because I am neither a cytologist nor a developmental biologist nor biochemist, just a "fly pusher" of the traditional *Drosophila* school, I can be completely objective in evaluating Beermann's research contributions. Furthermore, having been Beermann's friend for almost 25 years, having published with him and worked in *Abteilung* Beermann, I can provide some personal insights into the research environment which he cultivated. And, finally, since I am Beermann's senior by almost 5 years and have enjoyed the emeritus status for almost 4 years, I believe I can provide him with worthwhile advice as he embarks on his new career.

In the synopsis which follows I shall try and provide for you a personal view of Wolfgang Beermann's research accomplishments to date, the research environment he generated, the impact he and his students and associates have had in the realm of eukaryote chromosome structure and function and the lessons we can learn. I will close with some friendly advice!

The eminent bacteriologist and Nobel laureate Paul Ehrlich attributed scientific discovery to four elements: *Geld* (money), *Geduld* (patience), *Geschick* (skill or talent), and *Glück* (luck). If I may be so bold, I would add two more elements: *Gehirn* (brains) and *Gedanke* (idea). *Glück,* if translated as chance not luck, has a role in research only, as Pasteur emphasized, if one is prepared for the chance event.

How does Beermann's research fit into the remaining elements? *Geld,* it seems to me has played only a trivial role. His research employed a compound microscope, slides and coverglasses, stains, a few dissecting tools, and a place to raise *Chironomus*. All in all, by current standards these are an inexpensive array. Even his reprint costs were not excessive because the sum total of his publications spanning more than 3 decades is 45! (This seeming dearth of publications inparts an

[1] University of California, Davis, California 95616, USA.

Results and Problems in Cell Differentiation 14
Structure and Function of Eukaryotic Chromosomes
Edited by W. Hennig

important lesson on scientific research. Quality, not quantity, bespeaks Beermann's published work. When he had something to say, he published. He was immune from the "publish or perish" syndrome which generates too many premature and/or redundant publications and overburdens the libraries of the world!)

Geduld has been a necessary element in carrying out the painstaking, precise polytene chromosome cytology which Beermann produced; and the laboratory breeding of *Chironomus* species with their few annual generations manifestly demanded patience.

The remaining three elements, viz. *Geschick, Gehirn,* and *Gedanke,* taken together, I believe are necessary elements in the dictum of the renowned biologist and Nobel laureate, Szent-Györgi's on the nature of scientific research viz. "research is to see what everyone else has seen and to think what no one else has thought." Here, I shall briefly cite three Beermann publications which explicitly and elegantly fulfill Szent-Györgi's dictum.

First, *Chromomerenkonstanz und spezifische Modifikationen der Chromosomenstruktur in der Entwicklung und Organdifferenzierung von Chironomus tentans* (Chromomere constancy and specific modifications of chromosome structure in the development and organ differentiation of *C. tentans*), published in 1952, is a benchmark paper in the annals of research on chromosome structure and developmental biology. Here, Beermann sees in the polytene chromosomes what others have seen – Balbiani rings by Balbiani, puffs first seen in *D. melanogaster* and so named by Bridges – demonstrates the differential occurrence of puffs in different organs of *C. tentans* and thinks what no one else has thought before, viz. that the puffs must represent the visual manifestation of gene action! Their differential occurrence in different organs is the consequence of differential gene action. Subsequent biochemical research confirmed the correctness of this deduction.

In a second paper *Geschlechtsbestimmung und Evolution der genetischen Y-Chromosomen bei Chironomus* (Sex determination and the evolution of the genetic Y chromosomes of *Chironomus*) Beermann demonstrates that with patience and keen powers of observation, it is possible to carry out cytogenetic experiments with *Chironomus*. Here he establishes that in *Chironomus,* lacking heteromorphic sex chromosomes, males are the heterogametic sex and maleness is determined by a cytogenetically delimited "dominant" sex differentiator linked to one chromosome.

Finally, Beermann's striking powers of observation and deduction are manifest in a paper entitled *Ein Balbiani-Ring als Locus einer Speicheldrüsenmutation* (A Balbiani ring as the locus of a salivary gland mutation). Here he observes that in the salivary glands of *C. tentans,* four specific cells lack secretory granules not found in the equivalent four cells of *C. pallidivitatus*. By making crosses between the two species he demonstrates the granules to be inherited as a unitary, recessive Mendelian character, and linked to one particular chromosome. A puff specific to the four cells of *C. pallidivitatus* and absent in the same cells of *C. tentans* associates the secretory granules with the puff and affirms the conclusion that gene expression and puff formation are inexorably linked!

As I implied at the outset, an individual's research contributions are made in two ways: in part by personal research efforts and in part through the influence a researcher exerts on his students and associates, the role model one exemplifies. In his relationship with students and associates, Beermann followed, I do believe, a long-standing, but slowly disappearing German tradition which is illustrated in the following excerpt taken from the biography of the distinguished German biochemist and Nobel laureate, Otto Warburg, written by his equally distinguished Nobel laureate student, Hans Krebs. Krebs described Warburg's graduate education as follows: "In 1901, Warburg began his studies of chemistry at the University of Freiburg. As was customary in central Europe, he later moved to another university, Berlin, where he completed his studies with a doctoral thesis under Emil Fischer in 1906 ... Fischer himself, though a head of a large institute with many teaching commitments, spent most of his working time at the bench, side by side with his research students. He set an example by his style of working, his high standards of reliability and personal integrity on which Warburg modeled himself throughout his life."

In the foregoing, by substituting the name Beermann for Fischer, an accurate and apt description of Beermann's research style and role model is achieved. He worked and continues to work at the bench with the patience and the enjoyment of a researcher whose reliability and integrity are impeccable! His influence on his students and associates has generated an array of original and significant research projects which gained for *Abteilung* Beermann a rightly deserved international reputation. The scope of these investigations goes beyond the cytology of *Chironomus,* thereby attesting to the great breadth of Beermann's interests and knowledge in the areas of cytology, genetics, and development. I will cite here only a few studies to emphasize the scope of the research, e.g., the cytogenetics and differentiation of the Y chromosome in *Drosophila hydei* carried out by Hess and Meyer and by Hennig; chromosome diminution in *Cyclops* studied by Sigrid Beermann; the biochemistry of *Chironomus* larval salivary gland secretion initiated by Grossbach. Consistent with a long-standing German research tradition, all these studies – some of them doctoral dissertations – bear the imprint of Beermann's guidance, advice, and attention to detail. None of the publications derived from dissertations bears his name! (There is yet another important lesson in scientific research to be learned from the foregoing. Only at the bench can a researcher see the unusual, the unexpected. For it is upon those "cherished exceptions" that much scientific progress depends. How much is the current breed of researchers missing in discoveries and excitement by sitting behind a desk instead of at the bench or by indulging in what Otto Warburg disparagingly called "academic tourism?")

It is fitting and appropriate that a number of the chapters of this present book will bring up-to-date research themes pioneered by Beermann and his students and associates. Accordingly, Edström discusses the Balbiani ring system, Taylor considers the replication of eukaryote chromosomes, Schweizer examines the nature of polytene chromosome banding, Korge describes the current status of salivary gland secretions, Judd considers the organization of the white eye color locus in *D. melanogaster,* and Nöthiger reviews the problem of sex determination in eukaryotes.

To round out the theme of this book, Callan discusses the behavior of chromosomes during meiosis, Scheer considers the contributions of the Miller spreading technique to chromosome research, Saumweber discusses the chromosomes in the cell nucleus, Spradling evaluates gene amplification and eukaryote chromosome structure, Sippel considers DNA-protein interactions in eukaryote chromosomes, Schwarz-Sommer describes the place of transposable elements in eukaryote chromosomes, Rae considers basic chromosome research in biotechnology and medicine, and Wright describes gene clusters in *D. melanogaster*.

Finally, let me offer some friendly advice to a colleague gradually approaching emeritus status whose friendship I esteem and whose intellect I respect and admire. Emeritus status is not the end of a career, it is only a time for altering the course of a career. At long last you have earned freedom from administrative chores, commitments to students and associates and any other impedimenta which keep you from what you enjoy most. Now, you can continue at the bench pursuing whatever research strikes your fancy with unimpeded pleasure and personal satisfaction. For the dedicated researcher this is the best of an otherwise imperfect world and I am confident you will make the most of it.

In keeping with a long-time German tradition, I end these brief introductory remarks with a quotation from another eminent and distinguished Wolfgang, Johann Wolfgang von Goethe:

> *„Was glänzt, ist für den Augenblick geboren.*
> *Das Echte bleibt der Nachwelt unverloren.“*
> ("What glitters fills an instant and is gone.
> The genuine will for posterity live on.")

Lampbrush Chromosomes as Seen in Historical Perspective

H. G. Callan [1]

Lampbrush chromosomes were first seen by the founder/father of cytology, Walther Flemming, in 1878, but Flemming was not convinced that they were chromosomes, nor did he know them by the name that has now become so familiar. Flemming and his student Wiebe were studying the development of oocytes of Amphibia and fish, and in stained sections through the oocytes of the Mexican axolotl they noticed elongate objects, apparently consisting of thin fibres normally arranged to the axes of these structures, in germinal vesicle nuclei. A drawing of these objects (Fig. 1) was published by Flemming in 1882 in his classical monograph on cells and cell division. Similar structures were seen by Rabl (1885) in oocytes of *Proteus,* and by Holl (1890) in those of the chicken.

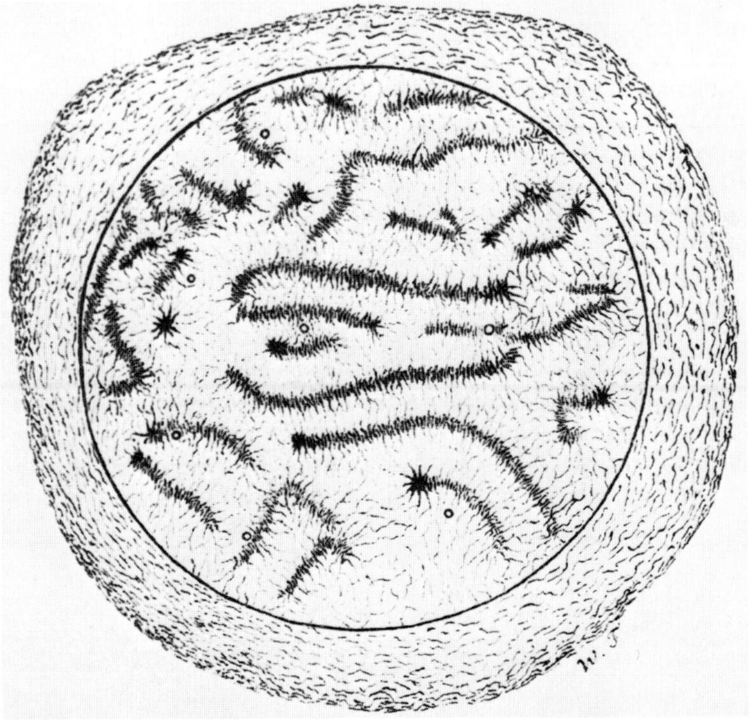

Fig. 1. Flemming's (1882) drawing of a stained section through a young oocyte nucleus of *Ambystoma mexicanum*

[1] Gatty Marine Laboratory, The University, St. Andrews, KY16 8LB, Scotland.

Results and Problems in Cell Differentiation 14
Structure and Function of Eukaryotic Chromosomes
Edited by W. Hennig

However, the real discoverer of lampbrush chromosomes was Rückert, who in 1892 published a fascinating and novel account of what happens to the chromosomes during the growth of elasmobranch oocytes. Rückert's studies were made at a time when chromosomes were coming to be a focus of attention because many cytologists, though by no means all, thought that they carried hereditary factors; and there was consequent lively debate as to precisely what occurs to the chromosomes during the development and maturation of germ cells. In 1887 Schultze had claimed that the "chromatin" present in young oocyte nuclei of Amphibia disappears during the course of oocyte growth, and only reappears, as chromosomes, on the first polar body spindle of the egg at maturation. If true, this clearly ruled out the possibility that the chromosomes are directly responsible for inheritance. Flemming's reluctance to identify as chromosomes the structures that he had seen in axolotl oocytes had a similar origin, for he recognized that, unlike interphase chromatin and ordinary chromosomes during cell division, they did not stain intensely with basic dyes.

By first studying stained sections through *Pristiurus* ovaries Rückert established that the nuclei of the smallest oocytes contain chromosomes similar to those of mitotic cells at prophase, and not ill-defined chromatin like that present in interphase nuclei. As oocytes grow to reach about 2 mm diameter, so too do their nuclei; thereafter the oocytes continue to enlarge with the deposition of yolk, but not their nuclei. During the growth of the oocyte nucleus its chromosomes grow longer and thicker, become more widely spaced and lose their stainability. For a while their axes remain defined by rows of stainable chromomeres that progressively increase in number and diminish in size, essentially to the point of vanishing altogether. At this stage, in sections, those regions of the nucleus that are occupied by the bloated chromosomes are scarcely distinguishable from neighbouring regions containing nuclear sap only, and the regions are in any case without sharp boundaries. However, as oocytes grow a little larger, to some 0.75–1 mm diameter, differentiation between chromosomes and nuclear sap improves, and Rückert was able to describe and draw (Fig. 2) the chromosomes at this stage; each portion consists of intertwined threads that loop back and forth across a major axis, thus resembling a „*Lampenzylinderputzer*" or lampbrush. Nowadays a rather closer comparison might be drawn to a steel brush of continental European design, of the kind used for removing deposit from shotgun barrels, and in which the steel wire, instead of having a multiplicity of free ends distal from the axis, is formed into a series of loops that project, from a stiffened twisted axis. Whatever analogy one prefers, the main point I wish to stress is Rückert's clear

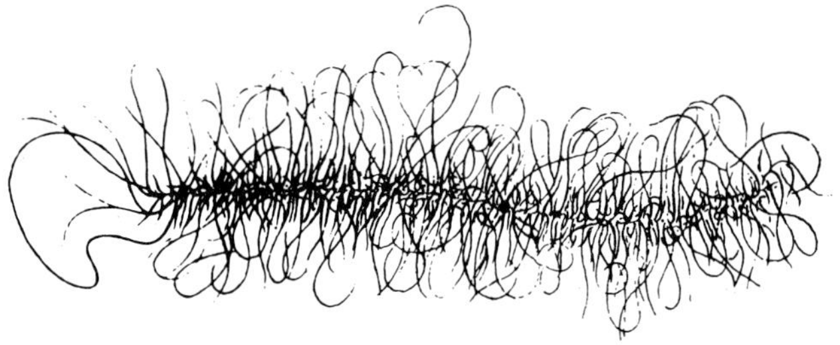

Fig. 2. Rückert's (1892) drawing of part of a lampbrush chromosome of *Pristiurus*

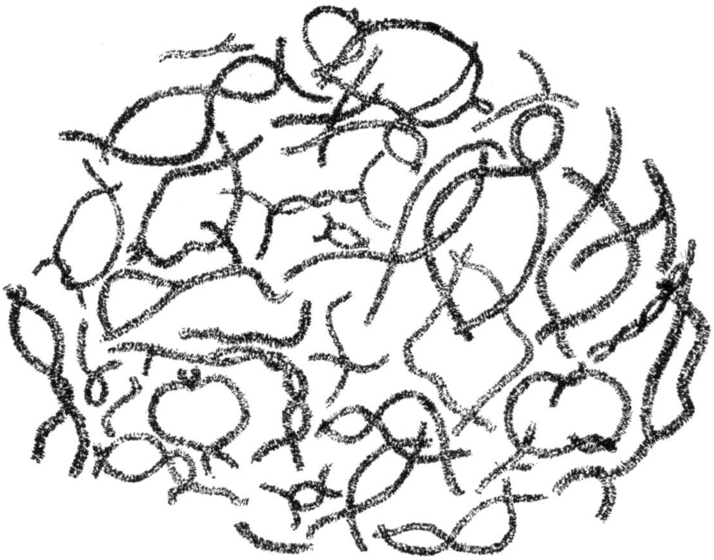

Fig. 3. Rückert's (1892) drawing of the entire lampbrush chromosome complement of *Pristiurus* from an isolated, fixed and stained germinal vesicle, when the chromosomes have reached their maximum size

recognition of laterally projecting *loops,* and his own happy choice of an arresting descriptive term is firmly established in cytological literature. Rückert was unsure whether chromomeres continue to occupy the chromosome axis when the lateral loops are maximally extended, but he was in no doubt that as the loops develop they extend from axial chromomeres, and as they regress, chromomeres reappear.

Because sections through *Pristiurus* oocytes only provided views of small portions of lampbrush chromosomes, and since Rückert now wanted to see the chromosomes in their entirety, and count them, he isolated germinal vesicles freehand using a dissecting lens, fixed them in acetic-sublimate, stained for several days in borax-carmine, and differentiated in acid alcohol until the nuclear sap was just colourless. Figure 3 shows his drawing of the lampbrush complement from an oocyte of 3 mm diameter. Rückert recognized that each lampbrush chromosome was one of a pair attached to each other by „*Ueberkreuzungen*" (crossovers!), producing the shapes that are so characteristic of diplotene bivalents. He counted between 30 and 36 per nucleus, most of the chromosomes being about 80 μm long, the longest some 120 μm. He found that they remain constant in number, but progressively diminish in size during further oocyte growth, and by the time an oocyte has reached 12 mm diameter the lateral loops have retracted completely. From now on the erstwhile lampbrush chromosomes looked and stained like familiar chromosomes, and in this state they proceeded on to the division spindle that produces the first polar body.

Rückert counted some 18 chromosomes at meiotic metaphase, a number that must be wrong because it should correspond with the number of bivalents at diplotene; doubtlessly this error occurred because the metaphase bivalents are small and tightly packed together on the division spindle of *Pristiurus*. It led Rückert to accept the proposal of Hertwig, that the chromosome number doubles very early on in oocyte development, and is reduced to one-quarter during the elimination of the two polar bodies. It is a minor blemish in an otherwise remarkable study.

Rückert's claim for chromosome continuity throughout oocyte development was speedily confirmed by Born (1892, 1894) working on the urodele *"Triton taeniatus"* (*Triturus vulgaris*). The illustration from Born's preliminary note, which was published in the same volume of the *Anatomische Anzeiger* as that containing Rückert's paper, shows all the lateral strands as loops leaving from and returning to the chromosome axis, and Born made the prescient point when describing these objects that the whole chromosome may well consist of a single, uninterrupted strand, many times woven back on itself to form the lateral loops.

At the turn of the century Carnoy and Lebrun (1897, 1898, 1899) published three papers on the development of amphibian oocytes, the first two of which must be included in a survey of early work on lampbrush chromosomes. The texts of these two papers are hard to evaluate, for the authors do not recognize chromosomes as such. They speak of the disappearance of a *"filament nucleinien primitif"* early in oocyte growth, and they claim that the remarkable structures found in oocyte nuclei midway through the period of growth sprout out as filaments from nucleoli, these filaments in turn sprout lateral filaments which at the peak of their development take the form of granular loops, giving rise to *"goupillons"* (bottle-brushes), and that thereafter the loops and filaments disintegrate. This description is so bizarre, and so much at variance with the observations of the more authoritative German cytologists of the time, that Carnoy and Lebrum's papers could be disregarded were it not for the beauty and accuracy of their illustrations, one of which is reproduced in Fig. 4. Clearly they worked with excellent preparations, and drew what they saw with precision. What they saw evidently included chromomeres, lateral loops of great morphological variety and recognizable bivalents with chiasmata. The third paper of the series (Carnoy and Lebrun 1899), which is concerned with the origin of the polar bodies in *Triturus*, starts off by denying continuity of the *"element nucleinien"* during oogenesis, claims de novo origin for the division chromosomes (this is the first occasion when Carnoy and Lebrun use this term) from a coalescence of nucleolar products, is equally well illustrated and the haploid chromosome number is correctly given as 12!

Evidence for and against the structural continuity of chromosomes during oocyte development was actively sought during the first few years of the 20th century, and supporters of Carnoy and Lebrun's views included Goldschmidt (1902) and Hartmann (1902), biologists who were later to make important positive contributions to the study of heredity and its material basis. An early and formidable opponent was Janssens, soon after to achieve fame by his "chiasma-type" interpretation of bivalent chromosome structure. In a brief, but cogent paper, without illustrations, Janssens (1904) emphasized the identity of the synaptic process in spermatocytes and oocytes of *Triturus*, and stated that he was able to follow the metamorphoses of the 12 bivalent chromosomes throughout oocyte growth and right up to the first maturation division.

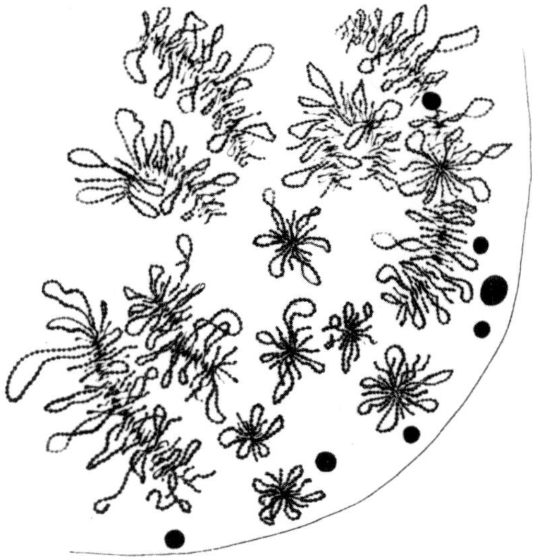

Fig. 4. Carnoy and Lebrun's (1897) drawing of parts of lampbrush chromosomes in a stained section through the germinal vesicle of an oocyte of *Salamandra*

The arguments of the time were reviewed by Maréchal (1907) who also came down firmly in favour of continuity despite his having worked at the Institut Carnoy in Louvain. Maréchal worked with stained sections through elasmobranch ovaries, and his observations gave full support to Rückert's account of the genesis of lampbrush chromosomes during the early stages of oocyte growth, their appearance at full development and the retraction of lateral loops and contraction of chromosome axes as these chromosomes give origin to normal condensed meiotic bivalents; the nucleoli do not intervene directly in any of these metamorphoses.

Around the turn of the century cytologists paid much attention to a certain "duality" of the constituents of cell nuclei, how to distinguish between them and their functional significance. Heidenhain (1894) introduced the terms basichromatin and oxychromatin to designate materials in fixed cell nuclei which stain with basic and acidic dyes respectively. A few years later Lubosch (1902) introduced the terms idiochromatin and trophochromatin, the former designating "stable" hereditary material, the latter being "metabolic" or "trophic". To a first approximation basichromatin = idiochromatin, and can be equated with DNA/histone in modern terminology, while oxychromatin = trophochromatin, and can be equated with nuclear proteins, some of which are associated with RNA. In the controversies surrounding the use of these terms the nature and behaviour of the constituents of germinal vesicles figured largely.

The transformation of an oocyte's chromosomes from basiphil (prior to oocyte growth) to acidophil (during oocyte growth) and back to basiphil (towards the end of oocyte growth in Amphibia, but much earlier in elasmobranchs and birds) led Rückert and most of his successors to the view that the predominantly acidophil lampbrush stage is a period when the chromosomes are metabolically active, i.e. generating trophochromatin according to the terminology of Lubosch, actively synthesizing RNA in modern parlance. Thus, Jörgensen (1913), in the course of a morphological study of oogenesis in the urodele *Proteus anguineus,* the cave-dwelling olm of Carpathia, showed that if sections of ovary are digested in the presence of trypsin or pepsin, the chromosomes when basiphil resist digestion, whereas the acidophil lampbrush chromosomes are completely destroyed. Jörgensen concluded from these observations that the protein backbone of the chromosomes is protected from hydrolysis by the nucleic acid present during the basiphil stages, and that when the chromosomes are acidophil and digestible no nucleic acid is present.

The famous Feulgen technique for specifically and unambiguously staining DNA, or thymonucleic acid as this substance was called at the time, was described by Feulgen and Rossenbeck in 1924, and a few years later Brachet (1929) applied the technique to sections of oocytes of various animals including *Rana, Triturus* and *Salamandra.* Brachet came to the conclusion that when they are maximally extended the lampbrush chromosomes of all three species are Feulgen-negative. Brachet's observations were at the time generally taken as evidence that thymonucleic acid cannot be the primary and persistent genetic material of chromosomes.

Brachet's observations were supported by those of several authors cited by him in a later paper (1940), including in particular Koltzoff (1938) who studied oocytes of *Triturus,* the hen and the pigeon by various techniques, including Feulgen. Koltzoff made a contrast between the stable genotype of the chromosome and its phenotype, essentially a resuscitation of Lubosch's distinction between idiochromatin and trophochromatin. Koltzoff visualized the stable material in a lampbrush chromosome as a gigantic chain molecule or "genonema" linking and including elementary chromomeres, possibly to be considered as genes, which reproduce by a conservative template copying mechanism and which, by a similar mechanism, induce the formation of maternal gene products in the form of granular lampbrush strands and loops; the transversely disposed lampbrush material ultimately detaches from the chromosomes' axes, and enters the egg cytoplasm at maturation. In the English summary of his paper Koltzoff states without qualification that "chromatin, i.e. thymonucleic acid completely disappears from the chromosomes during the second period of oogenesis and reappears again only in the fifth period; the chromatin should, therefore, by no means be included into the genotype of the chromosome and, hence, it cannot be regarded as an ingredient of the genes. Chromatin is primarily a solid protective sheath of the chromosome which at some stages isolates the genonema from the caryoplasm and determines the definite form of the chromosomes which is convenient for caryokinetic movements".

Surprising though this may now appear, Koltzoff's view of chromosome organization was a very fair statement of cytological opinion prevailing at the time. Thus Caspersson (1936), on the basis of microphotometric observations with uv light of the changing nucleic acid content of cell nuclei and chromosomes during mitosis, and supported by the apparently total disorganization of *Chironomus* salivary gland chromosomes produced by tryptic digestion, held the view that chromosomes have a protein framework to which nucleic acids become attached in preparation for mitosis, the nucleic acids dispersing once more to the nuclear sap as nuclei pass from telophase to interphase. Caspersson's view was championed by Darlington and La Cour (1940) in a paper where chromosomes or parts of chromosomes were stated to be charged or over-charged or starved of nucleic acid according to their staining behaviour, and this "nucleic acid charging" hypothesis was maintained by Darlington in several subsequent papers until as late as 1947.

The question as to whether chromosomes really contain no thymonucleic acid when in the lampbrush phase was evidently of crucial import in the late 1930s because of its bearing on the discussions then in progress concerning the nature of the primary genetic material; the problem was taken up again by Brachet. Working with the oocytes of various Amphibia Brachet (1940) now showed that, despite previous statements to the contrary, the chromomeres of the lampbrush chromosomes of *Triturus* remain Feulgen-positive throughout oogenesis. Brachet was still unable to demonstrate Feulgen-positive chromomeres in *Rana* oocytes at the stage when their lampbrush chromosomes are maximally extended, but he argued that anuran chromosomes are unlikely to be constructed in a fundamentally different way from those of urodeles, they are merely much smaller, and that consequently the chromomeres of *Rana* are too small for the Feulgen reaction to be detectable in the light microscope.

In his 1940 paper Brachet also described how Feulgen-positive granules are associated with free nucleoli in the small oocytes of *Rana fusca*. This finding was independently confirmed by Painter and Taylor (1942) in a study of the distribution of nucleic acids in oocytes of the toad *Bufo valliceps,* their paper including the first description of nucleolar DNA amplification, a phenomenon that was to be overlooked for another 25 years. Painter and Taylor were also able to establish that the chromomeres of toad lampbrush chromosomes are Feulgen-positive throughout oocyte development. Earlier Painter (1940), evidently with thoughts of *Drosophila* polytene chromosomes in mind, had speculated that lampbrush chromosomes are "... the result of some sort of reduplication process". However, Painter and Taylor corrected this error 2 years later with the statement "... that the germinal vesicle of the toad is highly polyploid in nucleolar organizers but otherwise lampbrush chromosomes are normal meiotic structures". Painter and Taylor's 1942 paper is important in another respect. Previous observers had generally viewed oogenesis as a process whereby material accumulates within the germinal vesicle, under some kind of chromosomal control, and that this "excess nuclear material" only becomes available to the developing embryo when the germinal vesicle wall breaks down. On the contrary, Painter and Taylor were able to show, by methyl green/pyronine staining with and without prior ribonuclease digestion, and in further confirmation of Brachet (1940), that a more dynamic situation exists, "... that there is a specific cytological mechanism in the toad (and frog) which begins to deposit ribonucleic acid in the cytoplasm of the oocyte soon after it is differentiated and continues to function through the months required to build up the mature egg. While large amounts of nuclear material are set free in the cytoplasm when the germinal vesicle breaks down, this appears to contain very little nucleic acid of either the ribose or desoxyribose type".

The last paper to consider in this context is one by Dodson (1948), who made a morphological and cytochemical study of the lampbrush chromosomes of the large American salamander ("Congo eel") *Amphiuma means*. Dodson's morphological observations on lampbrush chromosomes agreed with "... the classic picture as reported by Rückert and most subsequent writers". Using the Feulgen technique on *Amphiuma* Dodson confirmed once again that the chromomeres remain Feulgen-positive throughout oocyte development; after digestion with thymo-nuclease the chromomeres were no longer Feulgen-positive, though the morphology of the chromosomes remained unchanged. Staining with Unna's methyl green/pyronine mixture resulted in faint pyronine colouration of the lateral loops, but after digestion with thymo-nuclease all capacity to stain was lost; from this Dodson concluded that his nuclease was contaminated with ribonuclease, and that the lateral loops must contain RNA.

From the results of his studies taken overall, Dodson reached the following conclusion:

"The structural skeleton of the chromosome appears to be histone, while the nucleic acids are attached to this skeleton through protamine. In the main axis the nucleic acid is of the thymonucleic type, while in the loops this is rapidly converted to the ribonucleic type ... Functionally, the lampbrush chromosomes are regarded as agents for the synthesis of ribonucleic acid and enzymes, or other type of developmental agent, for use in the cytoplasm. ... In these chromosomes there is a functional separation of idiochromatin (the main axis) and trophochromatin (the side loops)".

It is appropriate at this point to turn to another approach to the study of lampbrush chromosomes, a method that was destined to lead to novel and far-reaching conclusions. Beginning his work in 1936 in Prof. Baltzer's laboratory in Bern, the American Duryee discovered how easy it is to isolate the germinal vesicles of amphibian and fish oocytes in saline media, and to remove the nuclear membrane, both these operations being done freehand, and thereafter to handle the lampbrush chromosomes with the aid of a micromanipulator. Duryee published the first account of his observations in 1937, and one of his photographs (of a bivalent lampbrush chromosome of *Triturus pyrrhogaster*) was reproduced in Waddington's (1939) textbook on genetics. Duryee's crucial contribution was his discovery that lampbrush chromosomes must not be exposed to Ringer's amphibian saline if they are to be handled and examined in a life-like state, and that for this deleterious action the calcium ions at the concentration present in Ringer's saline ($CaCl_2$ 1 mM) are responsible. Moreover, because the germinal vesicle membrane is permeable to calcium, oocyte nuclei must from the start be isolated and cleaned of cytoplasm in calcium-free Ringer. In Ringer containing calcium the chromosomes quickly become brittle and contracted, loops as well as main axes, whereas in calcium-free Ringer they remain extended as in life, and furthermore show remarkable elastic extensibility when stretched by micromanipulation. When the chromosomes are stretched, the bases of the loops do not open up, as they would if the loops were merely the lateral projections of a loose coil; instead, the lateral projections remain as loops, with their bases close together and associated with chromomeres visible in the stretched main chromosome axis.

Figure 5 shows Duryee's schematic diagram of the morphology of a lampbrush bivalent as published in 1941. It is erroneous in several respects, and in part

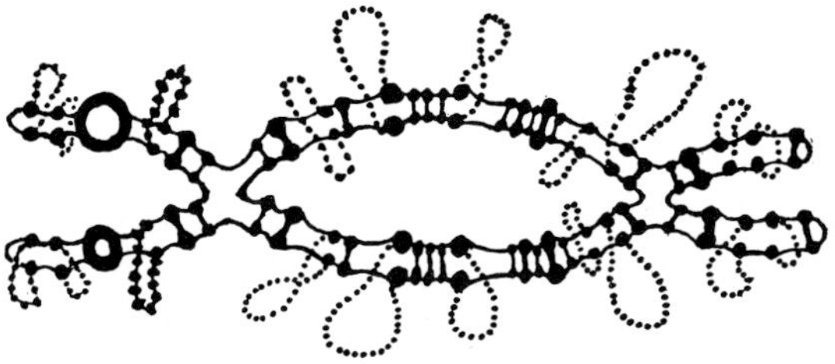

Fig. 5. Duryee's (1941) schematic diagram of the structure of a lampbrush bivalent

these errors must be attributed to diffraction artefacts; unfixed and unstained lampbrush chromosomes are exceedingly difficult to see, let alone to observe critically under a light microscope using ordinary illumination, and indeed Duryee stated that in order to take his photographs he was forced to increase the refractility of the chromosomes by adjusting the saline pH to 4.5.

Duryee's view of lampbrush chromosome organization was that the main axis consists of a "single plastic cylinder" with granules, chromomeres, embedded in this cylinder. The chromomeres are present in pairs, or are dumbbell-shaped, *transverse* to the chromosome axis, and where loops are present these are *single,* running from a granule on one side to its partner on the other. Duryee's conception of the loops was that they are the products of "molecular templates" which remain in the chromosome's axis cylinder, the loops starting as transverse bars between a chromomere pair (so giving rise to the dumbbell shape), then intermittently growing and buckling out from the main axis, then fragmenting into granules which come to lie free in the nuclear sap and which ultimately enter the egg cytoplasm at maturation. Duryee identified the junctions between homologous chromosomes as chiasmata, and he was unable to break these junctions by micromanipulation.

Considered in detail, there are a host of mistakes in Duryee's account of the structure of lampbrush chromosomes, but they should not distract the reader from appreciation of his main achievement, which was to show that lampbrush chromosomes can be easily isolated freehand and unfixed provided calcium-free saline is used as the medium for isolation.

In a paper primarily concerned with the structural organization of meiotic prophase chromosomes of grasshoppers, Ris (1945) considered the evidence for a "chromonemal" (i.e. continuous thread) as opposed to a "chromomeric" (i.e. beads on a string) type of organization, and came down strongly in favour of the former. Ris claimed that the chromomeres seen in leptotene by so many of the earlier cytologists are in reality regions of a coiled chromonema of uniform thickness where the gyres of the spiral are more tightly packed together than they are in neighbouring "interchromomeric" regions. The reason for referring to Ris's paper is that he included a photograph of a portion of a lampbrush chromosome from "... a frog oocyte, smeared in aceto-orcein", and stated that: "In contrast to Duryee, Koltzoff and Painter, it is here suggested that 'lamp-brush chromosomes' are typical diplotene chromosomes which differ from other diplotene chromosomes only in the tremendous longitudinal growth of the chromonemata.

The loops are then the major coils of the laterally separated chromonemata, the chromomeres are simply overlaps of the strands just as in diplotene chromosomes of the grasshopper". This opinion flew straight against the evidence from micromanipulation that had been given by Duryee, and was indisputably wrong, nevertheless as we now know it concealed a half-truth. There was, moreover, another half-truth in Ris's 1945 paper, for which the evidence was only provided several years later. Unlike Duryee and Koltzoff, both of whom regarded the lateral loops as gene products which are periodically sloughed off in their entirety from a primary chromosome axis, Ris claimed that "... the reduction in (lampbrush) chromosome size, just before the meiotic divisions, is accomplished ... not by throwing off parts of the chromosome ... but by the elimination of material on a submicroscopic level".

It was in 1947 that I first came to realize the potentialities which amphibian oocytes present for experimentation, but this realization did not stem from my having read Duryee's papers. Instead, it arose from a misconception of the process of oogenesis! I wished to compare first meiotic metaphase chromosomes of newt (*Triturus cristatus carnifex*) oocytes with those of newt spermatocytes, being already familiar with the latter (Callan 1942). For this purpose I fixed newt ovaries in Zenker's fluid, and used the Feulgen technique on the largest oocytes with the expectation of finding, under a dissecting binocular, a small purple object – the group of meiotic chromosomes – lying in an unstained mass of yolky cytoplasm. Instead, I found that entire oocytes stained bright purple, a consequence of the plasmal reaction, and when broken open these oocytes each revealed a large white sphere inside, the germinal vesicle. It was only thereafter that Duryee's photograph of a lampbrush bivalent in Waddington's (1939) textbook of genetics came to my attention and I then read Duryee's paper.

With the help of S. G. Tomlin, who operated an early model of the Siemens electron microscope in the Physics Department of King's College, London, I then attempted the freehand isolation of *Triturus* lampbrush chromosomes under a dissecting binocular, attaching parts of moderately stretched chromosomes to a film of metalized formvar, the film thereafter being mounted on an EM grid, shadowed and examined. These preparations showed that the axis of a newt lampbrush chromosome consists of a single filament of relatively even diameter, about 20 nm, to which lateral loops are attached at intervals. The preservation of the lateral loops left much to be desired, but the finding of a single axial filament of so little width (Fig. 6) came as something of a surprise, for even in the light microscope the dual nature of the axes of chromosomes at diplotene is readily observable in suitable material, such as grasshopper spermatocytes. This preliminary work was published in 1951.

While manipulating the chromosomes of *Triturus* for EM study it soon became apparent that there is considerable morphological variety amongst the objects attached to lampbrush chromosome axes; this was evident even under a dissecting binocular. I therefore attempted to examine in phase contrast unfixed preparations of lampbrush chromosomes which had been isolated in saline and then placed between a conventional coverslip and slide. This latter operation proved to be fraught with difficulty; indeed, it was at once clear that complete undamaged chromosome complements could not be prepared in this manner. I

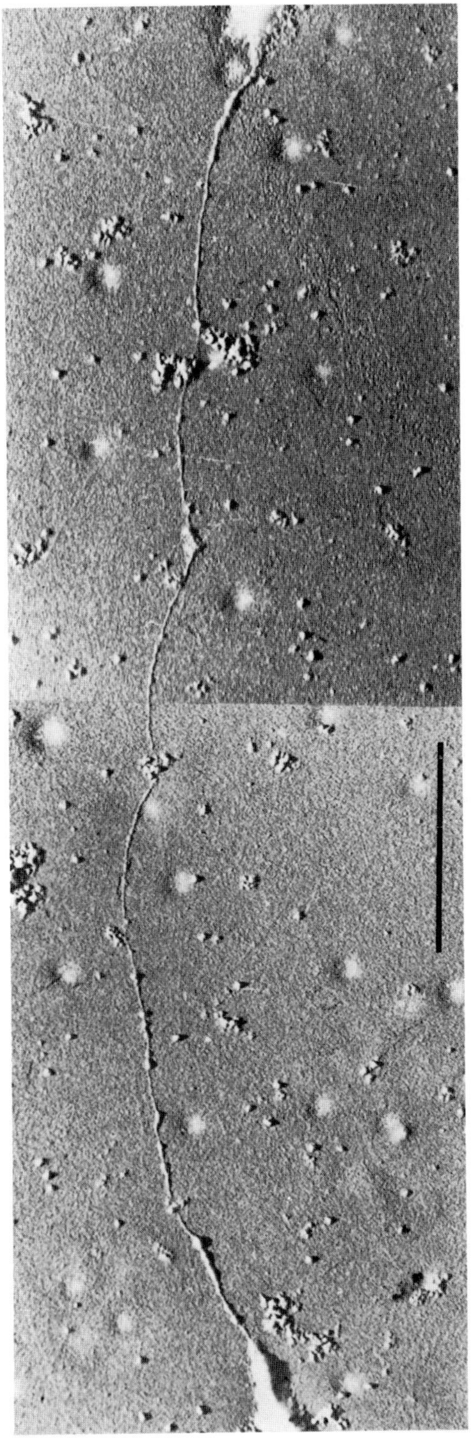

Fig. 6. Electron micrograph showing the single axial filament running between two chromomeres in part of a stretched chromosome of *Triturus cristatus*. *Bar* = 2 μm. (Tomlin and Callan 1951)

started to experiment using depression slides, and it was at this juncture that J. G. Gall, then a postgraduate student, first wrote to me from Yale University, giving at once invaluable advice on how to proceed.

Gall had been studying the lampbrush chromosomes of *Triturus* (*Notophthalmus*) *viridescens,* and his first account of this work was published in 1952. In it he described how preparations can be made which permit protracted study under phase contrast, and with immersion objectives resolving to the limit achievable by the light microscope. A germinal vesicle is isolated and cleaned of adhering yolk in calcium-free saline, then transferred to a large drop of saline on a coverslip, where its membrane is removed. As the nuclear sap disperses, nucleoli and chromosomes both become more and more clearly visible, and they fall to the coverslip surface. The coverslip preparation is placed for an hour in a moist chamber saturated with osmic or formaldehyde vapour, after which time the chromosomes generally come to adhere firmly to the glass. The coverslip is now inverted over a thin depression slide filled with immersion oil, and is finally ringed with paraffin wax. Gall had established that the method of fixation by vapour causes little morphological alteration of the chromosomes other than a measure of contraction which enhances their refractility, so that such preparations conserve a life-like appearance yet are essentially permanent.

Gall observed and photographed parts of the *N. viridescens* complement that were regularly identifiable from preparation to preparation. He saw chromomeres of various dimensions, the patterns of which were similar, though not always identical, in homologous region of bivalents, and he identified the centromeres (kinetochores), which in *N. viridescens* are elongate chromomeres without lateral loops. Gall confirmed earlier demonstrations that the chromomeres, but not the lateral loops, are Feulgen-positive, and as regards the latter Gall stated that "... the lateral loops are dissolved away if the chromosomes remain an hour or more in distilled water". He further confirmed that the interchromomeric fibril, which he termed the chromonema, is a single strand.

Gall's detailed account of the lampbrush chromosomes of *N. viridescens* was published in 1954. In the material and techniques section of this paper he described a novel method for examining lampbrush chromosomes with an inverted microscope, the use of which was soon to lead to a better understanding of the organization of these objects. Gall identified all 11 bivalents of *N. viridescens* on the basis of their lengths (which at maximum extension range from about 350 to 800 μm), centromere positions, the positions of loops which can be regularly recognized because of their exceptional lengths (some up to 200 μm) or exceptional textures and the positions of various other laterally attached objects, particularly "knobs" and nucleoli. As regards the latter, Gall compared the place where a nucleolus, morphologically similar to the free nucleoli, is attached to one lampbrush chromosome of *Amblystoma tigrinum,* with the position of the nucleolus organizer constriction in somatic chromosome complements of this organism, and demonstrated their correspondence. He thereby established that the hundreds of unattached nucleoli in amphibian germinal vesicle nuclei really are homologous to the nucleoli of somatic cells, which up until this time had been merely an assumption.

Gall confirmed Dodson's observation that the lateral loops and nucleoli (but not the "knobs") contain RNA. He also tested the reactions of unfixed chromosomes to digestion with trypsin and pepsin. Unlike Dodson, who had worked throughout with paraffin wax sections of oocytes, he found that trypsin, although it rapidly dissolves the lateral loops and causes the chromomeres to coalesce and lose optical contrast, does not destroy the linear integrity of lampbrush chromosomes. A few years later the full significance of this important observation was to become apparent.

Beginning their observations in 1949, Guyénot and Danon in Geneva carried out a parallel cytological study of oogenesis in *Triturus cristatus* and *Rana temporaria,* and published a long paper in 1953. Their work is not easy to evaluate, partly because several of the objects which they described from stained sections are certainly misidentified. Thus, for example, their *«macronucléoles»* and *«filaments nucléoplasmiques»* are particularly large and peculiar lateral loops that are characteristic of *T. cristatus* chromosomes. Guyénot and Danon's theory concerning the origin and development of lateral loops is bizarre; these are said to grow out from the chromosome axis as long bristles (*"poils"*), which are split lengthwise into two except where they are inserted into the chromosome axis (and presumably also at their distal extremities) thereby forming loops (*"boucles"*). They assert that the DNA content of the lampbrush chromosomes is minimal when oocytes are growing most rapidly, when the lateral loops are maximally extended, and that the chromosomes reacquire a coating of DNA as maturity approaches; this is a revival of the nucleic acid charging hypothesis of Darlington. Their paper could be largely disregarded were it not for an observation made in phase contrast of lampbrush chromosomes isolated in *«eau salée acétifiée»* (to enhance contrast). They noted *«... la fissuration du chromosome en deux chromatides, dont les grains sombres (chromioles) se correspondent deux à deux. C'est sur ces chromioles jumelés que sont insérés les poils et boucles qui sont disposés par pairs»*. This is the first mention in the literature of lateral loops occurring *in pairs*.

My own first studies of lampbrush chromosomes with the light microscope, using Gall's observation chamber and an inverted microscope, began in 1950. I started to work with oocytes of *Triturus vulgaris, T. helveticus* and *T. marmoratus,* and published a preliminary review of this work in 1955. An early and provocative observation was that if a freshly isolated lampbrush chromosome is stretched with a micromanipulator, not only does extension primarily occur in regions between chromomeres, without affecting the close juxtaposition of individual loops' two insertions in the chromosome axis (thus confirming Duryee's earlier observation), but further, occasional chromomeres split transversely when under tension; when they break in this way, the gap in the chromosome's axis *is invariably spanned by a pair of lateral loops* (Fig. 7). It later transpired that such "double-loop bridges" may be generated accidentally during the isolation of a germinal vesicle and the removal of its nuclear membrane (Fig. 8 a), more frequently in some newt species than in others. Still later I found that one particular pair of loops (the "giant fusing loops") on the smallest chromosome of *Triturus cristatus cristatus* often occur in double-bridge form, i.e. with their axial insertions separate from one another, in the natural state (Fig. 8 b).

The fact that mechanically produced or natural loop bridges occur at all, and are double, and that particularly large and easily recognizable loops on the smallest chromosome of *T. marmoratus* are always present in pairs (Fig. 8 c), prompted

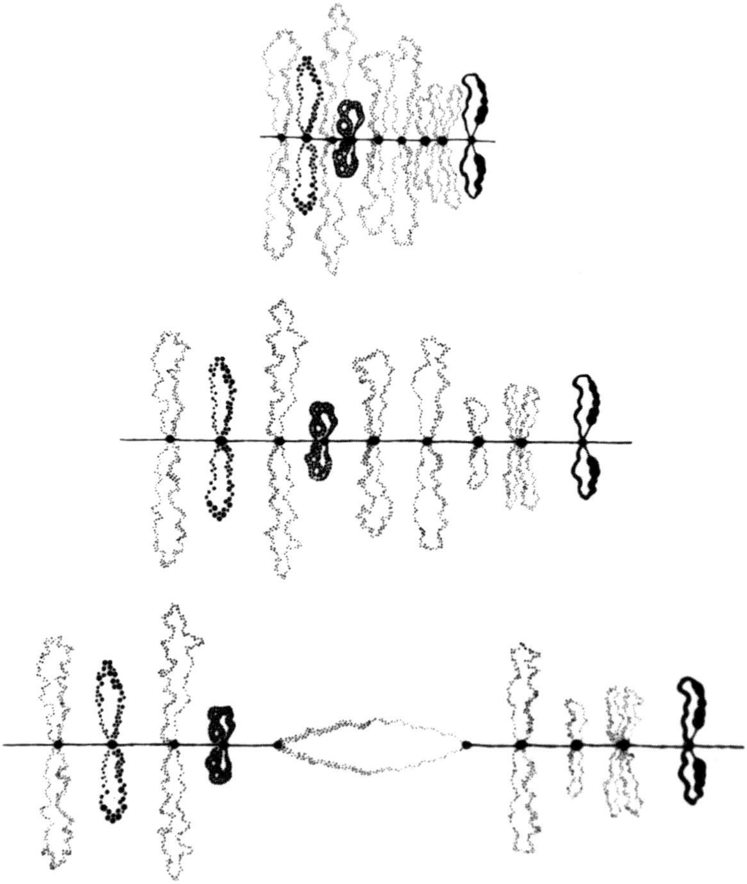

Fig. 7. Diagrams to illustrate what occurs when part of a lampbrush chromosome is stretched and a chromomere cleaves transversely, so producing a "double-loop bridge". (Callan 1963)

the speculation that there is a general duality of lateral loops (Fig. 9), indicating that each lampbrush chromosome consists of two chromatids, that each loop contains a fibrillar "core" running throughout its length and that these loop cores are reflected extended portions of the chromatids' axis, each being surrounded, asymmetrically, by the accumulated products of synthesis. This view was reinforced by the observation that along the smallest chromosome of *T. marmoratus* there are two short regions, one near the middle and the other at one end, where the chro-

→

Fig. 8 a–c. Phase contrast micrographs. **a** Double bridge accidentally produced by a pair of lampbrush loops of *Triturus cristatus carnifex*. *Bar* = 20 μm. **b** Naturally occurring double bridge formed by the pair of giant fusing loops on chromosome XII of *T. c. cristatus*. *Bar* = 20 μm. **c** Bivalent XII of *T. marmoratus* dispersed in dilute saline; *arrows* point to the two homologous pairs of giant fusing loops which have unravelled from their naturally compact state, while *arrowheads* point to the double axis regions. *Bar* = 50 μm. (**a,** From Callan unpublished; **b,** Callan and Lloyd 1960; **c,** Callan 1955)

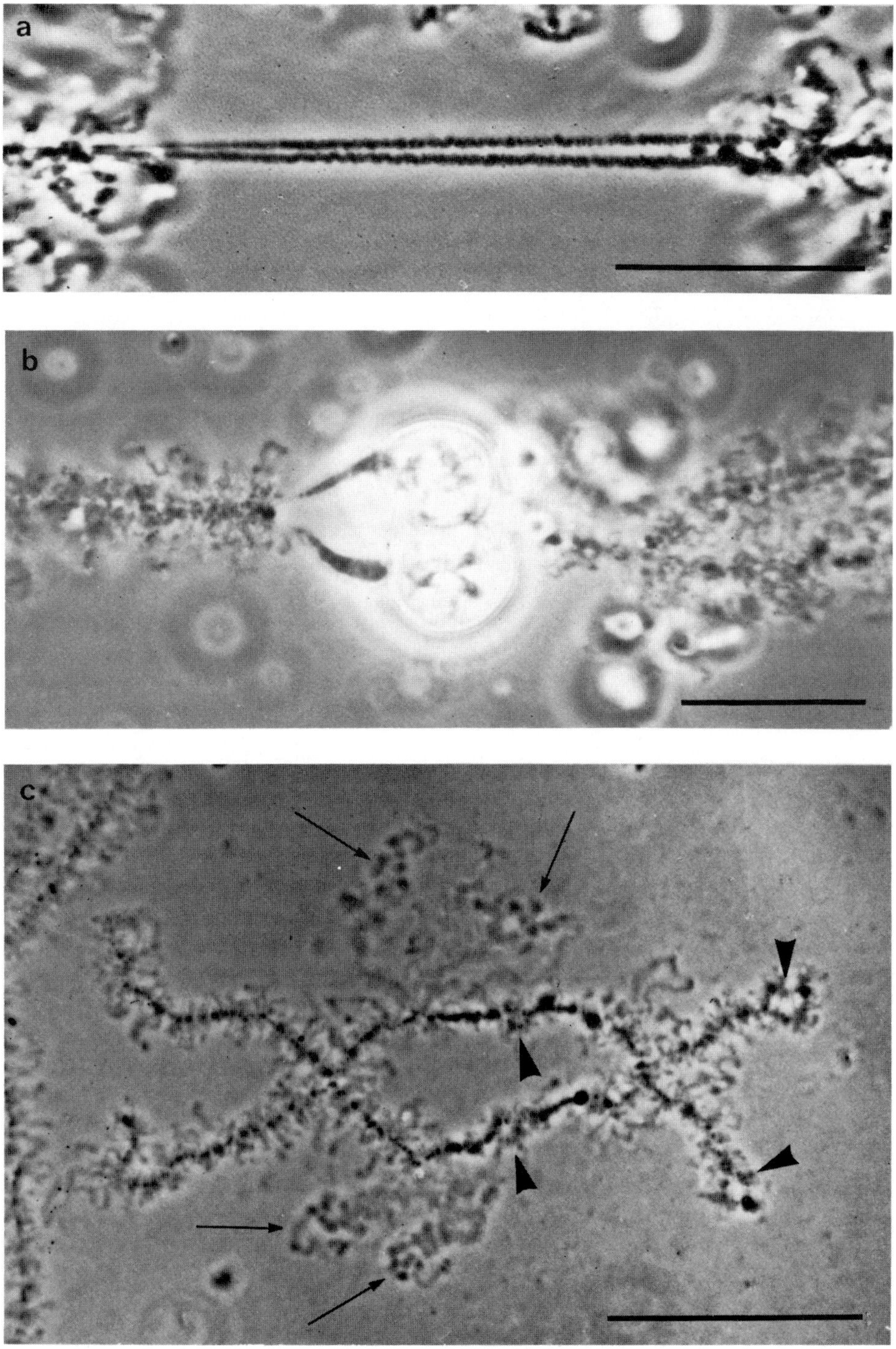
a
b
c

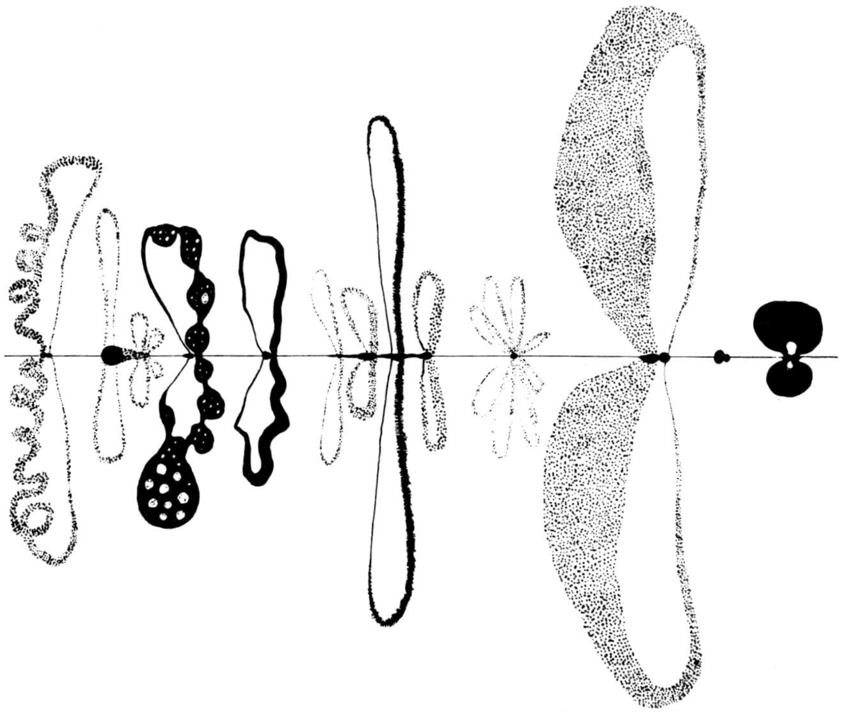

Fig. 9. Some characteristic types of newt lampbrush loops and the chromomeres with which they are associated. (Callan 1955)

mosome axis in its entirety is double (Fig. 8 c), with matching pairs of chromomeres on each axis, and each axis generating single loop bridges when mechanically broken. Thus, the relationship between lateral loops and the chromosome axis was put in rather a new light, betwixt and between the Duryee and Ris interpretations. Chromomeres are a reality, but the lateral loops are not simply "gene products" periodically shed from chromomeres, nor are they merely portions of a continuous chromonema; instead, embedded within each lateral loop, is a *portion* of the chromonema.

The interpretation of the organization of lampbrush chromosomes was discussed by Gall in 1955, and within a year the existence of a loop axis, resistant to digestion with pepsin and in this respect unlike the bulky surrounding material, had been demonstrated by electron microscopy (Gall 1956). For this paper Gall drew a diagram showing the structure postulated, which is reproduced here in Fig. 10. At the close of his 1956 paper Gall calculated the total length of DNA present in a lampbrush chromosome complement having the 4C weight of DNA, and from this concluded that a single averaged chromatid of *Notophthalmus viridescens* contains about 90 cm DNA. He also computed to a first approximation the total length of the lateral loops associated with such a lampbrush chromatid as 5 cm. He then went on to consider the possibility that if DNA is found uniformly throughout the chromatid (i.e. in the interchromomeric fibril and in the

Fig. 10. The organization of a chromomere and of its suggested relationship to a pair of lateral loops. (Gall 1956)

loop axis as well as in chromomeres) then one can predict an upper limit to the number of DNA strands per chromatid (18) and the further likelihood, because of the known compaction of DNA in chromomeres, that the number of strands per chromatid may be considerably less than this figure, perhaps only one. So far as I am aware, this was the first tentative step towards the proposition of uninemy. In the face of Gall's evidence for the existence of a single loop axis distinct from the surrounding products of synthesis, Lafontaine and Ris (1958) continued to assert, on the basis of EM studies of whole mounts and sections of lampbrush chromosomes, that each loop is in its entirety a portion of a continuous chromonema, and that it consists of bundles of microfibrils about 20 nm wide. These microfibrils we can now recognize as the RNP transcripts visible in Miller spreads of lampbrush chromosomes.

By 1955 I had decided to concentrate my attention on lampbrush chromosomes of newts of the multiracial species *Triturus cristatus,* and for three reasons. One of these was that there is a remarkable wealth of morphological diversity amongst the objects, loops and other structures, projecting laterally from the chromosomes of these animals; this permits the rapid and precise identification of particular chromosome regions (Callan and Lloyd 1960). A second reason lay in the fact that one of the races of *T. cristatus,* the Italian *T. c. carnifex,* is easy to maintain and breed in the laboratory, there being no obligatory terrestrial stage in its life history, and both sexes coming repeatedly into annual cycles of reproductive activity without any special treatment; the males of many other species of newt, including the American *N. viridescens* and the British race of *T. cristatus, T. c. cristatus,* are recalcitrant in this regard. The third reason was that I was already familiar with the male meiosis of four of the races of *T. cristatus,* and of several of its interracial hybrids, F_1 and others (Callan and Spurway 1951). I had many interracial hybrids still alive in the laboratory and I hoped by studying female interracial hybrids to find evidence for the Mendelian inheritance of "phenotypic" chromosomal characters, i.e. lateral loop morphologies, differentiating the races from one another.

The construction and examination of interracial hybrids in order to demonstrate the Mendelian inheritance of lateral loop characters later proved to be, at least in part, unnecessary; indeed, already in 1956 Callan and Lloyd published a preliminary note in which it was shown that individual females of *T. c. carnifex* may differ from one another with respect to the morphologies of loops at specific chromosomal loci, including spectacular heterozygosities that are consistent in all the oocytes of certain individuals, therefore providing strong presumptive evidence that loop morphologies are directly determined by "local" genetic constitution.

For a short period in 1956 I turned my attention to the possibility of studying the oocyte chromosomes of other organisms, particularly invertebrates, by the techniques already applied to newts, and found (Callan 1957) that there are genuine lampbrush chromosomes, as defined by the criteria that the chromosomes bear lateral loops, and that axial breaks are regularly spanned by double-loop bridges, in the cephalopod mollusc *Sepia* and the isopod crustacean *Anilocra*. These findings, and a perusal of the already voluminous literature on oogenesis, prompted the speculation that all animal oocytes may contain lampbrush chromosomes at some stage during oogenesis. Speculative generalizations are often unwise, and this one was no exception. If I had read more carefully Loyez's 1906 paper on the development of oocytes that store a large quantity of yolk, I would have learned that in the oocytes of some snakes and lizards the chromosomes are compact and metabolically inert throughout oogenesis, their original synthetic roles having been taken over by special nurse cells. A similar state of affairs occurs in many insects, notably the Diptera (Bier 1963a, b) and including *Drosophila* (King and Burnett 1959; Sirlin and Jacob 1960; Mahowald and Tiefert 1970), but whereas the nurse cells of reptiles are somatic in origin, those of insects are cells from the germ line deflected so as to fulfil a supportive function. In the oocytes of mammals too it now appears unlikely, despite the claim made by Baker and Franchi (1967a, b), that the chromosomes take on a lampbrush form at any stage of oogenesis. However, an historical perspective would be seriously distorted if I were to pursue this topic here; it is discussed, with the associated problem of how to define a lampbrush chromosome, in Callan (1986).

Returning to newts in 1957, the possibility presented itself that, by applying enzymes directly to unfixed lampbrush chromosomes floating freely in saline, the resolution of an urgent and contentious problem of the time might be in sight: how is DNA distributed along the length of a chromosome, is it interrupted, and if so, what material maintains a chromosome's linear integrity? The great advantage of this experimental system lay in the fact that isolated lampbrush chromosomes surrounded by saline are in violent Brownian movement, yet they remain coherent entities; any disruption of their linear integrity at once becomes manifest because portions then move independently and drift apart.

Macgregor and I applied the enzymes trypsin, pepsin, ribonuclease and deoxyribonuclease to freshly isolated lampbrush chromosomes (Fig. 11). We found that: "Pepsin, trypsin and ribonuclease all bring about the solution of loop matrix materials, together with certain other structures attached laterally to the axes of the chromosomes, but these enzymes do not destroy the linear integrity of the chromosomes. The action of deoxyribonuclease, however, is dramatically different. The submicroscopic fibrils connecting adjacent chromomeres snap, and the lateral loops fragment into smaller and smaller pieces, each piece however conserving for a time at least the finescale morphological peculiarities of the loop from which it originated. A drastic disruption of the chromosomes is already visible within a few minutes of the beginning of enzyme treatment, and in half an hour Brownian movement has so scattered the fragments that no recognizable trace of the original forms of the chromosomes remains. In our experience no

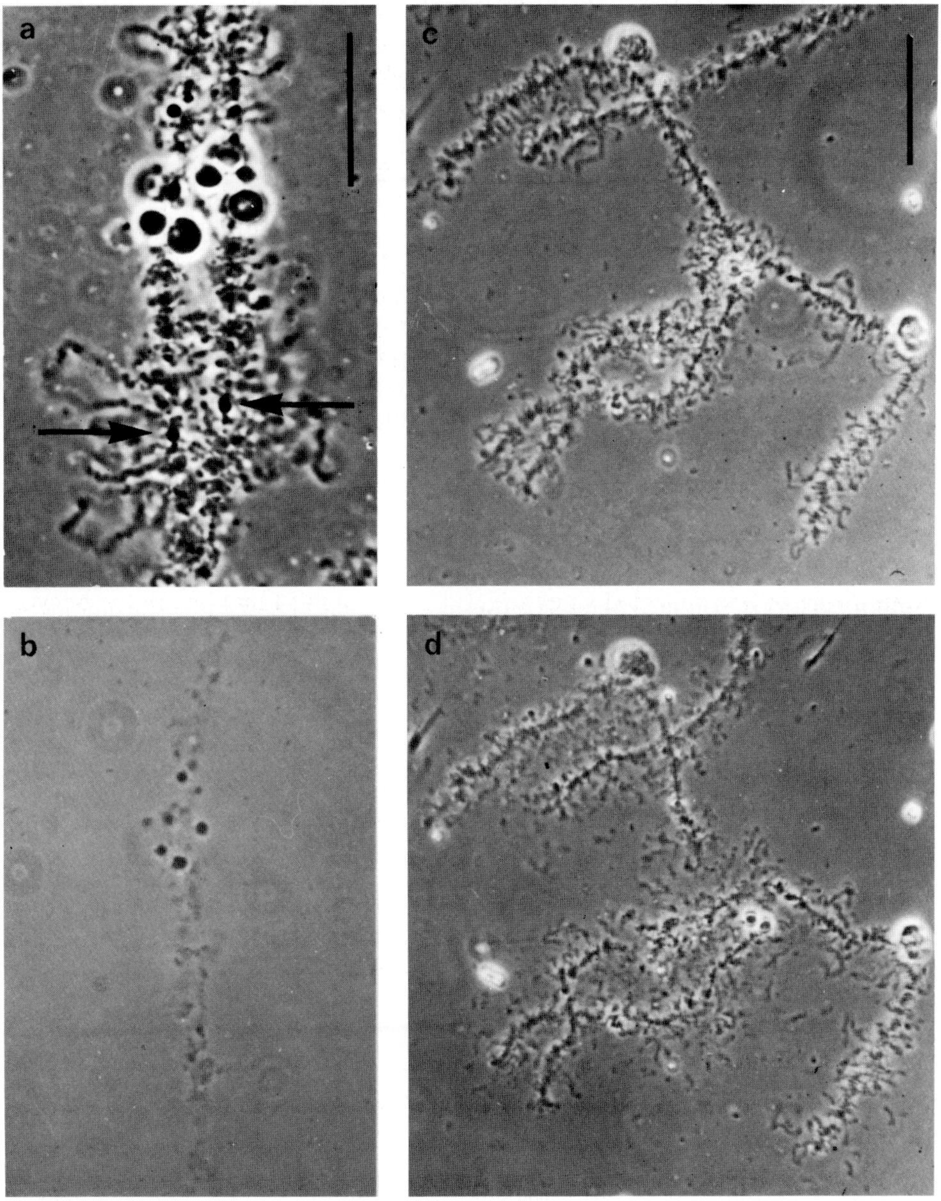

Fig. 11 a–d. Phase contrast micrographs. **a** The middle region of lampbrush bivalent II of *T. c. carnifex; arrows* point to the centromeres; the bulky refractile objects are lumpy loops, regular features near the centromeres of chromosome II. *Bar* = 20 μm. **b** As **a**, but after 40 min digestion with trypsin at pH 7.8. **c** Bivalent X of *T. c. carnifex*, entire. *Bar* = 50 μm. **d,** As **c,** but after 3 min digestion with pancreatic DNase at pH 6.3. (Macgregor and Callan 1962)

other chemical agent has been found capable of breaking lampbrush chromosomes in this fashion, and the conclusion is inescapable that deoxyribonucleic acid runs throughout the lengths of lampbrush chromosomes, including the axes of their lateral loops" (Callan and Macgregor 1958; Macgregor and Callan 1962).

Following up these observations Gall (1963) studied the kinetics of breakage of unfixed lampbrush chromosomes when digested by pancreatic DNase I. This enzyme was already known to attack the two polynucleotide chains of DNA in solution independently and at random, and that scission of the double helical molecule occurs only when breaks in the two chains occur within a few nucleotide pairs of each other; thus breaks in DNA in solution accumulate as a function of time raised to the power 2. Gall found that breaks of single loops (the giant loops near the centromeres of *Notophthalmus viridescens* were chosen because of their considerable lengths) accumulated as a function of time raised to the power 2.6 ± 0.2. Breaks of interchromomeric main axes accumulated as a function of time raised to the power 4.8 ± 0.4. Under the conditions of Gall's experiments, and bearing in mind that DNA is associated with histones in lampbrush chromosomes, both these figures are likely to be overestimates. Gall concluded that the simplest model consistent with his results is that a chromatid, as seen in single lampbrush loops, contains one DNA double helix, and that the interchromomeric strand in a lampbrush chromosome contains two running side by side. These conclusions have now been substantiated by experiments of other kinds, and the unineme nature of chromatids generally thereby established.

I have chosen to terminate my historical survey when studies on lampbrush chromosomes had reached this stage. Other speakers will handle the many more recent lampbrush discoveries, such as the Y-loop story in *Drosophila* begun by Meyer et al. (1961), the nature of transcription complexes begun by Miller and Beatty (1969a, b), the study of nucleic acid sequences that are transcribed into RNA by in situ hybridization begun by Pukkila (1975) and the immunolocalization of proteins associated with transcription begun by Sommerville (1973). In all these fields advantage has been taken of the "magnification factor" provided by lampbrush chromosomes, the many transcription units with closely packed transcripts, which is their most distinctive feature. Whether transcription and its control in oocytes differs in scale only from what habitually occurs in ordinary somatic cell nuclei remains to be demonstrated, but my guess is that it will be. If so, lampbrush chromosomes will continue to be "in fashion" for a good many more years.

References

Baker TG, Franchi LL (1967a) The fine structure of oogonia and oocytes in human ovaries. J Cell Sci 2:213–224

Baker TG, Franchi LL (1967b) The structure of the chromosomes in human primordial oocytes. Chromosoma 22:358–377

Bier K (1963a) Synthese, interzellulärer Transport und Abbau von Ribonukleinsäure im Ovar der Stubenfliege *Musca domestica*. J Cell Biol 16:436–440

Bier K (1963 b) Autoradiographische Untersuchungen über die Leistungen des Follikelepithels und der Nährzellen bei der Dotterbildung und Eiweißsynthese im Fliegenovar. Arch Entwicklungsmech Org 154:552–575
Born G (1892) Die Reifung des Amphibieneies und die Befruchtung unreifer Eier bei *Triton taeniatus*. Anat Anz 7:772–781 and 803–811
Born G (1894) Die Struktur des Keimbläschens im Ovarialei von *Triton taeniatus*. Arch Mikrosk Anat Entwicklungsmech 43:1–79
Brachet J (1929) Recherches sur le comportement de l'acide thymonucléinique au cours de l'oogenèse chez diverses espèces animales. Arch Biol 39:677–697
Brachet J (1940) La localization de l'acide thymonucléique pendant l'oogenèse et la maturation chez les Amphibiens. Arch Biol 51:151–165
Callan HG (1942) Heterochromatin in *Triton*. Proc R Soc Lond Ser B 130:324–335
Callan HG (1955) Recent work on the structure of cell nuclei. In: Symp 8th Congr Cell Biol. IUBS Ser B21:89–109 Noordhoff, Groningen
Callan HG (1957) The lampbrush chromosomes of *Sepia officinalis* L., *Anilocra physodes* L. and *Scyllium catulus* Cuv. and their structural relationship to the lampbrush chromosomes of Amphibia. Publ Stn Zool Napoli 29:329–346
Callan HG (1963) The nature of lampbrush chromosomes. Int Rev Cytol 15:1–34
Callan HG (1986) Lampbrush Chromosomes. Springer, Berlin Heidelberg New York Tokyo
Callan HG, Lloyd L (1956) Visual demonstration of allelic differences within cell nuclei. Nature 178:355–357
Callan HG, Lloyd L (1960) Lampbrush chromosomes of crested newts *Triturus cristatus* (Laurenti). Philos Trans R Soc Lond Ser B 243:135–219
Callan HG, Macgregor HC (1958) Action of deoxyribonuclease on lampbrush chromosomes. Nature 181:1479–1480
Callan HG, Spurway H (1951) A study of meiosis in interracial hybrids of the newt, *Triturus cristatus*. J Genet 50:235–249
Carnoy JB, Lebrun H (1897) La vésicule germinative et les globules polaires chez les batraciens. I. Salamandre et Pleurodèle. Cellule 12:191–295
Carnoy JB, Lebrun H (1898) La vésicule germinative et les globules polaires chez les batraciens II. Axolottl et Tritons. Cellule 14:113–200
Carnoy JB, Lebrun H (1899) La vésicule germinative et les globules polaires chez les batraciens III. Les globules polaires des urodèles. Cellule 16:304–391
Caspersson T (1936) Über den chemischen Aufbau der Strukturen des Zellkernes. Skand Arch Physiol 73:suppl 8
Darlington CD (1947) Nucleic acid and the chromosomes. Symp Soc Exp Biol 1:252–269
Darlington CD, La Cour L (1940) Nucleic acid starvation of chromosomes in *Trillium*. J Genet 40:185–213
Dodson EO (1948) A morphological and biochemical study of lampbrush chromosomes of vertebrates. Univ Calif Publ Zool 53:281–314
Duryee WR (1937) Isolation of nuclei and non-mitotic chromosome pairs from frog eggs. Arch Exp Zellforsch 19:171–176
Duryee WR (1941) The chromosomes of the amphibian nucleus. In: Cytology, genetics and evolution. Univ Pennsylvania Press, Philadelphia, pp 129–141
Feulgen R, Rossenbeck H (1924) Mikroskopisch-chemischer Nachweis einer Nukleinsäure vom Typus der Thymonukleinsäure. Z Physiol Chem 135:213–248
Flemming W (1882) Zellsubstanz, Kern- und Zelltheilung. Vogel, Leipzig
Gall JG (1952) The lampbrush chromosomes of *Triturus viridescens*. Expl Cell Res Suppl 2:95–102
Gall JG (1954) Lampbrush chromosomes from oocyte nuclei of the newt. J Morphol 94:283–352
Gall JG (1955) Problems of structure and function in the amphibian oocyte nucleus. Symp Soc Exp Biol 9:358–370
Gall JG (1956) On the submicroscopic structure of chromosomes. Brookhaven Symp Biol 8:17–32
Gall JG (1963) Kinetics of deoxyribonuclease action on chromosomes. Nature 198:36–38

Goldschmidt R (1902) Untersuchungen über die Eireifung, Befruchtung und Zellteilung bei *Polystomum integerrimum*. Rud Z Wiss Zool 71:397–444
Guyénot E, Danon M (1953) Chromosomes et ovocytes de batraciens. Étude cytologique et au microscope électronique. Rev Suisse Zool 60:1–129
Hartmann M (1902) Studien am tierischen Ei. I. Ovarialei und Eireifung von *Asterias glacialis*. Zool Jahrb (Anat) 15:793–812
Heidenhain M (1894) Neue Untersuchungen über die Centralkörper und ihre Beziehungen zum Kern- und Zellenprotoplasma. Arch Mikrosk Anat Entwicklungsmech 43:423–758
Holl M (1890) Über die Reifung der Eizelle des Huhnes. Sitzungsber Akad Wiss Wien 99:311–370
Janssens FA (1904) Das chromatische Element während der Entwicklung des Ovocyts von Triton. Anat Anz 24:648–651
Jörgensen M (1913) Zellenstudien I. Morphologische Beiträge zum Problem des Eiwachstums. Arch Zellforsch 10:1–126
King RC, Burnett RG (1959) Autoradiographic study of uptake of tritiated glycine, thymidine and uridine by fruitfly ovaries. Science 129:1674–1675
Koltzoff NK (1938) The structure of the chromosomes and their participation in cell-metabolism. Biol Zh 7:44–46
Lafontaine JG, Ris H (1958) An electron microscope study of lampbrush chromosomes. J Biophys Biochem Cytol 4:99–105
Loyez M (1906) Recherches sur le développement ovarien des oeufs méroblastiques à vitellus nutritif abondant. Arch Anat Microsc Morphol Exp 8:69–397
Lubosch W (1902) Über die Nucleolarsubstanz des reifenden Tritoneneies nebst Betrachtungen über das Wesen der Eireifung. Jena Z Naturwiss 37:217–296
Macgregor HC, Callan HG (1962) The actions of enzymes on lampbrush chromosomes. Q J Microsc Sci 103:173–203
Mahowald AP, Tiefert M (1970) Fine structural changes in the *Drosophila* oocyte nucleus during a short period of RNA synthesis. Arch Entwicklungsmech Org 165:8–25
Maréchal J (1907) Sur l'ovogénese des sélaciens et de quelques autres chordates. Cellule 24:5–239
Meyer GF, Hess O, Beermann W (1961) Phasenspezifische Funktionsstrukturen in Spermatocytenkernen von *Drosophila melanogaster* und ihre Abhängigkeit vom Y-Chromosom. Chromosoma 12:676–716
Miller OL, Beatty BR (1969 a) Visualization of nucleolar genes. Science 164:955–957
Miller OL, Beatty BR (1969 b) Portrait of a gene. J Cell Physiol 74:suppl 1, 225–232
Painter TS (1940) On the synthesis of cleavage chromosomes. Proc Natl Acad Sci USA 26:95–100
Painter TS, Taylor AN (1942) Nucleic acid storage in the toad's egg. Proc Natl Acad Sci USA 28:311–317
Pukkila PJ (1975) Identification of the lampbrush loops which transcribe 5S ribosomal RNA in *Notophthalmus (Triturus) viridescens*. Chromosoma 53:71–89
Rabl C (1885) Ueber Zelltheilung. Morphol Jahrb 10:214–330
Ris H (1945) The structure of meiotic chromosomes in the grasshopper and its bearing on the nature of "chromomeres" and "lampbrush chromosomes". Biol Bull 89:242–257
Rückert J (1892) Zur Entwicklungsgeschichte des Ovarialeies bei Selachiern. Anat Anz 7:107–158
Schultze O (1887) Untersuchungen über Reifung und Befruchtung des Amphibieneies. Z Wiss Zool 45:178–226
Sirlin JL, Jacob I (1960) Cell function in the ovary of *Drosophila*. Exp Cell Res 20:283–293
Sommerville J (1973) Ribonucleoprotein particles derived from the lampbrush chromosomes of newt oocytes. J Molec Biol 78:487–503
Tomlin SG, Callan HG (1951) Preliminary account of an electron microscope study of chromosomes from newt oocytes. Q J Microsc Sci 92:221–224
Waddington CH (1939) An introduction to modern genetics. Allen & Unwin, London

Polytene Chromosomes

G. Korge [1]

1 Introduction

The chromosomal nature of the „Kernschleifen" in Diptera had been established in the 1930's (Heitz and Bauer 1933; Painter 1933, 1934; King and Beams 1934; Koltzoff 1934). During the subsequent 10 years these giant chromosomes or polytene chromosomes were intensively used to answer classical cytogenetic and genetic questions. At the beginning of the 1950's, a new era of polytene chromosome research was inaugurated by W. Beermann's fundamental work on chromosome structure and function (Bauer and Beermann 1952; Beermann 1950, 1952a, b, 1955). A tremendous number of publications appeared during the following decades, dealing with various aspects of polytene chromosomes in different tissues of different organisms (Beermann 1962, 1972a; Ashburner 1970; Crick 1971; Lefevre 1974a, b; Sorsa 1974a, b, 1976; Zhimulev 1975a; Ashburner and Berendes 1978; Spradling and Rubin 1981; Zhimulev et al. 1983; Eissenberg et al. 1985; Lefevre and Watkins 1986).

Polytene chromosomes are used as a handy model for studying structural and functional organization of eukaryote chromosomes. Great efforts have been made to understand chromomere organization, correspondence between genes and chromomeres, chromosome replication, and transcription-related chromatin decondensation or puffing (Beermann 1972b). New molecular techniques of gene analysis and in situ hybridization offer new ways of experimental approach. Most of the work reviewed in the following has been carried out on polytene chromosomes of Diptera and the discussion is focused on three topics:

1. Structure of polytene chromosomes.
2. Gene organization in polytene chromosomes.
3. Gene activation-related chromosome puffing.

2 Structure of Polytene Chromosomes

2.1 Disproportionate Replication

Polytene chromosomes are known to exist in nuclei as different as the macronucleus anlagen of ciliated protozoa (Ammermann 1964; Ammermann et al.

[1] Institut für Allg. Genetik der Freien Universität Berlin, Arnimallee 5–7, 1000 Berlin 33, FRG.

Results and Problems in Cell Differentiation 14
Structure and Function of Eukaryotic Chromosomes
Edited by W. Hennig

1974) or from suspensor cells of plant seeds, particularly of beans (Nagl 1962, 1974, 1981; Tschermak-Woess 1963), salivary gland cells of Collembola (Prabhoo 1961; Cassagnau 1971, 1976; Deharveng 1982), and different tissues of Diptera (Beermann 1962). Although originating from taxonomically very different organisms, they show the same characteristic features in general: they are cable-like chromatin structures of varying diameter, exhibiting specific cross-striping patterns produced by compact bands of highly condensed DNA (chromomeres) which are separated by light regions of low DNA content (interchromomeres).

Polytene chromosomes arise, according to the "polyteny hypothesis" (Heitz and Bauer 1933; Painter 1933; Koltzoff 1934) by repeated cycles of endoreplication of the chromatids, during which, in Diptera, the homologs remain in close contact by somatic pairing. Analysis of DNA replication and determination of DNA content of nuclei using autoradiographic and cytophotometric methods (Kurnick and Herskowitz 1952; Swift and Rasch 1954; Welch 1956; Swift 1962; Rodman 1967; Rudkin 1972; Hartmann-Goldstein and Goldstein 1979) have shown that the euchromatic parts of the chromatids replicate synchronously, causing geometric doublings of the DNA content, and finally reach levels of polyteny ranging from 1024 in *D. melanogaster* salivary glands through 8192 in suspensor cell nuclei of beans (Nagl 1974, 1981) to 16,384 in *Chironomus* or even 32,768 in *Rhynchosciara* (Beermann 1962). These values are only correct under the assumption that all euchromatic DNA sequences are equally replicated.

That not all chromosome segments polytenize to the same extent had already been known since the rediscovery of giant chromosomes during the 1930's (Bauer 1935; Heitz 1933, 1934; Hinton 1942). Thus, for instance, in *Drosophila melanogaster* only euchromatic parts of the X chromosomes and some segments of heterochromatin seem to polytenize. It has also been shown by cytogenetic, cytophotometric and autoradiographic studies that heterochromatin remains at a 4C level or undergoes only a few replication cycles (Rudkin and Schultz 1961; Keyl and Pelling 1963; Rudkin 1969; Berendes and Keyl 1967; Mulder et al. 1968). Underreplication is demonstrated most strikingly in the case of the Y chromosome in *Drosophila melanogaster*. In diploid cells the Y chromosome is heterochromatic and approximately of the same size as the X chromosome (Fig. 1). In nuclei of larval salivary glands, however, it is completely undetectable and is hidden in the chromocenter, which also contains the heterochromatic centromeric regions of the other chromosomes (Fig. 1). Out of the great number of other known conspicuous differences in the properties of chromosomes between diploid cells and polytene chromosomes (Beermann 1962), the absence of chromosome arms (Zacharias 1979) or even the absence of entire chromosomes in the polytene chromosome complement (Southern et al. 1973) can only be mentioned here.

Investigations of DNA sequences by density gradient centrifugation, filter and in situ hybridization, and renaturation kinetics have shown that highly repetitive DNA sequences, the vast majority of which constitutes heterochromatin, are excluded from polytenization (Rudkin 1969; Hennig et al. 1970; Jones and Robertson 1970; Rae 1970; Botchan et al. 1971; Gall et al. 1971; Dickson et al. 1971; Hennig 1972; Peacock et al. 1978; Redfern 1981 a). Gall and coworkers presented evidence that only centromeric α-heterochromatin does not replicate dur-

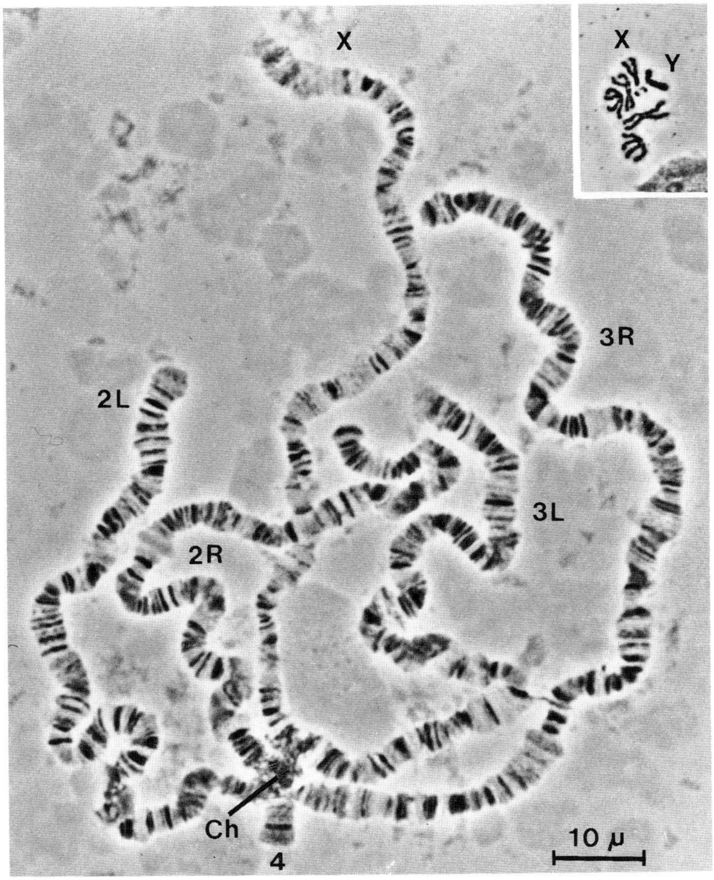

Fig. 1. Salivary gland chromosomes from a male larva of *Drosophila melanogaster* wild type. The chromosomes are indicated. The Y chromosome is hidden in the chromocenter, *Ch*. *Insert* the metaphase chromosomes from the ganglion of a male larva, same magnification. X and Y chromosomes are marked. The Y chromosome is heterochromatic. Phase contrast

ing polytenization, but that some or all of the repetitive sequences in the *β*-heterochromatin replicate like euchromatin (Gall et al. 1971).

In addition to the centromeric regions, also telomeres contain heterochromatin consisting of complex sets of repeated sequences (Rubin 1978; Young et al. 1983; Renkawitz-Pohl and Bialojan 1984). Some but not all of the members of these sets are underreplicated during polytenization in *Drosophila melanogaster* (Young et al. 1983).

In *Drosophila melanogaster* the genes for ribosomal 18S and 28S RNA's are located in the heterochromatic part of the X chromosome and in the short arm of the Y chromosome (Pardue et al. 1970). The replication of these genes seems to be more complicated than that of highly repetitive simple sequences of the heterochromatic regions. The rRNA genes are clustered at two loci, each of which contains 200–250 copies of the tandemly repeated genes. From comparison of gene copy numbers from diploid cells and polytene chromosomes the following picture emerges:

1. The genes for 18S and 28S rRNA are underreplicated in polytene chromosomes compared to the bulk of the euchromatic DNA. While euchromatic sequences complete nine replication cycles, rRNA genes complete only seven, which causes a reduction to 30% of the amount expected in the case of normal polytenization (Hennig and Meer 1971; Spear and Gall 1973; Spear 1974; Lifschytz 1983).

2. The rDNA of only one of the two X chromosomes in females and only that of the X chromosome in males was found to polytenize (Endow and Glover 1979; Endow 1980). These data were obtained by using *Drosophila* strains carrying rDNA clusters with different spacers. Since rDNA clusters of *Drosophila* generally are heterogenous, the origin from one or the other homolog of an amplified sequence can hardly be recognized. Therefore, those results may have been overinterpreted by the authors. Further experiments are needed to answer the question whether rDNA genes of only one X chromosome are amplified, and if so, whether rDNA sequences of only the maternal or of the paternal X chromosome are involved in amplification.

3. Some gene families of the rDNA clusters seem to be favored for replication and others undergo no or only a few replication cycles (Endow and Glover 1979).

How the replication of rDNA during polytenization is regulated specifically is so far unknown. Its obvious consequence is a more or less constant number of rRNA genes in polytenized chromosomes independent of the number of nucleolus organizers in the diploid genome. Underreplication of rDNA during polytenization could be caused by surrounding heterochromatin. Selective amplification of specific rDNA families, on the other hand, requires specific regulation.

The genes for rRNA in *Chironomus tentans* are located in the euchromatic region of chromosome 2 and are proportionately represented in relation to the bulk of the genome (Hollenberg 1976). Therefore, Hollenberg (1976) suggested that the polytene chromosomes of salivary glands in *Chironomus* are formed by an equal number of replications of the entire genome.

From cytodensitometrical (Rudkin et al. 1956) and biochemical data, Beermann (1972b) estimated the average DNA content of individual chromomeres in polytene chromosome bands of *D. melanogaster* to be 30,000 base pairs (30 kilo bases, 30 kb) ranging from less than 5 kb to about 100 kb, and to be 2 kb in interchromomeres (interbands). The calculation of the values is based on the hypothetical chromosome model of continuous DNA fibers running along the chromosome. Values obtained from densitometric measurements of the total DNA mass of individual band-interband units (Rudkin et al. 1956; Rudkin 1969) were related to single chromomeres and interchromomeres on the assumption that all DNA fibers are completely and synchronously polytenized. An alternative chromosome model of disproportionate replication was created on the basis of DNA values obtained for bands and interbands by dry-mass determinations (Laird 1980) and of electron micrographed thin sections of whole-mounted polytene chromosomes (Sorsa 1974b). The authors state that in large chromosome bands there is a significantly higher level of polyteny than in small bands or interbands.

More precise analyses are now possible by testing restriction fragments of cloned DNA in Southern blots comparing DNA's from tissues of diploid and of polytene genomes. Equal amounts of genomic DNA's digested by restriction endonucleases are loaded on gels. After gel electrophoresis and Southern transfer to nitrocellulose filters, defined labeled DNA fragments are hybridized to the filter-bound DNA. Identical intensities of hybridization signals are interpreted as identical representation of the DNA sequences in the genomes. In addition to the variations of hybridization and densitometry, the most critical point of the procedure seems to be the exact determination of quantities of DNA's which have different qualities. Nevertheless, it should be possible to recognize a twofold difference in the quantities of DNA sequences. This method opens the way for precise molecular analyses of continuous long DNA sequences covering more than one chromosome band. Using this technique, Lifschytz (1983) and A. Spierer and P. Spierer (1984) analyzed the representation of various DNA fragments in polytene chromosomes of *Drosophila melanogaster*. In addition to the DNA sequences of heterochromatic regions a significant underreplication of DNA of the histone gene complex (Lifschytz 1983) and of the *bithorax* locus (Spierer 1984) was found in polytene chromosomes, but no significant differences in the level of polyteny were observed for 11 unique sequences and 3 multisite moderate repetitive DNA sequences (Lifschytz 1983), as well as for DNA which spans 315 kb of genomic DNA of the *rosy* locus (Bender et al. 1983a; Spierer 1984; Spierer and Spierer 1984). In the latter experiments A. Spierer and P. Spierer (1984) tested the hybridization efficiencies of 84 restriction fragments (each about 2 kb long) covering the chromosome region 87D5/6 to 87E5/6 (Fig. 2) which contains 15 chromosome bands. Very faint bands were found to contain about 3 kb, faint bands 7 kb and the largest band 87E1/2, a doublet according to Bridges' map (1941), 160 kb of the 315 kb DNA tested (Bender et al. 1983a; Spierer 1984). These values agree well with those measured and estimated by Rudkin et al. (1956) and Beermann (1972b). As was expected from the polytene chromosome model (Beermann 1972b; DuPraw and Rae 1966), all fragments showed identical hybridization efficiencies signifying equal levels of polyteny in interbands, faint bands, and even in the center of the largest band 87E1/2. Different amounts of DNA in interbands and bands in the *rosy* locus, therefore, must result from corresponding differences of DNA packing ratios which were estimated at 23 for very faint bands and at about 180 for the heavy band (Bender et al. 1983a; Spierer 1984). These results strictly contradict those of dry mass determinations of whole-mounted polytene chromosomes (Laird 1980) and of electron micrographs of thin sections of polytene chromosomes (Sorsa 1974b). It seems likely that this contradiction is not due to different chromosome regions analyzed in the different tests, but rather seems to reflect techniqual problems in dry mass determination and electron micrograph methods and/or interpretation of the data. Nevertheless, to prove different levels of polyteny within different bands, more detailed molecular analyses are needed to measure the abundance of specific DNA sequences covering more than one band. The hybridization experiments using cloned DNA allow direct testing of the abundance of each DNA fragment in the polytene chromosome, but one

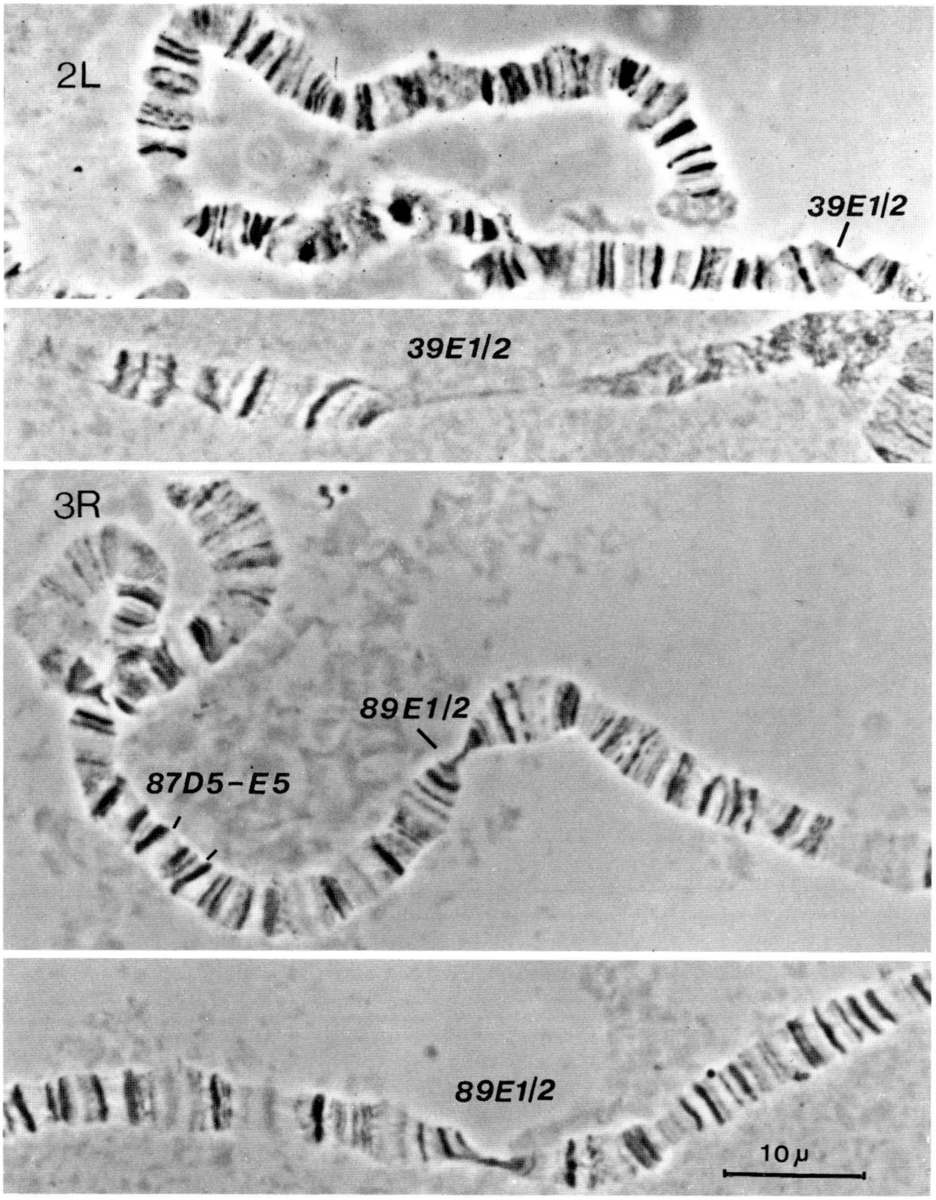

Fig. 2. Weak points in the polytene chromosomes 2L and 3R of *D. melanogaster* indicating under-replication. 39E1/2, the locus of histone gene clusters. 89E1/2, the locus of the *bithorax* gene complex. 87D5–E5, the *rosy* region which has been analyzed by Southern blotting, details see text. Phase contrast

should keep in mind that molecular analyses, compared to cytological observations, have the disadvantage of yielding mean values only and that the situation in various individual nuclei can be rather different. A detailed analysis of the same chromosome segments using both principally different techniques might help to clarify real situations.

Spierer's (1984) data for the *rosy* region indicate that for the majority of euchromatin Beermann's (1972b) model of polytene chromosomes can be accepted; however, disproportionate replication during polytenization seems to occur in euchromatic regions as well as in heterochromatin. Up to now the question for the reasons of underreplication cannot be answered, but it seems to be that underreplicated sequences either are constituents of heterochromatin or are located within or adjacent to heterochromatic sequences. This is obviously also relevant for the histone gene complex and for DNA of the *bithorax* locus. Both were found to lie in chromosomal weak point loci. The histone genes lie at the basis of chromosome arm 2 L in the region 39D3 through 39E1/2 (Pardue et al. 1977) close to the chromocenter (Fig. 2a). This locus very often shows ectopic pairing. The gene complex of *bithorax* is located in 89E1/2 of chromosome arm 3 R (Fig. 2b) (Lewis 1978; Spierer 1984; Spierer and Spierer 1984; Bender et al. 1983b).

As has been shown by Arcos-Terán and Beermann (1968; Arcos-Teran 1972), DNA of weak point loci in the X chromosome of *Drosophila melanogaster* exhibit late replication. This and ectopic pairing (Hennig et al. 1970; Ananiev et al. 1978) are characteristic properties of heterochromatin. Although heterochromatic sequences have not been demonstrated so far, at the histone and *bithorax* loci they may exist and cause underreplication of the neighboring sequences.

In some cases euchromatic regions which in chromosome rearrangements are translocated to chromocentric heterochromatin were reported to be underreplicated (Hartmann-Goldstein 1967; Ananiev and Gvozdev 1974; Hartmann-Goldstein and Cowell 1976; Cowell and Hartmann-Goldstein 1980), whereas in another case the heat shock gene locus 87C was found to be normally present even in the case of gene inactivation caused by position-effect variegation (Henikoff 1981). The techniques used in those experiments – microdensitometrical measuring of Feulgen-DNA content or silver grain counting in autoradiographs after ^{3}H-thymidin incorporation or in situ hybridization with labeled DNA, respectively – might not be sensitive enough to show slight alterations in replication levels.

In the translocation *T(1;4)*$w^{m258\text{-}11}$ (Cowell and Hartmann-Goldstein 1980) the translocated chromosome carries the gene *Sgs-4*, i.e., the gene for larval secretion protein 4 (Korge 1975), which is now juxtaposed to heterochromatin. Using cloned *Sgs-4* DNA for Southern blot analyses, the gene in the translocation polytene chromosome was found to be increasingly underreplicated, the lower the breeding temperature was (Kornher and Kauffman 1986). Underreplication was correlated with decreased gene expression and therefore seems to be the reason for position effect variegation, at least in this case.

So far, the replication behavior of chromosomal regions along the longitudinal chromosome axis has been investigated in complete polytene chromosomes or in chromosome rearrangements. To answer the question whether there exist

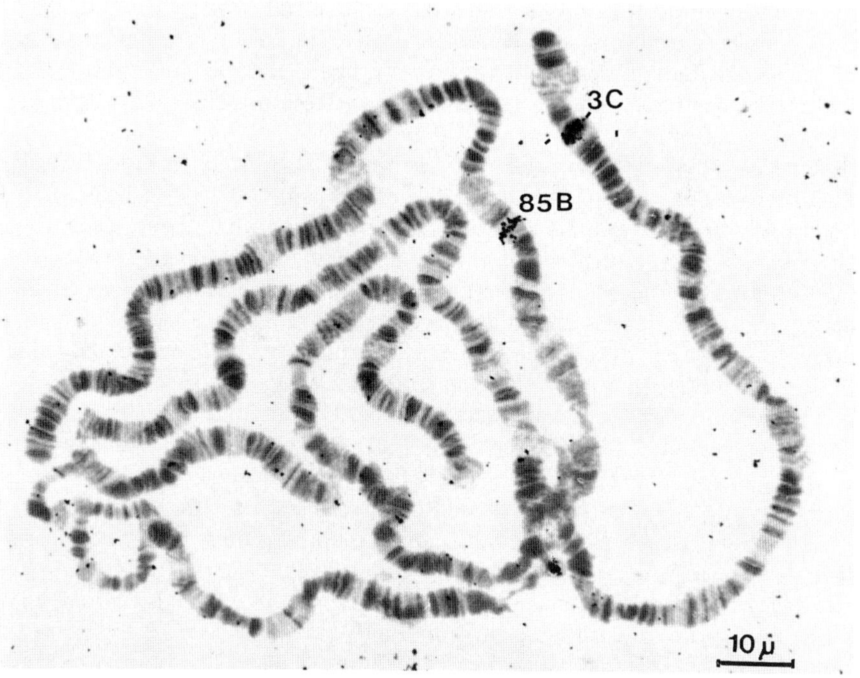

Fig. 3. Autoradiography of salivary gland chromosomes from the transformant K24-1 (Krumm et al. 1985) after in situ hybridization with ^{3}H-labeled *Sgs-4* DNA. 3C, the locus of the endogenous *Sgs-4* gene of the recipient strain. 85B, the integration locus of the transposed *Sgs-4* gene. Giemsa staining. DNA analysis by Southern blotting revealed identical polytenization levels for both genes

specific DNA sequences responsible for specific polytenization levels, defined DNA fragments should be tested with respect to their capability in regulating replication.

To identify such "replication-controlling elements" it is now possible to test normal and underreplicated DNA sequences by transposition with the help of the excellent P-element-mediated transformation system established by Allan Spradling and Gerald Rubin (Spradling and Rubin 1982; Rubin and Spradling 1982). In our laboratory we started to find out whether transposed *Sgs-4* genes replicate in transformed larvae as they do at their original locus, or whether they acquire the replicative properties of the chromosomal region into which they are inserted (Fig. 3). To avoid the methodical difficulties of determinations of DNA amounts mentioned above, we compared DNA fragments of transposed genes directly to different-sized fragments of endogenous genes in the same lane of Southern blots. Similar techniques were used in a recently published work (Kornher and Kauffman 1986). First results from transpositions into euchromatic regions indicate normal replication of the integrated *Sgs-4* fragment (Krumm et al. 1985). It will be interesting to test fragments which are integrated close to or into weak point loci, to see whether they acquire new replicative properties and therefore underreplicate.

The existence of replication-controlling elements is demanded, since DNA puffs in salivary glands of *Rhynchosciara* (Breuer and Pavan 1955) and *Sciara* (Gabrusewycz-Garcia 1964, 1971; Crouse and Keyl 1968) are known and DNA amplification in these puffs has been proven (Glover et al. 1982). Replication-controlling elements are now most clearly identified for chorion protein genes of *D. melanogaster*. Two clusters of these genes and their flanking sequences undergo differential amplification in the ovarian follicle cells (Spradling and Mahowald 1980; Spradling 1981) even in cases of transposition of DNA fragments containing the "amplification-controlling element", which is only 300 base pairs long in the case of the gene cluster in chromosome 3 (De Cicco and Spradling 1984; Levine and Spradling 1985), but amplification is reduced or fails completely in the inversion mutant *ocelliless* which has the breakpoint near the gene cluster (Spradling and Mahowald 1981).

2.2 Constancy of Banding Patterns

Polytene chromosomes exhibit species-specific banding patterns. Beermann (1952a, 1955, 1962, 1972b) discussed in detail all facts known at that time concerning "constancy of banding patterns" and tissue-specific variations. Apparent differences in banding of chromosomes from different tissues, like presence or absence of faint bands, singlet-doublet differences, differential spacing and tissue-specific changes in staining intensity of bands and interbands could be interpreted as being caused by functional modifications or changes in DNA contents or packing. Beermann stated that none of the limited data available at that time on banding differences were in conflict with his important generalization: "The number and sequence of bands and interbands are constant in all tested somatic tissues" (Beermann 1972b). This statement was well founded upon Beermann's own work on polytene chromosomes from different tissues of *Chironomus* (Beermann 1952a, b), and, with the acceptance of local variations (Beermann 1972b), still holds true although differences were observed between chromosomes from salivary glands and from Malpighian tubules in *Drosophila hydei* (Berendes 1966) and between chromosomes from salivary glands and fat body of *D. melanogaster* (Richards 1980).

A striking similarity in banding patterns also exists in polytene chromosomes from tissues as different as larval salivary glands and adult ovarian nurse cells of *Anopheles* (Coluzzi et al. 1970; Redfern 1981b; Mahmood and Sakai 1985) or larval salivary glands and pseudo nurse cells (PNC) in the female sterile mutant *fs231* of *D. melanogaster* (King 1970; King et al. 1981). Ovary nurse cells in *Anopheles* normally have well-structured polytene chromosomes, in *Drosophila* wild types, however, only oligotene chromatid bundles appear, which later during oogenesis separate, forming a reticular network of about 1024 chromatids (King 1970). In the female sterile mutant *fs231,* 10 to 12 cycles of DNA replications are performed. Therefore, the amount of DNA in PNC often is equal to or greater than that in wild-type nurse cells or in salivary gland cells (King et al. 1981).

Ovary nurse cells of *Calliphora* polyploidize similarly to those of *Drosophila* wild types, but in some inbred lines and under artificial breeding conditions well-

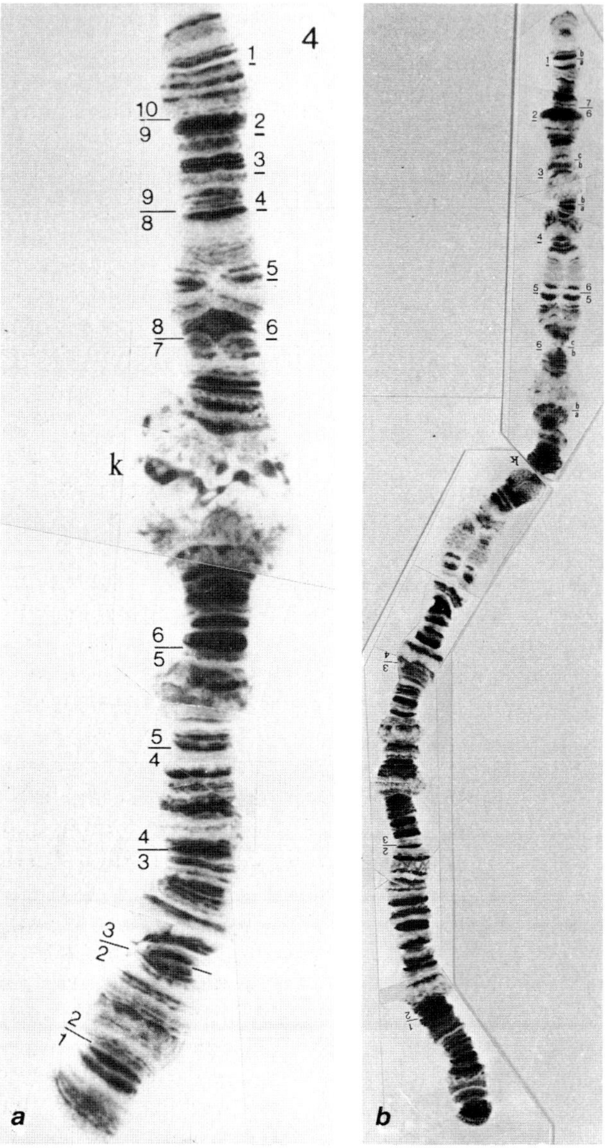

Fig. 4. Polytene chromosome 4 experimentally induced in nurse cells (**a**) and the homolog from pupal bristle forming cells (**b**) of *Calliphora erythrocephala*. (Ribbert 1979). The experimentally induced polytene chromosome shows uniform width and has no constrictions, unlike the homologous chromosome from bristle-forming cells

structured polytene chromosomes of 2^5–2^7 polyteny values arise (Fig. 4; Bier 1957, 1960; Ribbert 1979). Comparing polytene chromosomes of trichogen-forming cells (TC) (Ribbert 1967) with those induced in ovary nurse cells (NC) Ribbert (1979) made a surprising discovery: "while both chromosome complements exhibit a constant banding pattern, it is not possible to homologize the two tissue-specific patterns by identifying homologous band-sequences" (Fig. 4). Although it was not difficult to homologize polytene chromosomes from *Calliphora* footpad with those of *Calliphora* and *Lucilia* TC, "any attempt at homologizing chromosomes of NC and TC within the same species fails completely" (Ribbert 1979). The most striking differences between both types of chromosome are: the number of bands in TC chromosomes exceeds that of NC by a factor 1.7, and the characteristic properties "weak points", "constrictions" and "sites of ectopic pairing", which are typical for TC chromosomes and indicate underreplication, are totally absent in NC polytene chromosomes (Fig. 4). Since ovary nurse cells are germ line cells, the patterns and intensity of gene activity should very much differ from that of somatic cells. Puffing in NC chromosomes really was found "to be unusual in its extent as well as that it remains constant during long periods of oogenesis" and in this behavior NC polytene chromosomes resemble lampbrush chromosomes.

In a recent study of satellite DNA, Ribbert and Dover (in prep.) demonstrated by analytical centrifugation experiments that DNA from polyploid ovarian nurse cell nuclei (NC) of *Calliphora,* both with reticular chromatin arrangement and experimentally induced polytene chromosomes, shows the same satellite pattern as DNA from diploid genomes, and that certain satellites in somatic polytene chromosomes are underreplicated. The results strongly support the assumption that "weak points" in the somatic polytene chromosomes (TC) contain underreplicated repetitive DNA sequences which in ovary nurse cells are proportionately replicated, producing chromosomes of uniform diameter and lacking "constrictions", "weak points" and "sites of ectopic pairing". Proportional replication of highly repetitive DNA sequences (constituents of heterochromatin) and an altogether intensive transcriptional activity of genes in NC might explain some of the conspicuous morphological differences between NC and TC polytene chromosomes. The NC polytene chromosomes demonstrate that underreplication of heterochromatin is not an obligatory consequence of polytenization but is specific to the situation in endomitotic active nuclei of somatic tissues. Nevertheless, underreplication of satellite DNA does not include all satellite fractions of all species. So, for instance, AT-rich satellite DNAs of some Chironomids and in euchromatin of the *Drosophila virilis* group (Cohen and Bowman 1979) replicate during polytenization to the same extent as euchromatin (Walter 1973; Steinemann 1978; Schmidt 1980).

3 Gene Organization in Polytene Chromosomes

Constancy of banding patterns in polytene chromosomes stands for constancy in DNA packing and is variegated, as was shown above, by differential replication or modified by differential gene activation. Since polytene chromosomes have been known to be chromosomes, the question has arisen as to what the relationship between bands/interbands (chromomeres/interchromomeres) on the one hand and genetical units on the other hand might be. The species specificity of banding patterns and peculiar similarities of chromosome features between related species was a temptation to strictly correlate chromomere/interchromomere units to genetic units. Beermann (1972b) presented a detailed analysis of the problems concerning correlations between genes and chromomeres, and since that time some further reviews dealing with the same subject have been published (Lefevre 1974a, b; Sorsa 1974a, b, 1976; Zhimulev 1975a; Laird 1980; Lewin 1980; Zhimulev et al. 1983; Lefevre and Watkins 1986). The general questions to which an answer was sought are: (1) How many bands exist in polytene chromosomes or in defined chromosome divisions? (2) How many complementation groups or genes are located within an exactly determined number of band/interband units?

New techniques in electron microscope analyses of surface spread polytene chromosomes (Kalisch 1982; Kalisch et al. 1986) and of thin sections from whole-mounted chromosomes (Berendes 1970; Sorsa et al. 1984*; Zhimulev and Belayeva 1975a, Zhimulev et al. 1983; Ananiev and Barsky 1985) allow an insight into elementary structures of the chromosomes and in the numbers of bands. There is no doubt that Beermann's (1952a, 1962, 1972b) chromosome model of continuous chromatin fibers running along the longitudinal axis of the chromosomes is correct in general. Local modifications of underreplicated or possibly amplified DNA are mentioned above. Depending on the methods used, in particular with respect to the applied fixation and spreading technique, different numbers of chromosome bands have been counted. In comparison to the light microscope observation of polytene chromosomes, mostly more bands are found in electron microscope observations. So, for instance, in surface-spread chromosomes of *D. hydei*, the band number was found to be 34% higher than in light microscope observations (Kalisch et al. 1986) and in whole mount sections of an X chromosome division of *D. melanogaster* 8.5% more bands are revealed (Sorsa 1982) as compared with the light microscope map of Bridges (1938). For *D. melanogaster* the entire chromosome set was found by Bridges to contain at least 5000, bands which could be corrected by EM analyses to approximately 5500.

The genetic content and capacity of a chromomere/interchromomere (band/interband) unit can principally be investigated in two ways: (1) Genetically by determination of the number of complementation groups or genes contained in a morphological unit, (2) molecularly by using new DNA cloning and hybridization techniques.

1. The number of complementation groups which lies within specific chromomere/interchromomere units has been tried to be found by extensive saturation experiments. From earlier work reviewed by Beermann (1972b) and later by Lefevre (1974a, b) a one-to-one correspondence of polytene chromosome bands and complementation groups was strongly suggested especially by data obtained for X chromosome segments (Lefevre 1974b; Judd et al. 1972; Lefevre and Green 1972; Lifschytz and Falk 1968, 1969) and for chromosome 4 (Hochman 1971, 1973).

Saturation studies determine the number of independent loci in a limited interval containing a known number of bands. Mutations, particularly recessive lethal mutations induced by X-rays or by chemical mutagens such as EMS are the material for gene counting. Since saturation studies use mutants identified by their readily detectable effects on viability or phenotype, many other kinds of mutions producing no easily identified alteration obviously escape the usual screen for mutants. These critical points were the cause for a more detailed re-examination of the X chromosome segment 3A-3C7 (Young and Judd 1978). In this study complementation groups have been discovered which have no lethal alleles and are "without any genetically observable effect". The authors state "that division 3B contains more genes (i.e., complementation groups) than polytene chromosome bands, while portions of 3A and 3C seem to have no functional significance. Accordingly, many polytene chromosome bands may be composites of several complementing functional units. This investigation also indicates that there are chromosomal segments that are seemingly dispensible and thus function in a manner that is difficult or impossible to define with available methods" (Young and Judd 1978). Since then, however, further studies have been published supporting the one band-one gene correspondence and are cited and discussed in Lefevre and Watkins (1986), but for the region 9/10 of the X chromosome some bands were found to contain more than one gene (Zhimulev et al. 1983).

Most of the studies are based on the assumption that all genes mutate with equal probability following a Poisson distribution. That this might be incorrect has long been supposed (Morgan 1935) and was critically reviewed by Lefevre and Watkins (1986). Keeping in mind data of detailed complementation and gene-counting studies available at present, Lefevre and Watkins (1986) concluded: "Unfortunately, we cannot estimate the total numbers of nonvital loci, loci with undetectable phenotypes and loci having extremely low mutabilities. In any event, our estimate of the total vital gene number was far short of the total number of bands in the analyzed regions; yet, in several short intervals, we have found more vital genes than bands, in other intervals, fewer. We conclude that the one-band, one-gene hypothesis, in its literal sense, is not true; furthermore, it is difficult to support, even approximately".

2\. Earlier discussions about genetic capacities and organization of chromomeres and interchromomeres in polytene chromosomes are based on the assumption that genes on the average need about 1 kb DNA for protein coding, that a chromomere contains 30–40 kb DNA, an interchromomere 2 kb on average, and that only one gene is located in a chromomere/interchromomere unit (Crick 1971; Beermann 1972b; Paul 1972). The great excess of obviously noncoding sequences was assumed to fulfill structural and/or regulatory functions or to be composed of silent, duplicated genes or nonsense sequences. Rather different chromosome models have been created and are briefly given in the following: (1) Interbands contain coding sequences, bands contain control elements (Crick 1971). (2) Interbands contain "address sites" (polymerase binding sites), bands contain transcribed sequences (Paul 1972). (3) "The essential portion of each chromomere-interchromomere complex, at least in the case of heavier bands, is restricted to a small "initial" fraction at the edge of the band or, alternatively, only occupies the interband portion" (Beermann 1972b). (4) Interbands contain active housekeeping genes, bands contain temporarily or constantly inactive genes (Zhimulev et al. 1983).

Direct tests of gene contents of band/interband DNA have been possible since recombinant DNA techniques have been available. Using cloned overlapping DNA fragments wich cover more than one chromosome band, transcripts arising from the DNA at any time during development can principally be identified by DNA/RNA hybridization. Most extensive studies have been performed in this way for the *rosy* region 87D,E in chromosome 3 of *D. melanogaster*. The *rosy* locus, carrying the structural information for xanthine dehydrogenase (XDH) has been extensively studied genetically and cytogenetically (Chovnick et al. 1977; Hilliker et al. 1980). Within a segment containing 23 or 24 bands 21 complementation groups were observed which was interpreted as close correspondence between complementation group number and band number (Hilliker et al. 1980; Gausz et al. 1986). From a genomic DNA library clones containing a total of 315 kb DNA, which constitutes about 14 bands in polytene chromosomes of the *rosy* locus and contains essential sequences for at least 12 complementation groups (Bender et al. 1983 a; Spierer et al. 1983), were hybridized to polyadenylated RNA (Hall et al. 1983). From various developmental stages 20 discrete polyadenylated RNAs were found in total. The coding sequences of some genes were located within chromosome bands. Two large bands contained several transcription units (Hall et al. 1983). The reason for several transcripts arising from the same band could be that they indeed originate from several transcription units or that they originate from one and the same gene but result from different transcription starts, as in the case of alcohol dehydrogenase (Benyajati et al. 1983), or from different transcription termination or processing of the RNAs, as in several other cases (Henikoff et al. 1986 a, b; Schneuwly et al. 1986).

Further analyses are necessary to clarify the situation at the *rosy* region. Nevertheless, at first glance the *rosy* region seems to demonstrate a one-to-one correspondence between number of bands, number of complementation groups and number of transcripts. However, in addition to the exceptions mentioned above and to the already discussed difficulty of identifying complementation groups, one has to consider that only polyadenylated RNA was used in the hybridization tests. Moreover, the sensitivity of hybridization tests should be extremely high in order to ensure that RNA's are detected which originate in very low amounts from weakly expressed genes or from those which are expressed only in a small number of cells for a short time.

Discussing the distribution of genes in chromosome substructures one should also take into account the physical extensions of genes (Spradling and Rubin 1981).

a) Whereas the length of mature RNA sequences may range between 0.12 kb as for 5S rRNA (Artavanis-Tsakonas et al. 1977) and 7.3 kb as for *rudimentary* (Segraves et al. 1983), some unusually large genes like *Antennapedia* have exons of more than 100 kb in total (Schneuwly et al. 1986). The sizes of unprocessed transcription products differ enormously, depending on the numbers and the sizes of intervening sequences. Extremely large genes like those of the *bithorax* complex (Bender et al. 1983 b), the *Antennapedia* complex (Scott et al. 1983; Schneuwly et al. 1986) or of the *Notch* locus (Artavanis-Tsakonas et al. 1983; Kidd et al. 1983) contain introns of more than 60 kb long which amount to more than 100 kb in *Antennapedia* (Schneuwly et al. 1986). To date, such extremely large transcription units have been considered to be exceptions, but the number of these exceptions steadily increases with the number of genes analyzed. The size of entire functional units, i.e., transcription unit plus adjacent sequences of regulatory function, might also differ between genes. A rough estimate of the size of functional units with relatively small structural genes can be obtained from P-element-mediated transformation experiments (Spradling and Rubin 1982; Rubin and Spradling 1982). The lowest limits of functional DNA sizes are not definitely known, but mostly 3 to 10 kb DNA were sufficient for a proper function of the transposed genes.

b) The density of RNA coding sequences was found to be much higher than would have been expected from the one gene-one band hypothesis (Spradling and Rubin 1981). For the DOPA decarboxylase clusters Wright (this vol.) reported that within a region of 8 to 12 chromosome bands containing 160 kb DNA 18 complementation groups plus at least two unmutated transcriptional units were found and that two subclusters exist of nine transcriptional units within 25 kb and five within 15.5 kb DNA. In another case four genes for larval cuticle proteins are clustered within 7.9 kb DNA (Snyder et al. 1981) and the transcribed sequence of the gland-specific gene *gsg* and the secretion gene *Sgs-4* are separated only by less than 1 kb DNA (Hofmann and Korge, in press).

c) Different transcripts arise from the same gene, resulting from differences in transcription start or termination or RNA processing or polyadenylation (Benyajati et al. 1983; Henikoff et al. 1986a; Schneuwly et al. 1986). Different starts and terminations of transcription could yield more than one complementation group in saturation tests.

d) Different genes share the same enhancer or regulatory sequences. This seems to be true for the larval secretion protein genes *Sgs-7* and *Sgs-8* (Garfinkel et al. 1983) and may also exist in the case of the gland-specific gene *gsg* and the secretion protein gene *Sgs-4* (Hofmann and Korge, in press).

e) Genes are coded on opposite DNA strands at the same genetical locus or nearby, as is the case of *gsg* and *Sgs-4* (Hofmann and Korge, in press). In its extreme, a gene is situated in the intron of another gene, as is the case for a pupal cuticle protein gene at the GART locus (Henikoff et al. 1986b).

The presented examples of gene arrangements and sizes of functional units suggest that for some genes, perhaps for the majority, there might be space enough in bands as well as in interbands. However, large genes on the one hand must cover more than one chromosome band/interband unit (a), while in cases of densely arranged genes (b), especially those genes which share enhancers (d) or in extreme cases are interlocked (e), more than one cistron should occupy one band/interband unit. Therefore, one cannot but state that the genetic content of band/interband units can differ from unit to unit and, therefore, a generalization as is implied from the one band-one gene hypothesis is no longer justified.

More detailed information about the structural organization of specific DNA sequences in polytene chromosomes should be obtained by using combined methods of chromosome spreading or sectioning, in situ hybridization, and electron microscopic observation. So, for instance, Kress et al. (1985) used gold-labeled biotinylated DNA probes (Wu and Davidson 1981; Hutchinson et al. 1982; Langer-Sofer et al. 1982; Manuelidis et al. 1982) for in situ hybridization on surface-spread polytene chromosomes (Kalisch 1982). The electron microscope analysis yielded resolution of ca. 10 kb DNA and transcription sites of polyadenylated RNA could be localized in bands, interbands, and in puffs (Kress et al. 1985). Further development of these techniques and possibly also using whole mount sections of chromosomes (Sorsa et al. 1984*; Semeshin et al. 1986*) to increase resolution without loss of chromosome structure should allow localizing defined DNA sequences of regulatory or coding function in transcriptionally active and inactive chromosome sites.

4 Gene Activation-Related Chromosome Puffing

In his fundamental work on *Chironomus* polytene chromosomes Beermann (1952a) reported stage- and tissue-specific local modifications of the chromosome structure, i.e., inflated loci or chromosome puffs, in their extreme forms in Chironomids named Balbiani rings. Beermann interpreted puffs as functional modifications of the chromosome structure indicating high metabolic activity. Stage- and tissue-specificity of puff and Balbiani ring patterns were simultaneously found in *Rhynchosciara* (Pavan and Breuer 1952). During the following years, Beermann's observation that puffs and Balbiani rings represent loci of genes and are metabolically highly active has been corroborated for plants (Nagl 1981) as well as for Collembola (Cassagnau 1971) and Diptera, particularly *D. melanogaster,* by demonstrating stage- and tissue-specific puffing patterns (Becker 1959, 1962; Ashburner 1972*; Ashburner and Berendes 1978*), specific induction or repression of puffs or Balbiani rings (Beermann 1961, 1973; Clever 1963; Ashburner 1971; Kress 1973; Ashburner and Berendes 1978*), uridine incorporation (Pelling 1964) and correlation of puffs with the synthesis of specific proteins (Beermann 1961; Grossbach 1969, 1977*; Korge 1975).

Since chromosome puffs are known to be related to transcriptionally active genes and represent light-microscopically visible differential gene expression, this fascinating phenomenon has been used as a model for studying eukaryotic gene expression (Ashburner and Berendes 1978*). Among the factors which are involved in puff formation upon induction, the accumulation of acidic proteins seems to be the first process at the chromosomal level (Berendes 1968; Berendes and Helmsing 1974). Despiralization of DNA within the puff region as a precondition for RNA synthesis and the storage of newly synthesized RNA simultaneously covered with protein, are thought to be the main features of puffing (Beermann 1965). In his autoradiographic study Pelling (1964) stated "that RNA in puffed regions is synthesized in situ" and that "a general relationship exists between the rate of synthesis and the size of a puff". Uridine incorporation was only found in "structurally modified loci" (Pelling 1964; Beermann 1965). Puff formation and puff size, therefore, seem to be strictly correlated with RNA synthesis. RNA synthesis in puffs was evidenced by other workers too, but also unpuffed loci were found to be transcribed (Ashburner and Berendes 1978*). Zhimulev and

Belyaeva (1975b) even claimed that all interbands are transcriptionally active sites. Therefore, transcription in polytene chromosomes is not generally restricted to puffs, but is also correlated with chromatin decondensation in restricted areas of interbands.

Since puffing in wild-type chromosomes obviously always corresponds with intensive transcription of the puffed DNA, puffing could be a precondition for high-level transcription or a result of it. Data which seem to favor the second alternative came from P-element-mediated transformation experiments using hybrid genes containing the *hps70* heat shock gene promoter and various structural genes. If the complete *hsp70* promoter was used, fused genes were activated and puffs were induced on heat shock at the site of integration. The puffs appeared nd regressed at the same time as the puff at the endogenous *hsp70* locus and attained similar sizes (Lis et al. 1983). Using shortened *hsp70* 5′upstream sequences, i.e., deficient *hsp70* promoter sequences the activity level of the genes and the sizes of the puffs were reduced. If a fused gene remained inactive no puff was formed at the integration site (Bonner et al. 1984; Cohen and Meselson 1984; Dudler and Travers 1984; Simon et al. 1985). In addition to this general correlation between the level of gene activity and puff size, a dependence of puff sizes on the length of the transcripts was observed. An internal deletion of the fused gene that reduced the transcript size from 9 kb to 0.8 kb resulted in a dramatic reduction in puff size (Simon et al. 1985).

The result of the transformation experiments using heat shock promoter hybrid genes indicate that promoter strength or transcript length directly influence puffing, but the statement that "these results ... indicate that puff formation requires transcription" (Cohen and Meselson 1984) is not substantiated.

Further information about puffing was obtained from analyses of larval secretion protein gene expression. In *D. melanogaster* eight genes are known (*Sgs-1–8*) to code for larval secretion proteins. These genes are transcribed only in larval salivary glands from the middle of the third larval instar to about 5 h prior to prepupa formation (Korge 1975; Beckendorf and Kafatos 1976; Muskavitch and Hogness 1980; Velissariou and Ashburner 1980, 1981; Meyerowitz and Hogness 1982; Guild and Shore 1984). The *Sgs* gene activities are strictly correlated with puffing at the corresponding gene loci (Korge 1977a). In larvae which carry a deletion in the promoter region of the secretion gene *Sgs-4*, causing underproduction of sgs-4 mRNA and of protein, the *Sgs-4* gene locus remains unpuffed (Korge 1977b; Muskavitch and Hogness 1980). These results again indicate a clear correspondence between puffing and RNA synthesis, but the causality of the processes remains unclear. Uncoupling of transcription from puffing, particularly puffing without transcription, should help to answer the question.

First indications that transcription and puffing might be separable came from Berendes' experiments on *Drosophila hydei* which were performed in Beermanns' laboratory. Chromosome puffs could be induced at appropriate sites by heat shock or by ecdysone, althouth RNA synthesis was inhibited by actinomycin D treatment (Berendes 1968). The induced puffs, however, were smaller than normal. Even if very weak RNA synthesis could not be completely excluded, the data pointed to separability of puffing from transcription processes. RNA synthesis within the puff seems to be necessary for the maximal extension of a puff.

In the temperature-sensitive *suppressor of forked* mutant *l(1)su(f)*^{ts67g} of *D. melanogaster*, larvae fail to molt if the breeding temperature is shifted from permissive 25 °C to nonpermissive 30 °C, which is due to a failure in the rise of the concentration of the hormone ecdysone. If mutant larvae were reared at 25 °C until just before the molt of second to third instar larvae and then the breeding temperature was shifted to nonpermissive 30 °C, neither secretion proteins sgs-3,4,7, and 8 nor their corresponding mRNAs were synthesized, but, surprisingly, chromosome puffs were formed at the loci of the corresponding *Sgs* genes (Hansson et al. 1981; Hansson and Lambertson 1983). Injection of ecdysone rescued the larvae and induced secretion synthesis and prepupariation.

More detailed analyses were performed by Meyerowitz et al. (1985), using the recessive lethal mutant *l(1)npr-1* of *D. melanogaster*, and testing its influence on the expression of the *Sgs-3,7,8* gene cluster located at 68C of chromosome 3 (Meyerowitz and Hogness 1982). Larvae that are homozygous for the mutation *l(1)npr-1* never pupariate but die as third instar larvae as a result of a failure in the response to ecdysone (Kiss et al. 1978). In salivary glands of such mutant third instar larvae the *Sgs-3,7,8* gene cluster locus 68C is puffed, but as in *l(1)su(f)*^{ts67g} no concomitant accumulation of the corresponding RNAs is recognizable. After pulse-labeling of RNA no radioactivity was incorporated into the 68C saliva RNAs (Crowley et al. 1984; Meyerowitz et al. 1985).

These results again demonstrate uncoupling of puff formation and transcription. Therefore, in addition to the specific induction of transcription a specific induction of puffing or chromatin decondensation must be supposed. Using P-element-mediated transformation it should be possible to delimit and to identify regions which are responsible for the one or the other function.

In transformants in which the introduced 6 kb or 7 kb DNA included only *Sgs-3*, the largest of the three secretion protein genes of 68C, the transformed gene was abundantly expressed and appropriately regulated. Formation of a new puff at the site of insertion, however, was not observed (Richards et al. 1983; Meyerowitz et al. 1985; Crosby and Meyerowitz 1986). Electron microscope analysis of polytene chromosomes of transformants which contained an 8 kb DNA fragment harboring *Sgs-3* and *Sgs-7* (Richards et al. 1983) demonstrated wide interbands or small puffs at the integration sites (Semeshin et al. 1986). Only if the introduced DNA (8.1 kb) from 68C contained all three genes (*Sgs-3,7,8*) a light microscopically visible new intermolt puff at the chromosomal location of the insert was formed.

The analyses of *Sgs-3* expression in mutants and in transformants have shown clearly that puffing is neither necessary nor sufficient for abundant *Sgs-3* transcription, and that high level transcription is neither necessary nor sufficient for puffing. The results of these transformation experiments – *Sgs-3* transcription without puffing, and puff formation only if all three *Sgs* genes 3, 7, and 8 are transposed – could have several reasons: (1) The *Sgs-3* transcript is less than 1.2 kb in length (Crosby and Meyerowitz 1986) and could be too short to cause or stabilize puffing. (2) All three transcribing *Sgs* genes are needed to gain the number or total length of transcripts necessary for puffing. (3) The transposed DNA fragment needs a special size to retain its puff-forming capacity. (4) There is a specific DNA sequence responsible for puff formation. This DNA sequence could be located within or adjacent to the structural gene.

If the latter is true and if the puff-forming sequence solely is transformed, a puff should appear at the integration site provided that originally adjacent sequences are not needed for co-operation in puff formation, or that new surrounding sequences do not inhibit puff formation (position effect). Further transformation experiments are needed to decide what is true for the 68C puff.

In the case of the gene *Sgs-4* for larval secretion protein sgs-4 of *D. melanogaster* the above questions seem to be close to an answer. The gene *Sgs-4* lies in the X chromosome at position 3C11, and in larvae of most wild-type strains this locus is puffed during the sgs-4 RNA and protein synthesis phase (Fig. 5a; Korge 1975, 1977b). In some wild-type strains, however, neither sgs-4 RNA and protein are synthesized abundantly nor does a puff appear at 3C (Fig. 5b; Korge 1977b) due to a deletion in the 5′upstream region of *Sgs-4* (Muskavitch and Hogness 1980). Such a *Sgs-4* underproducer strain was used as a recipient for P-element-mediated transformation of DNA fragments containing 2.5 kb 5′upstream, 1.0 kb coding, and 1.3 kb 3′downstream *Sgs-4* sequences from a normal producer strain (Krumm et al. 1985). All transformants yielded significant, though low, *Sgs-4* expression, tested at the RNA and protein levels (Krumm et al. 1985). In some transformants puffs were formed at the integration site and regressed at the same time as the endogenous *Sgs* puffs at autosomal loci, when primary ecdysone-inducible puffs appeared (Fig. 6). The dimensions of the puffs differed between transformed lines, depending on the local situation at the integration site. The puffs seemed to be more extended if integration occurred into bands than into interbands.

In the presented example integration occured at 3A1–3 in the X chromosome (Fig. 6). This locus normally is unpuffed showing three or four compact thick DNA bands (doublets) during the phase of secretion synthesis in normal and in

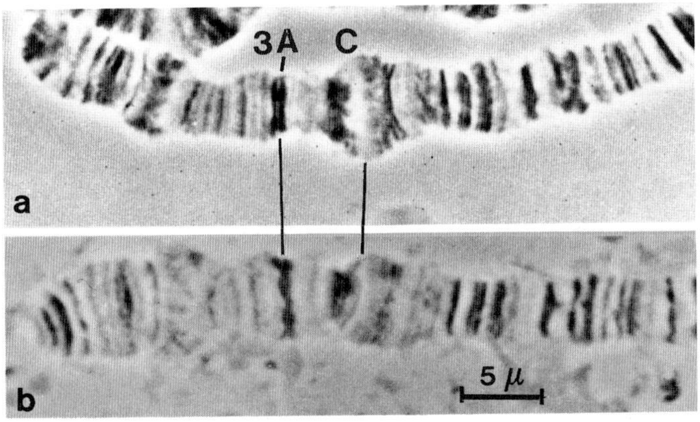

Fig. 5. Distal part of X chromosomes in female larvae of *D. melanogaster*. **a** wild-type Oregon-R. **b** the underproducer strain Kochi-R. 3C, the locus of the gene *Sgs-4* is puffed during the phase of secretion synthesis in Oregon-R but is unpuffed in Kochi-R. 3A1–4 is unpuffed in both strains. Phase contrast

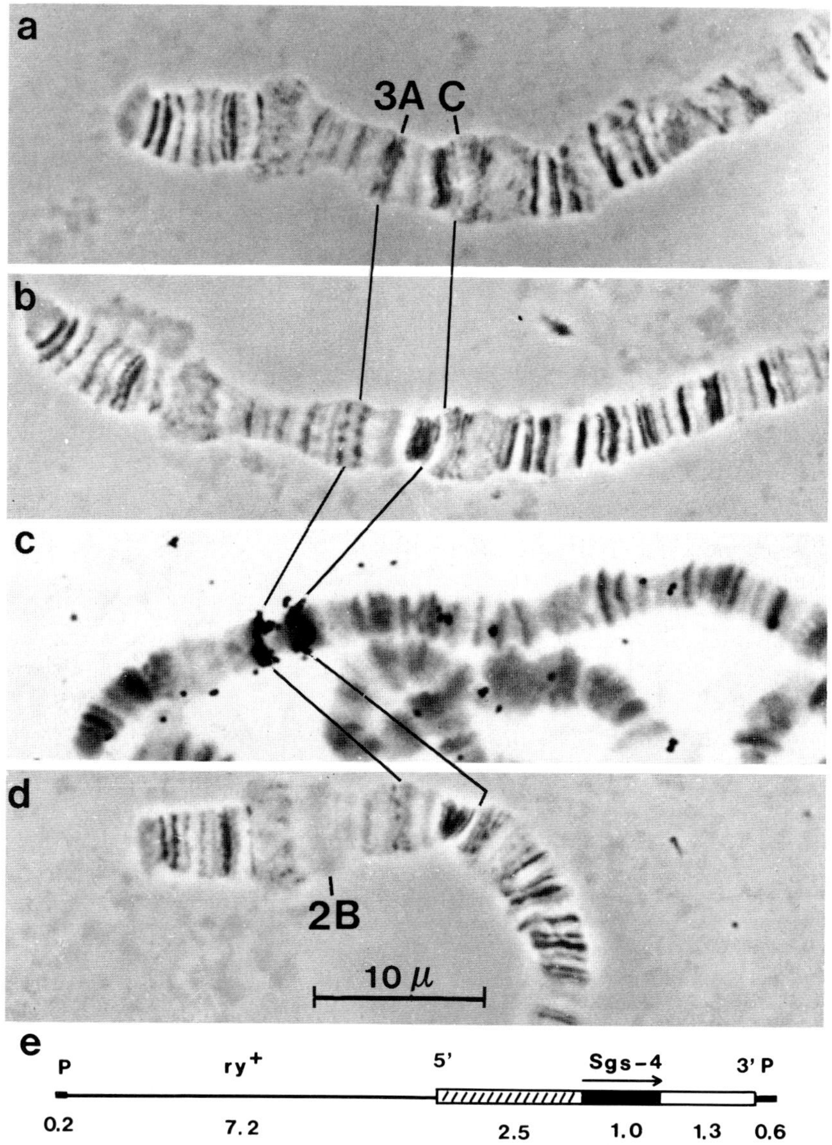

Fig. 6. Distal part of X chromosomes from female larvae of the *Sgs-4* underproducer recipient strain Kochi-R (**a**) and of the transformant K7-1 (**b–d**). *3A* the locus of DNA integration in K7-1. *3C* the unpuffed locus of the endogenous underproducing gene *Sgs-4* of the recipient. *3A* is unpuffed in **a**, but is puffed in **b** during the phase of secretion synthesis. **d** the puff in 3A is regressing and the ecdysone responsible puff at 2B appears. **c** autoradiogram after in situ hybridization with ^{3}H-labeled *Sgs-4* DNA. **a, b, d** phase contrast. **c** Giemsa staining. **e** the transposed DNA fragment. *P* fragment of the P-element. ry^+ sequence containing the $rosy^+$ gene for selection. Lengths of the fragments in kb are given below

Fig. 7. Monoclonal antibody staining of RNA polymerase II (ant pol II) (**a'**) and of a DNA binding protein (BX42) (**b', c'**) in polytene X chromosomes of the wild-type stock Oregon-R (**a, b**) and of the transformed line K7-1 (**c**). **a, b, c** Hoechst (DNA) staining. *3A* the locus of DNA in-

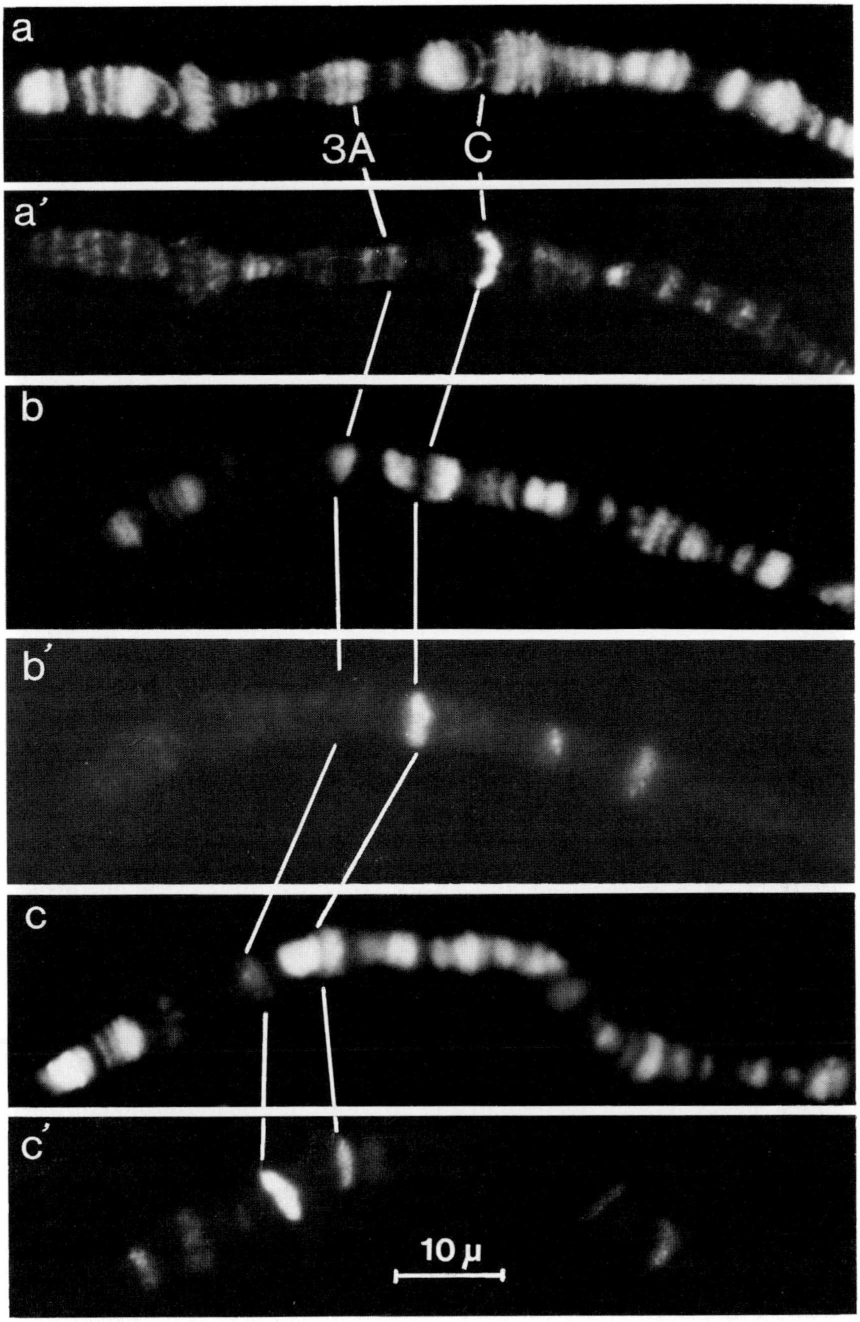

tegration in line K7-1 (**c**). *3C* the locus of the endogenous *Sgs-4* gene which has normal activity in the wild-type stock Oregon-R (**a, b**) and is weakly active in the recipient underproducer stock (**c**). Stage-specific ant pol II and BX42 staining in 3C of the wild-type stock (**a, b**), weak staining in 3C of the transformed line K7-1. No significant staining of 3A in the wild-type stock (**a, b**), strong and stage-specific staining of 3A with ant pol II (not shown) and BX42 in the line K7-1 (**c**) (Saumweber, unpublished) underproducer wild-type strains (Fig. 5). In the transformant the 3A1–3 locus is inflated or puffed and the chromosome bands 3A1 to 3A4 are decondensated (Fig. 6), and sgs-4 RNA and protein are produced (Krumm et al. 1985). Staining of RNA polymerase II in the polytene chromosomes with monoclonal antibodies (ant pol II) confirmed the results of RNA and protein analyses (Fig. 7). The endogenous *Sgs-4* locus of the underproducer recipient strain does not stain with ant pol II but at the integration site 3A1–3 ant pol II staining increases and decreases in correlation with secretion synthesis. A similar staining of DNA binding proteins was observed after applying the monoclonal antibody BX 42 (Frasch et al. 1985). In addition to some other loci these antibodies stained the locus 3C in wild-type chromosomes intensively only during the phase of secretion synthesis, but in the underproducer strains 3C failed to stain or stained very weakly (Fig. 7; Saumweber, unpublished).

In the underproducer strain, as well as in normal wild-type strains, the locus 3A1–3 neither stains with ant pol II nor with BX42 (Fig. 7). In the transformant the integration site 3A1–3 was intensively stained with both antibodies and its staining pattern was identical to that of the 3C locus in normal *Sgs-4* producer wild-type strains (Fig. 7). Monoclonal antibody staining, therefore, is an ideal cytological counterpart to biochemical RNA and protein determinations for identification of gene activities. A more detailed presentation of the data is in preparation.

The results of our transformation experiments, so far, differ from those of *Sgs-3* inasmuch as a relatively short DNA fragment of 4.8 kb exhibited gene activity-correlated puffing at the integration site but, as was found for *Sgs-3*, puffing still seems to correspond with transcription. The length of the transcript (1 kb) obviously is sufficient for puff formation.

For further transformation experiments a plasmid was constructed which contained 2.5 kb upstream sequences of *Sgs-4* connectec to the 1.3 kb coding sequence of the viral oncogene *mil* and a 0.7 kb fragment of the viral oncogene *myc* (Fig. 8), both originating from the avian virus Mill Hill 2 (Jansen et al. 1983). In transformants containing this DNA fragment RNA which might be derived from the *mil* or *myc* sequence was never detected. Accidentally, one of the transformants carries an integrated fragment at 3A1–3 (Fig. 8) close to the *Sgs-4* integration site described above (Fig. 6). This locus, surprisingly, exhibits conspicuous puffing strictly correlated with secretion synthesis, although only 2.5 kb upstream sequences of *Sgs-4* are present and no transcription of the linked viral coding sequences occurred. The transcriptional inactivity of the locus was also proven by ant pol II staining which in the puff was only at the back ground level

Fig. 8 a–f. Stage-specific chromatin decondensation induced by a transposed upstream sequence of *Sgs-4*. Distal parts of X chromosomes of female larvae of a transformed line from the stage of secretion synthesis (**a, c, d**) and from the stage after inhibition of secretion synthesis by ecdysone (**b, e**). **a, b** phase contrast. **c, d, e** autoradiography after in situ hybridization with ^{3}H-labeled upstream DNA of *Sgs-4*. *3A* the locus of DNA integration. *3C* the locus of the endogenous inactive *Sgs-4* gene of the recipient stock. The integration locus 3A is puffed only during the phase of secretion synthesis and stains with BX42 (not shown) but does not stain with ant pol II. The *mil* gene combined with the *Sgs-4* upstream sequence is not transcribed. *f* the transposed DNA fragment. *mil, myc,* avian virus oncogenes. Lengths of DNA sequences in kb are given below

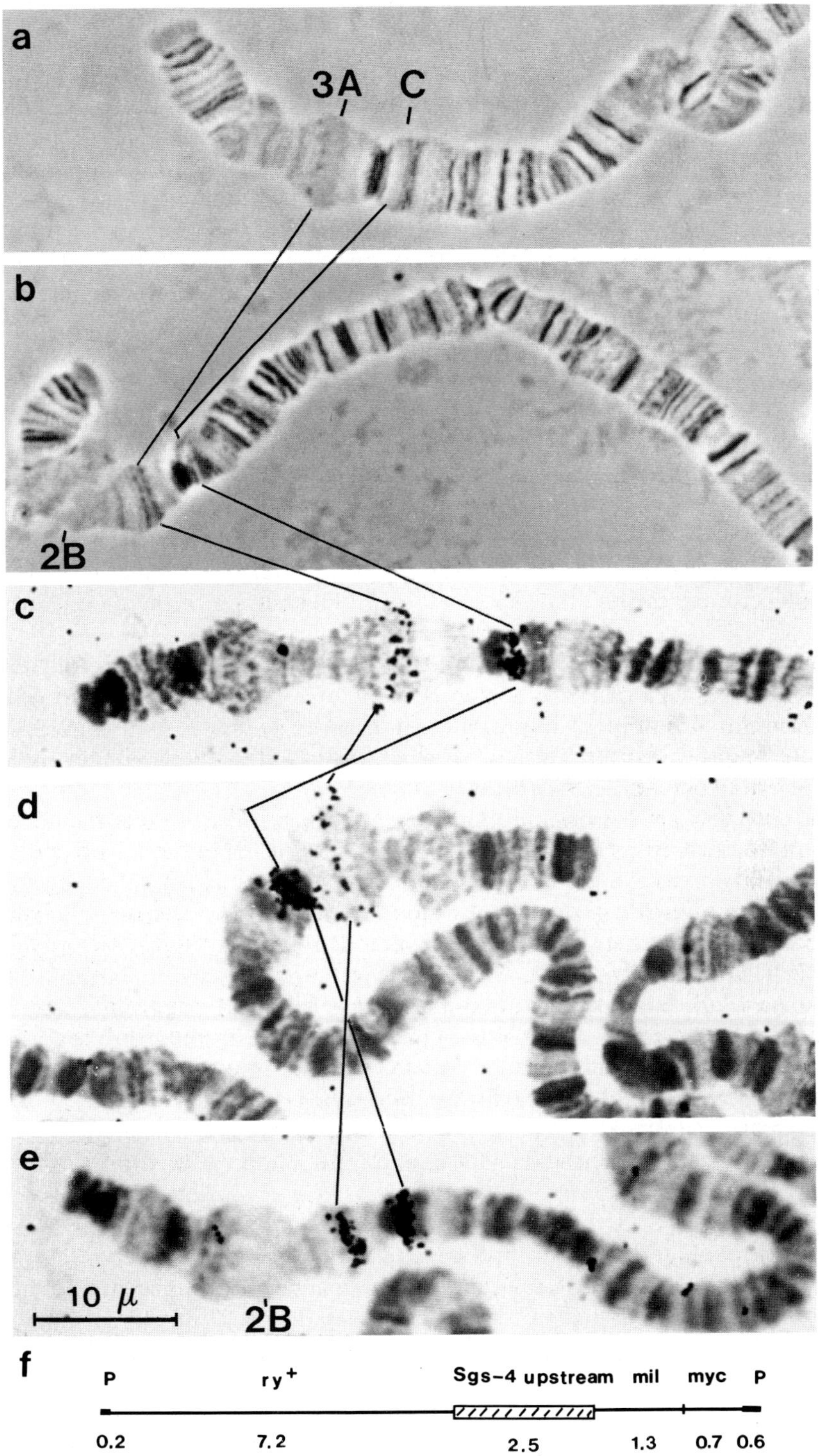
a
3A
C
b
2B
c
d
e
10 μ
2B
f
P
ry+
Sgs-4 upstream
mil
myc
P
0.2
7.2
2.5
1.3
0.7
0.6

Fig. 8a–f

(Saumweber, unpublished). In contrast, staining of a DNA-binding protein with the antibody BX42 was very intensive and stage-specific, as in the case of the original *Sgs-4* locus 3C in wild-type chromosomes (Saumweber, unpublished). This indicates specific binding of the antigen to BX42 to the 5′ upstream region of *Sgs-4*.

The results of our transformation experiments concerning puffing clearly demonstrate: (1) Puff formation, at least in the case of *Sgs-4*, does not depend on high level transcription or transcription at all within the puffed locus. (2) The 2.5 kb upstream sequence of *Sgs-4* contains the capacity and quality for stage-specific puff formation or chromatin decondensation. This upstream sequence seems to respond to a specific inducer of chromatin decondensation. The exact identification of this specific sequence is still open.

5 General Conclusions

Out of the wide field and the manifold facets of polytene chromosome research only a few aspects could be discussed in this review. On the basis of Beermann's work the attempt was made to focus on new points of view in chromosome structure and function-related morphological modifications. The polytene chromosome model of geometrically multiplied strictly paired chromatin fibers running along the longitudinal axis of the chromosomes is in general correct. So is, on the average, the one-to-one correspondence between the number of chromomere/interchromomere units and the number of genetic units within restricted chromosome regions.

The use of new molecular-biological techniques like DNA cloning, sequencing, in vitro recombination and hybridization, in combination with new cytogenetic and cytomorphological techniques like in situ hybridization and electron microscopic observation of chromosomes, opened new ways of chromosome research, and revealed several exceptions to the general chromosome model. So, for instance, within the heterochromatin, which was already known to be excluded from polytenization, sequences were found to be polytenized to various extents, and euchromatin was found to be interspersed with underreplicated sequences. Further enhancement of the sensitivity of techniques in chromatin analysis might yield increasing numbers of underreplicated and possibly amplificated sequences, as in the case of the DNA puffs in *Rhynchosciara*.

The genetic content and capacity of the structural units chromomere and interchromomere have been under discussion for some decades. Now, taking into account the new data about gene sizes and gene organization, one has to accept that the situation can differ between different chromomere/interchromomere units: extremely large and complex genes cover more than one unit, whereas in cases of a high density of small genes and "genes within genes", in particular, more genes occupy one structural unit. For the majority of the genes a one-to-one relationship between genes and chromomer/interchromomere units may exist.

Chromosome puffs and Balbiani rings are structural modifications of polytene chromosome loci which Beermann supposed to be morphological signals of high genetic activity. In wild-type chromosomes puffing always coincides with transcription in the puff in accordance with Beermann's interpretation; but in some mutants and under abnormal conditions, especially in cases of transposed in vitro recombinated genes, both processes, transcription and puffing, can be separated.

Altogether, polytene chromosomes represent dynamic structures, which in their features, although sometimes supposed to be exceptional, demonstrate characteristics of typical eukaryotic chromosomes. The many already known varieties in the structure of chromosomes and organization of genes complicate the design of a general chromosome model.

Acknowledgments. I wish to thank Drs. A. Hofmann, H. Kress and G. E. Roth for valuable discussion and help in preparation of the manuscript. I am also thankful to Drs. E. Meyerowitz, D. Ribbert and H. Saumweber who generously permitted me to cite their results before publication. I am indebted to Drs. P. Cassagnau, W. Nagl, V. Sorsa and J. F. Zhimulev for their kindness and help in providing me with data and photographic material of their work. The author's work, some of which is discussed here, is supported by the Deutsche Forschungsgemeinschaft (Ko 434/9-2).

References

Ammermann D (1964) Riesenchromosomen in der Makronukleusanlage des Ciliaten *Stylonychia spec.* Naturwissenschaften 51:249

Ammermann D, Steinbrück G, Berger L von, Hennig W (1974) The development of the macronucleus in the ciliated protozoan *Stylonychia mytilus.* Chromosoma 45:401–429

Ananiev EV, Barsky VE (1985) Elementary structures in polytene chromosomes of *Drosophila melanogaster.* Chromosoma 93:104–112

Ananiev EV, Gvozdev VA (1974) Changed pattern of transcription and replication in polytene chromosomes of *Drosophila melanogaster* resulting from eu-heterochromatin rearrangement. Chromosoma 45:173–191

Ananiev EV, Gvozdev VA, Ilyin YV, Tchurikov NA, Georgiev GP (1978) Reiterated genes with varying location in intercalary heterochromatin regions of *Drosophila melanogaster* polytene chromosomes. Chromosoma 70:1–18

Arcos-Terán L (1972) DNS-Replikation und die Natur der spät replizierenden Orte im X-Chromosom von *Drosophila melanogaster.* Chromosoma 37:233–296

Arcos-Terán L, Beermann W (1968) Changes of DNA replication behaviour associated with intragenic changes of the *white* region in *Drosophila melanogaster.* Chromosoma 25:377–391

Artavanis-Tsakonas S, Schedl P, Tschudi C, Pirotta V, Steward R, Gehring W (1977) The 5S genes of *Drosophila melanogaster.* Cell 12:1057–1067

Artavanis-Tsakonas S, Muskavitch MAT, Yedvobnick C (1983) Molecular cloning of *Notch,* a locus affecting neurogenesis in *Drosophila melanogaster.* Proc Natl Acad Sci USA 80:1977–1981

Ashburner M (1970) Function and structure of polytene chromosomes during insect development. Adv Insect Physiol 7:1–95

Ashburner M (1971) Induction of puffs in polytene chromosomes of in vitro cultured salivary glands in *Drosophila melanogaster* by ecdysone analogues. Nature New Biol 230:222–224

Ashburner M (1972) Puffing patterns in *Drosophila melanogaster* and related species. In: Beermann W (ed) Results and problems in cell differentiation, vol IV. Springer, Berlin Heidelberg New York, pp 102–151

Ashburner M, Berendes HD (1978) Puffing of polytene chromosomes. In: Ashburner M, Wright TRF (eds) The genetics and biology of *Drosophila,* vol 2b. Academic Press, London New York San Francisco, pp 316–395

Bauer H (1935) Der Aufbau der Chromosomen aus den Speicheldrüsen von *Chironomus thummi* Kiefer (Untersuchungen an den Riesenchromosomen der Dipteren I). Z Zellforsch Mikrosk Anat 23:280–313

Bauer H, Beermann W (1952) Die Polytänie der Riesenchromosomen. Chromosoma 4:630–648
Beckendorf SK, Kafatos FC (1976) Differentiation in the salivary glands of *Drosophila melanogaster:* characterization of the glue proteins and their developmental appearance. Cell 9:365–373
Becker HJ (1959) Die Puffs der Speicheldrüsenchromosomen von *Drosophila melanogaster* I. Beobachtungen zum Verhalten des Puffmusters im Normalstamm und bei zwei Mutanten, *giant* und *lethal-giant-larvae*. Chromosoma 10:654–678
Becker HJ (1962) Die Puffs der Speicheldrüsenchromosomen von *Drosophila melanogaster* II. Die Auslösung der Puffbildung, ihre Spezifität und ihre Beziehung zur Funktion der Ringdrüse. Chromosoma 13:341–384
Beermann W (1950) Chromomerenkonstanz bei *Chironomus*. Naturwissenschaften 37:543–544
Beermann W (1952a) Chromomerenkonstanz und spezifische Modifikationen der Chromosomenstruktur in der Entwicklung und Organdifferenzierung von *Chironomus tentans*. Chromosoma 5:139–198
Beermann W (1952b) Chromosomenstruktur und Zelldifferenzierung in der Speicheldrüse von *Trichocladius vitripennis*. Z Naturforsch 7b:237–242
Beermann W (1955) Cytologische Analyse eines *Camptochironomus* Artbastards. I. Kreuzungsergebnisse und die Evolution des Karyotyps. Chromosoma 7:198–259
Beermann W (1961) Ein Balbianiring als Locus einer Speicheldrüsenmutation. Chromosoma 12:1–25
Beermann W (1962) Riesenchromosomen. Protoplasmatologia VI
Beermann W (1965) Operative Gliederung der Chromosomen. Naturwissenschaften 52:365–375
Beermann W (1972a) Developmental Studies on Giant Chromosomes. In: Beermann W (ed) Results and problems in cell differentiation, vol IV. Springer, Berlin Heidelberg New York
Beermann W (1972b) Chromomeres and Genes. In: Beermann W (ed) Results and problems in cell differentiation, vol IV. Springer, Berlin Heidelberg New York, pp 1–33
Beermann W (1973) Directed changes in the pattern of Balbiani ring puffing in *Chironomus:* effects of a sugar treatment. Chromosoma 41:297–326
Bender W, Spierer P, Hogness DS (1983a) Chromosomal walking and jumping to isolate DNA from the *bithorax* complex and the *Ace* and *rosy* loci in *Drosophila melanogaster*. J Mol Biol 168:17–33
Bender W, Akam M, Karch F, Beachy PA, Peifer M, Spierer P, Lewis EB, Hogness DS (1983b) Molecular genetics of the *bithorax* complex in *Drosophila melanogaster*. Science 221:23–29
Benyajati Ch, Spoerel N, Haymerle H, Ashburner M (1983) The messenger RNA for alcohol dehydrogenase in *Drosophila melanogaster* differs in its 5 end in different developmental stages. Cell 33:125–133
Berendes HD (1966) Gene activities in the Malpighian tubules of *Drosophila hydei* at different developmental stages. J Exp Zool 162:209–218
Berendes HD (1968) Factors involved in the expression of gene activity in polytene chromosomes. Chromosoma 24:418–437
Berendes HD (1970) Polytene chromosome structure at the submicroscopic level. I. A map of region X, 1–4 E of *Drosophila melanogaster*. Chromosoma 29:118–130
Berendes HD, Helmsing PJ (1974) Non-histone proteins of Dipteran polytene nuclei. In: Cameron IL, Jeter JR Jr (eds) Acidic proteins of the nucleus. Academic Press, New York London
Berendes HD, Keyl HG (1967) Distribution of DNA in heterochromatin and euchromatin of polytene nuclei of *Drosophila hydei*. Genetics 57:1–13
Bier K (1957) Endomitose und Polytänie in den Nährzellkernen von *Calliphora erythrocephala*. Chromosoma 8:493–522
Bier K (1960) Der Karyotyp von *Calliphora erythrocephala* Meigen unter besonderer Berücksichtigung der Nährzellkernchromosomen im gebündelten und gepaarten Zustand. Chromosoma 11:335–364
Bonner JJ, Parks C, Parks-Thornburg J, Martin MA, Pelham HRB (1984) The use of promoter fusion in *Drosophila* genetics: isolation of mutants affecting the heat shock response. Cell 37:979–991

Botchan M, Kram R, Schmid CW, Hearst JE (1971) Isolation and chromosomal localization of highly repeated DNA sequences in *Drosophila melanogaster*. Proc Natl Acad Sci USA 68:1125–1129

Breuer ME, Pavan D (1955) Behavior of polytene chromosomes of *Rhynchosciara angelae* at different stages of larval development. Chromosoma 7:371–386

Bridges CB (1938) A revised map of the salivary gland X-chromosome of *Drosophila melanogaster*. J Hered 29:11–13

Bridges CB (1941) A revision of the salivary gland 3R-chromosome map of *Drosophila melanogaster*. J Hered 32:299–300

Cassagnau P (1971) Les chromosomes salivaires polytenes chez *Bilobella grassei* (Denis) (Collemboles: Neanuridae). Chromosoma 35:57–83

Cassagnau P (1976) La variabilite des chromosomes polytenes chez *Bilobella aurantiaca caroli* (Collembole Neanuridae) et ses rapports avec la biogeographie et l'ecologie de l'espece. Arch Zool Exp Gen 117:511–572

Chovnick A, Gelbart W, McCarron M (1977) Organization of the *rosy* locus in *Drosophila melanogaster*. Cell 11:1–10

Clever U (1963) Von der Ecdysonkonzentration abhängige Genaktivitätsmuster in den Speicheldrüsenchromosomen von *Chironomus tentans*. Dev Biol 6:73–98

Cohen EH, Bowman SC (1979) Detection and location of three simple sequence DNAs in polytene chromosomes from *virilis* group species of *Drosophila*. Chromosoma 73:327–355

Cohen RS, Meselson M (1984) Inducible transcription and puffing in *Drosophila melanogaster* transformed with *hsp70*-phage λ hybrid heat shock genes. Proc Natl Acad Sci USA 81:5509–5513

Coluzzi M, Cancrini G, DiDeco M (1970) The polytene chromosomes of *Anopheles superpictus* and relationship with *Anopheles stephensi*. Parasitologia 12:101–112

Cowell J, Hartmann-Goldstein IJ (1980) Contrasting response of euchromatin and heterochromatin to translocation in polytene chromosomes of *Drosophila melanogaster*. Chromosoma 79:329–340

Crick F (1971) General model for the chromosomes of higher organisms. Nature 234:25–27

Crosby MA, Meyerowitz EM (1986) *Drosophila* glue gene *Sgs-3:* Sequences required for puffing and transcriptional regulation. Dev Biol 118:593–607

Crouse HV, Keyl HG (1968) Extra replications in the "DNA puffs" of *Sciara coprophila*. Chromosoma 25:357–364

Crowley TE, Mathers PH, Meyerowitz EM (1984) A trans-acting regulatory product necessary for expression of the *Drosophila melanogaster* 68C glue gene cluster. Cell 39:149–156

DeCicco DV, Spradling AC (1984) Localization of a cis-acting element responsible for the developmentally regulated amplification of *Drosophila* chorion genes. Cell 38:45–55

Deharveng L (1982) Polymorphism of polytene chromosomes in *Bilobella aurantiaca* (Insecta: Collembola). Chromosoma 85:201–214

Dickson E, Boyd JB, Laird CD (1971) Sequence diversity of polytene chromosome DNA from *Drosophila hydei*. J Mol Biol 61:615–627

Dudler R, Travers AA (1984) Upstream elements necessary for optimal function of the hsp 70 promoter in transformed flies. Cell 38:391–398

DuPraw EJ, Rae PMM (1966) Polytene chromosome structure in relation to the folded fiber concept. Nature 212:598–600

Eissenberg JC, Cartwright JL, Thomas GH, Elgin SC (1985) Selected topics in chromatin structure. Annu Rev Genet 19:485–536

Endow SA (1980) On ribosomal gene compensation in *Drosophila*. Cell 22:149–155

Endow SA, Glover DM (1979) Differential replication of ribosomal gene repeats in polytene nuclei of *Drosophila*. Cell 17:597–605

Frasch M, Saumweber H, Symmons P, Dequin R (1985) Chromatin-associated antigens with specific distribution in embryonal tissues and on polytene chromosomes of *Drosophila melanogaster*. Hoppe-Seyler's Z Phys Chem 366:113–126

Gabrusewycz-Garcia N (1964) Cytological and autoradiographic studies in *Sciara coprophila* salivary gland chromosomes. Chromosoma 15:312–344

Gabrusewycz-Garcia N (1971) Studies in polytene chromosomes of *Sciarids*. I. The salivary chromosomes of *Sciara (Lycoriella) pauciseta*(II) Felt. Chromosoma 33:421–435

Gall JG, Cohen EH, Polan ML (1971) Repetitive DNA sequences in *Drosophila*. Chromosoma 33:319–344

Garfinkel MD, Pruitt RE, Meyerowitz EM (1983) DNA sequences, gene regulation and modular protein evolution in the *Drosophila* 68C glue gene cluster. J Mol Biol 168:765–789

Gausz J, Hall LMC, Spierer A, Spierer P (1986) Molecular genetics of the *rosy-Ace* region of *Drosophila melanogaster*. Genetics 112:65–78

Glover DM, Zaha A, Stocker AJ, Santelli RV, Pueyo MT, Toledo SM de, Lara FJS (1982) Gene amplification in *Rhynchosciara* salivary gland chromosomes. Proc Natl Acad Sci USA 79:2947–2591

Grossbach U (1969) Chromosomen-Aktivität und biochemische Zelldifferenzierung in den Speicheldrüsen von *Camptochironomus*. Chromosoma 28:136–187

Grossbach U (1977) The salivary gland of *Chironomus* (Diptera): A model system for the study of cell differentiation. In: Beermann W (ed) Biochemical differentiation in insect glands. Results and problems in cell differentiation, vol 8. Springer, Berlin Heidelberg New York, pp 147–196

Guild GM, Shore EM (1984) Larval salivary gland secretion proteins in *Drosophila:* identification and characterization of the *Sgs-5* structural gene. J Mol Biol 179:289–314

Hall CMC, Mason PJ, Spierer P (1983) Transcripts, genes and bands in 315,000 base-pairs of *Drosophila* DNA. J Mol Biol 169:83–96

Hansson L, Lambertsson A (1983) The role of su(f) gene function and ecdysterone in transcription of glue polypeptide mRNAs in *Drosophila melanogaster*. Mol Gen Genet 192:395–401

Hansson L, Lineruth K, Lambertsson A (1981) Effect of the $l(1)su(f)^{ts67g}$ mutation of *Drosophila melanogaster* on the glue protein synthesis. Wilhelm Roux Arch Dev Biol 190:308–312

Hartmann-Goldstein IJ (1967) On the relationship between heterochromatinization and variegation in *Drosophila* with special reference to temperature sensitive periods. Genet Res 10:143–159

Hartmann-Goldstein IJ, Cowell J (1976) Position effect on the DNA content of chromosome regions in *Drosophila*. In: Jones K, Brandhaus PE (eds) Current chromosome research. Elsevier, Amsterdam, pp 43–50

Hartmann-Goldstein IJ, Goldstein DJ (1979) Effect of temperature on morphology and DNA-content of polytene chromosomes in *Drosophila*. Chromosoma 71:333–346

Heitz E (1933) Die somatische Heteropyknose bei *Drosophila melanogaster* und ihre genetische Bedeutung. Z Zellforsch Mikrosk Anat 20:237–287

Heitz E (1934) Über α- und β-Heterochromatin, sowie Konstanz und Bau der Chromomeren bei *Drosophila*. Biol Zbl 54:588–609

Heitz E, Bauer H (1933) Beweise für die Chromosomen-Natur der Kernschleifen in den Knäuelkernen von *Bibio hortulanus*1. Z Zellforsch Mikrosk Anat 17:68–82

Henikoff S (1981) Position-effect variegation and chromosome structure of a heat shock puff in *Drosophila*. Chromosoma 83:381–393

Henikoff S, Keene MA, Sloan JS, Bleskan J, Hards R, Patterson D (1986a) Multiple purine pathway enzyme activities are encoded at a single genetic locus in *Drosophila*. Proc Natl Acad Sci USA 83:720–724

Henikoff S, Keene MA, Fechtel K, Fristrom JW (1986b) Gene within a gene: nested *Drosophila* genes encode unrelated proteins on opposite DNA strands. Cell 44:33–42

Hennig W (1972) Highly repetitive DNA sequences in the genome of *Drosophila hydei*. II. Occurence in polytene tissues. J Mol Biol 71:419–431

Hennig W, Meer B (1971) Reduced polyteny of ribosomal RNA cistrons in giant chromosomes of *Drosophila hydei*. Nature New Biol 233:70–72

Hennig W, Hennig J, Stein H (1970) Repeated sequences in the DNA of *Drosophila* and their localization in giant chromosomes. Chromosoma 32:31–63

Hilliker AJ, Clark SH, Chovnick A (1980) Cytogenetic analysis of the chromosomal region immediately adjacent to the *rosy* locus in *Drosophila melanogaster*. Genetics 95:95–110

Hinton T (1942) A comparative study of certain heterochromatic regions in the mitotic and salivary gland chromosomes of *Drosophila melanogaster*. Genetics 27:119–127

Hochman B (1971) Analysis of chromosome 4 in *Drosophila melanogaster*. II. Ethyl methane sulfonate induced lethals. Genetics 67:235–252

Hochmann B (1973) Analysis of a whole chromosome in *Drosophila*. Cold Spring Harbor Symp Quant Biol 38:581–589

Hofmann A, Korge G (1987) Upstream sequences of dosage compensated and non-compensated alleles of the larval secretion protein gene *Sgs-4* in *Drosophila*. Chromosome, in press

Hollenberg CP (1976) Proportionate representation of rDNA and Balbiani ring DNA in polytene chromosomes of *Chironomus tentans*. Chromosoma 57:185–197

Hutchinson NJ, Langer-Sofer PR, Ward DC, Hamkalo BA (1982) *In situ* hybridization at the electron microscopic level: hybrid detection by autoradiography and colloidal gold. J Cell Biol 95:609–618

Jansen HW, Patschinsky T, Bister K (1983) Avian oncovirus MH2: Molecular cloning of proviral DNA and structural analysis of viral RNA and protein. J Virol 48:61–73

Jones KW, Robertson FW (1970) Localization of reiterated nucleotide sequences in *Drosophila* and mouse by in situ hybridization of complementary RNA. Chromosoma 31:331–345

Judd BH, Shen MW, Kaufmann TC (1972) The anatomy and function of a segment of the X chromosome of *Drosophila melanogaster*. Genetics 71:139–156

Kalisch W-E (1982) EM chromosome mapping using surface-spread polytene chromosomes. Genetica 60:21–24

Kalisch W-E, Schwitalla G, Whitmore T (1986) Electron microscopic band-interband pattern of the X chromosome in *Drosophila hydei*. Chromosoma 93:387–392

Keyl H-G, Pelling C (1963) Differentielle DNS-Replikation in den Speicheldrüsen-Chromosomen von *Chironomus thummi*. Chromosoma 14:347–359

Kidd S, Lockett TJ, Young MW (1983) The *Notch* locus of *Drosophila melanogaster*. Cell 34:421–433

King RC (1970) Ovarian development in *Drosophila melanogaster*. Academic Press, London New York

King RC, Riley SF, Cassidy JD, White PE, Paik YK (1981) Giant polytene chromosomes from the ovaries of a *Drosophila* mutant. Science 212:441–443

King RL, Beams HW (1934) Somatic synapsis in *Chironomus*, with special reference to the individuality of the chromosomes. J Morphol 56:577–586

Kiss I, Szabad J, Major J (1978) Genetic and developmental analysis of puparium formation in *Drosophila*. Mol Gen Genet 164:77

Koltzoff N (1934) The structure of the chromosomes in the salivary glands of *Drosophila*. Science 80:312–313

Korge G (1975) Chromosome puff activity and protein synthesis in larval salivary glands of *Drosophila melanogaster*. Proc Natl Acad Sci USA 72:4550–4554

Korge G (1977a) Larval saliva in *Drosophila melanogaster:* production, composition and relationship to chromosome puffs. Dev Biol 58:339–355

Korge G (1977b) Direct correlation between a chromosome puff and the synthesis of a larval saliva protein in *Drosophila melanogaster*. Chromosoma 62:155–174

Kornher JS, Kauffmann SA (1986) Variegated expression of the *Sgs-4* locus in *Drosophila melanogaster*. Chromosoma 94:205–216

Kress H (1973) Specific repression of a puff in the salivary gland chromosomes of *Drosophila virilis* after injection of glucosamine. Chromosoma 40:379–386

Kress H, Meyerowitz EM, Davidson N (1985) High resolution mapping of in situ hybridized biotinylated DNA to surface-spread *Drosophila* polytene chromosomes. Chromosoma 93:113–122

Krumm A, Roth G, Korge G (1985) Transformation of salivary gland secretion protein gene *Sgs-4* in *Drosophila:* stage- and tissue-specific regulation, dosage compensation, and position effect. Proc Natl Acad Sci USA 82:5055–5059

Kurnick NB, Herskowitz JH (1952) The estimation of polyteny in *Drosophila* salivary gland nuclei based on determination of desoxyribonucleic acid content. J Cell Comp Physiol 39:281–300

Laird CD (1980) Structural paradox of polytene chromosomes. Cell 22:869–874

Langer-Sofer PR, Levine M, Ward DC (1982) Immunological method for mapping genes on *Drosophila* polytene chromosomes. Proc Natl Acad Sci USA 79:4381–4385

Lefevre GF (1974a) The relationship between genes and polytene chromosome bands. Annu Rev Genet 8:51–62

Lefevre GF (1974b) The one band-one gene hypothesis: Evidence from a cytogenetic analysis of mutant and nonmutant rearrangement breakpoints in *Drosophila melanogaster*. Cold Spring Harbor Symp Quant Biol 38:591–599

Lefevre GF, Green MM (1972) Genetic duplication in the white-split interval of the X chromosome in *Drosophila melanogaster*. Chromosoma 36:391–412

Lefevre GF, Watkins W (1986) The question of the total gene number in *Drosophila melanogaster*. Genetics 113:869–895

Levine J, Spradling A (1985) DNA sequence of a 3.8 kilobase region controlling *Drosophila* chorion gene amplification. Chromosoma 92:136–142

Lewin B (ed) (1980) Gene expression 2. Wiley & Sons, New York Chichester

Lewis EB (1978) A gene complex controlling segmentation in *Drosophila*. Nature 276:565–570

Lifschytz E (1983) Sequence replication and banding organization in the polytene chromosomes of *Drosophila melanogaster*. J Mol Biol 164:17–34

Lifschytz E, Falk R (1968) Fine structure analysis of a chromosome segment in *Drosophila melanogaster*. Analysis of X-ray induced lethals. Mutat Res 6:235–244

Lifschytz E, Falk R (1969) Fine structure analysis of a chromosome segment in *Drosophila melanogaster*. Analysis of ethyl methane sulfonate-induced lethals. Mutat Res 8:147–155

Lis JT, Simon JA, Sutton CA (1983) New heat shock puffs and β-galactosidase activity resulting from transformation of *Drosophila* with an hsp 70 – lac Z hybrid gene. Cell 35:403–410

Mahmood F, Sakai RK (1985) An ovarian chromosome map of *Anopheles stephensi*. Cytobios 43:79–86

Manuelidis L, Langer-Sofer PR, Ward DC (1982) High-resolution mapping of satellite DNA using biotin-labeled DNA probes. J Cell Biol 95:619–625

Meyerowitz EM, Hogness DS (1982) Molecular organization of a *Drosophila* puff site that responds to ecdysone. Cell 28:165–176

Meyerowitz EM, Crosby MA, Garfinkel MD, Martin CH, Mathers PH, Raghavan V (1985) The 68C glue puff of *Drosophila*. Cold Spring Harbor Symp Quant Biol 50:347–353

Morgan TH (1935) The relation of genetics to physiology and medicine. Sci Monthly 41:5–18

Mulder MP, Duijn P van, Gloor HJ (1968) The replicative organization of DNA in polytene chromosomes of *Drosophila hydei*. Genetica 39:385–428

Muskavitch MAT, Hogness DS (1980) Molecular analysis of a gene in a developmentally regulated puff of *Drosophila melanogaster*. Proc Natl Acad Sci USA 77:7362–7366

Nagl W (1962) 4096-ploidie und Riesenchromosomen im Suspensor von *Phaseolus collineus*. Naturwissenschaften 49:261–262

Nagl W (1974) The *Phaseolus* suspensor and its polytene chromosomes. Z Pflanzenphysiol 73:1–44

Nagl W (1981) Polytene chromosomes of plants. Int Rev Cytol 73:21–53

Painter TS (1933) A new method for the study of chromosome rearrangements and the plotting of chromosome maps. Science 78:585–586

Painter TS (1934) Salivary chromosomes and the attack on the gene. J Hered 25:456–476

Pardue ML, Gerbi SA, Eckhardt RA, Gall JG (1970) Cytological localization of DNA complementary to ribosomal RNA in polytene chromosomes of Diptera. Chromosoma 29:268–290

Pardue ML, Kedes LH, Weinberg ES, Birnstiel ML (1977) Localization of sequences coding for histone messenger RNA in the chromosomes of *Drosophila melanogaster*. Chromosoma 63:135–151

Paul J (1972) General theory of chromosome structure and gene activation in eukaryotes. Nature 238:444–446

Pavan C, Breuer ME (1952) Polytene chromosomes in different tissues of *Rhynchosciara*. J Hered 43:152–157

Peacock WJ, Lohe AR, Gerlach WL, Dunsmuir P, Dennis ES, Appels R (1978) Fine structure and evolution of DNA in heterochromatin. Cold Spring Harbor Symp Quant Biol 42:1121–1135

Pelling C (1964) Ribonukleinsäure-Synthese der Riesenchromosomen. Autoradiographische Untersuchungen an *Chironomus tentans*. Chromosoma 15:71–122

Prabhoo NR (1961) A note on giant chromosomes in the salivary glands of *Womersleya sp.* (Collembola, insecta). Bull Entomol 2:21–22

Rae P (1970) Chromosomal distribution of rapidly reannealing DNA in *Drosophila melanogaster*. Proc Natl Acad Sci USA 67:1018–1025
Redfern CPF (1981 a) Satellite DNA of *Anopheles stephensi* (Diptera: Culicidae): Chromosomal location and under-replication in polytene nuclei. Chromosoma 82:561–581
Redfern CPF (1981 b) Homologous banding patterns in the polytene chromosomes from the larval salivary glands and ovarian nurse cells of *Anopheles stephensi* Liston (Culicidae). Chromosoma 83:221–240
Renkawitz-Pohl R, Bialojan S (1984) A DNA sequence of *Drosophila melanogaster* with a differential telomeric distribution. Chromosoma 89:206–211
Ribbert D (1967) Die Polytänchromosomen der Borstenbildungszellen von *Calliphora erythrocephala* unter besonderer Berücksichtigung der geschlechtsgebundenen Heterozygotie und des Puffmusters während der Metamorphose. Chromosoma 21:296–344
Ribbert D (1979) Chromomeres and puffing in experimentally induced polytene chromosomes of *Calliphora erythrocephala*. Chromosoma 74:269–298
Ribbert D, Dover G (in prep) Reduplication of satellite DNA in experimentally induced polytene chromosomes of *Calliphora erythrocephala*
Richards G (1980) The polytene chromosomes in the fat body nuclei of *Drosophila melanogaster*. Chromosoma 79:241–250
Richards G, Caasab A, Bourouis M, Jarry B, Dissous C (1983) The normal developmental regulation of a cloned *Sgs-3* glue gene chromosomally integrated in *Drosophila melanogaster* by P element transformation. EMBO J 2:2137–2142
Rodman TC (1967) DNA replication in salivary gland nuclei of *Drosophila melanogaster* at successive larval and prepupal stages. Genetics 55:375–385
Rubin GM (1978) Isolation of a telomeric DNA sequence from *Drosophila melanogaster*. Cold Spring Harbor Symp Quant Biol 62:1041–1046
Rubin GM, Spradling AC (1982) Genetic transformation of *Drosophila* with transposable element vectors. Science 218:348–353
Rudkin GT (1969) Non replicating DNA in *Drosophila*. Genetics (Suppl) 61:227–238
Rudkin GT (1972) Replication in Polytene Chromosomes. In: Beermann W (ed) Results and problems in cell differentiation, vol IV. Springer, Berlin Heidelberg New York, pp 59–85
Rudkin GT, Schultz J (1961) Disproportional synthesis of DNA in polytene chromosome regions in *Drosophila melanogaster*. Genetics 46:893–894
Rudkin GT, Corlette SL, Schultz J (1956) The relations of the nucleic acid content in salivary gland chromosome bands. Genetics 41:657–658
Schmidt ER (1980) Two AT-rich satellite DNAs in the Chironomid *Glyptotendipes barbipes* (Staeger). Chromosoma 79:315–328
Schneuwly S, Kuroiwa A, Baumgartner P, Gehring WJ (1986) Structural organization and sequence of the homeotic gene *Antennapedia* of *Drosophila melanogaster*. EMBO J 5:733–739 (1986)
Scott MP, Weiner AJ, Hazelrigg TI, Polisky BA, Pirotta V, Scalenghe F, Kaufmann TC (1983) The molecular organization of the *Antennapedia* locus of *Drosophila*. Cell 35:763–776
Segraves WA, Louis C, Schedl P, Jarry BP (1983) Isolation of the *rudimentary* locus of *Drosophila melanogaster*. Mol Gen Genet 189:34–40
Semeshin VF, Belyaeva ES, Zhimulev IF, Lis JT, Richards G, Bourois M (1986) Electron microscopical analysis of *Drosophila* polytene chromosomes IV. Chromosoma 93:461–468
Simon JA, Sutton CA, Lobell RB, Glaser RL, Lis JT (1985) Determinants of heat shock-induced chromosome puffing. Cell 40:805–817
Snyder M, Hirsh J, Davidson N (1981) The cuticle genes of *Drosophila*: a developmentally regulated gene cluster. Cell 25:165–177 (1981)
Sorsa V (1974 a) Organization of chromomeres. Cold Spring Harbor Symp Quant Biol 38:601–608
Sorsa V (1974 b) Organization of replicative units in salivary gland chromosome bands. Hereditas 78:298–302
Sorsa V (1976) A cytological view of the functional organization of chromomeres. Hereditas 82:63–68
Sorsa V (1982) Electron microscopic map for the salivary gland chromosome X of *Drosophila melanogaster*. Divisions 1 5. In: Lakovaara S (ed) Advances in genetics, development and evolution of *Drosophila*. Plenum, New York London, pp 23–31

Sorsa V, Saura AO, Heino TJ (1984) Electron microscopic map of divisions 61, 62, and 63 of the salivary gland 3 L chromosome in *Drosophila melanogaster*. Chromosoma 90:177–184

Southern DJ, Pell PE, Craig-Cameron TA (1973) Polytene chromosomes of the Tsetse fly *Glossina morsitans morsitans*. Chromosoma 40:107–120

Spear BB (1974) The genes for ribosomal RNA in diploid and polytene chromosomes of *Drosophila melanogaster*. Chromosoma 48:159–179

Spear BB, Gall JG (1973) Independent control of ribosomal gene replication in polytene chromosomes of *Drosophila melanogaster*. Proc Natl Acad Sci USA 70:1359–1363

Spierer A, Spierer P (1984) Similar level of polyteny in bands and interbands of *Drosophila* giant chromosomes. Nature 307:176–178

Spierer P (1984) A molecular approach to chromosome organization. Dev Genet 4:333–339

Spierer P, Spierer A, Bender W, Hogness DS (1983) Molecular mapping of genetic and chromomeric units in *Drosophila melanogaster*. J Mol Biol 168:35–50

Spradling AC (1981) The organization and amplification of two chromosomal domains containing *Drosophila* chorion genes. Cell 27:193–201

Spradling AC, Mahowald AP (1980) Amplification of genes for chorion proteins during oogenesis in *Drosophila melanogaster*. Proc Natl Acad Sci USA 77:1096–1100

Spradling AC, Mahowald AP (1981) A chromosome inversion alters the pattern of specific DNA replication in *Drosophila* follicle cells. Cell 27:203–209

Spradling AC, Rubin GM (1981) *Drosophila* genome organization: conserved and dynamic aspects. Annu Rev Genet 15:219–264

Spradling AC, Rubin GM (1982) Transposition of cloned P elements into *Drosophila* germ line chromosomes. Science 218:341–347

Steinemann M (1978) Co-replication of satellite DNA of *Chironomus melanotus* with mainband DNA during polytenization. Chromosoma 66:127–139

Swift H (1962) Nucleic acids and cell morphology in dipteran salivary glands. In: Allen JM (ed) The molecular control of cellular activity. McGraw Hill, New York, pp 73–125

Swift H, Rasch EM (1954) Nucleoproteins in *Drosophila* polytene chromosomes. J Histochem Cytochem 2:456–457

Tschermak-Woess E (1963) Strukturtypen der Ruhekerne von Pflanzen und Tieren. Protoplasmatologia V:1

Velissariou V, Ashburner M (1980) The secretory proteins of the larval salivary glands of *Drosophila melanogaster*. Chromosoma 77:13–27

Velissariou Y, Ashburner M (1981) Cytogenetic and genetic mapping of a salivary gland secretion protein in *Drosophila melanogaster*. Chromosoma 84:173–185

Walter L (1973) Syntheseprozesse an den Riesenchromosomen von *Glyptotendipes*. Chromosoma 41:327–360

Welch RM (1956) A developmental analysis of the lethal mutant L(2)GL of *Drosophila melanogaster* based on cytophotometric determination of nuclear desoxyribonucleic acid (DNA) content. Genetics 42:544–559

Wu M, Davidson N (1981) Transmission electron microscopic method for gene mapping on polytene chromosomes by in situ hybridization. Proc Natl Acad Sci USA 78:7059–7063

Young BS, Pession A, Traverse KL, French C, Pardue ML (1983) Telomere regions in *Drosophila* share complex DNA sequences with pericentric heterochromatin. Cell 34:85–94

Young MW, Judd BH (1978) Nonessential sequences, genes and the polytene chromosome bands of *Drosophila melanogaster*. Genetics 88:723–742

Zacharias H (1979) Underreplication of a polytene chromosome arm in the Chironomid *Prodiamesa olivacea*. Chromosoma 72:23–51

Zhimulev JF, Belyaeva ES (1975a) Proposals to the problem of structural and functional organization of polytene chromosomes. Theor Appl Genet 45:335–340

Zhimulev JF, Belyaeva ES (1975b) ^{3}H-uridine labelling patterns in the *Drosophila melanogaster* salivary gland chromosomes X, 2R and 3L. Chromosoma 49:219–231

Zhimulev JF, Pokholkova GV, Bgatov AV, Semeshin VF, Umbetova GH, Belyaeva ES (1983) Genetic interpretation of polytene chromosomes banding pattern. Molec Biol Rep 9:19–23

Cited papers marked by an asterisk (*) in the text contain references to preceding publications of the same author(s).

Giant Chromosomes in Ciliates

D. AMMERMANN [1]

1 Introduction

Ciliates are characterized by the occurrence of two different cell nuclei. The DNA-rich *macronucleus* is the "somatic nucleus" of the cell and is responsible for almost the entire cellular RNA synthesis. However, during sexual reproduction, or conjugation, it is resorbed. The diploid *micronuclei* are the "germ line nuclei" of the cells. They are, at least in many species, dispensable during most of the cells' life cycle, which is characterized by asexual reproduction (division). During conjugation they undergo meiosis and form a synkaryon in each cell. One of its derivatives develops into a new macronucleus in the animals which, after conjugation, are now called exconjugants. During the development of the macronucleus anlage into a new macronucleus giant chromosomes are found in several species.

The history of the discovery of giant chromosomes is as curious in Ciliata as it is in Diptera. Balbiani (1890) was the first to identify them in *Loxophyllum meleagris,* and he already compared these structures with the dipteran giant chromosomes he had discovered some years earlier. We do not know, however, what Balbiani really observed. According to his figures (Fig. 1) the giant chromosomes are present in the macronuclei. Reexamination by Ruthmann (1963) and Ammermann (unpublished) revealed that no giant chromosomes are present in these nuclei. Until now giant chromosomes were only found as transient stages in macronucleus anlagen of exconjugants. Therefore it is probable that Balbiani saw the giant chromosomes in exconjugants, but erroneously thought that they are present in mature macronuclei. Because no one has yet succeeded in inducing conjugation in this species, this assumption cannot be proved.

Grassé (1952) mentioned the occurrence of giant chromosomes in *Nyctotherus cordiformis* (Fig. 2). This observation was later confirmed, and additional light and electron microscope studies were performed by Golikova (1964). However, many questions remained open, e.g., the further fate of the giant chromosomes and the molecular biology of the macronucleus. The reason is that this species is an endosymbiont in the intestine of tadpoles and frogs. Conjugation is rare and is apparently connected with the metamorphosis of its host, thus making it difficult to handle.

[1] Institut für Biologie III, Abt. Zellbiologie, Universität Tübingen, 7400 Tübingen, FRG.

Results and Problems in Cell Differentiation 14
Structure and Function of Eukaryotic Chromosomes
Edited by W. Hennig

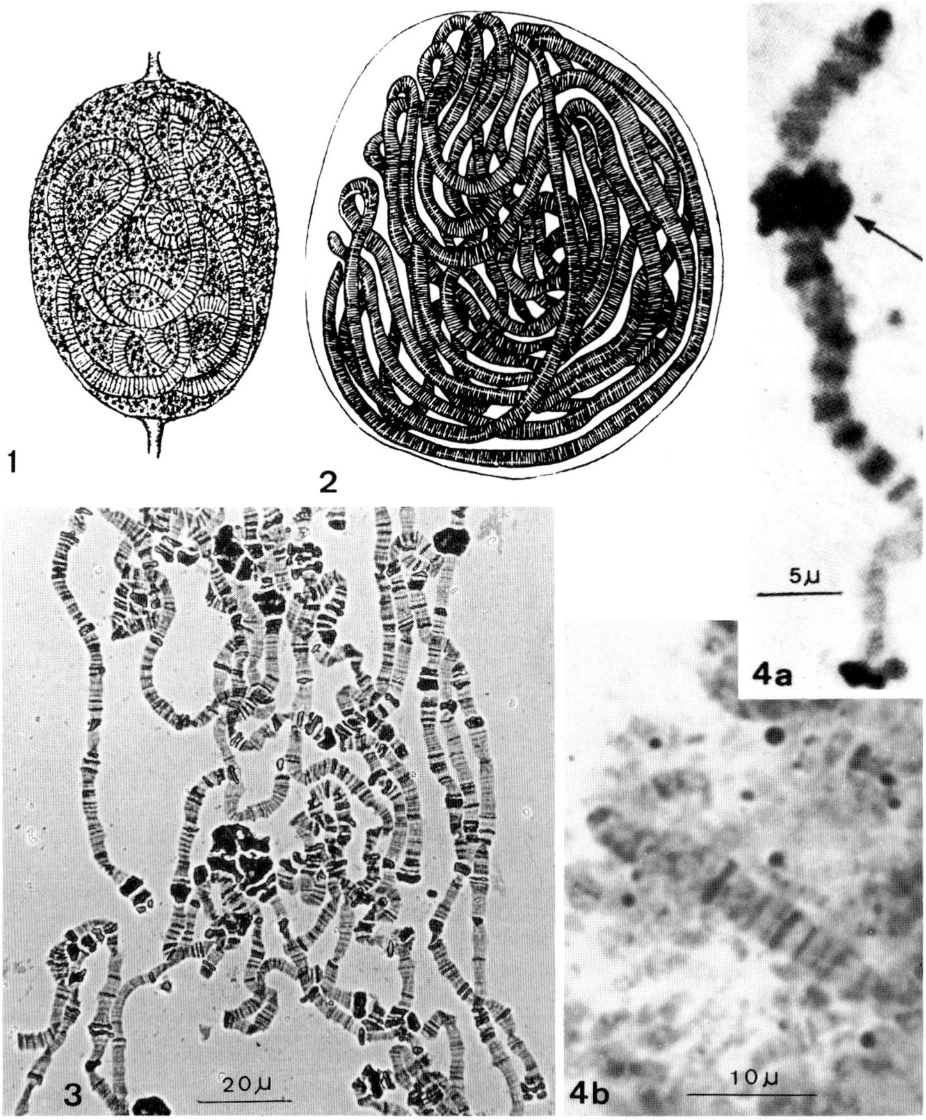

Figs. 1–4. Giant chromosomes of ciliates. **1** *Loxophyllum meleagris* (Balbiani 1890). **2** *Nyctotherus cordiformis* (Grassé 1952). **3** *Stylonychia lemnae*. **4a, b** *Chilodonella cucullulus*. (Radzikowski 1979 and unpublished)

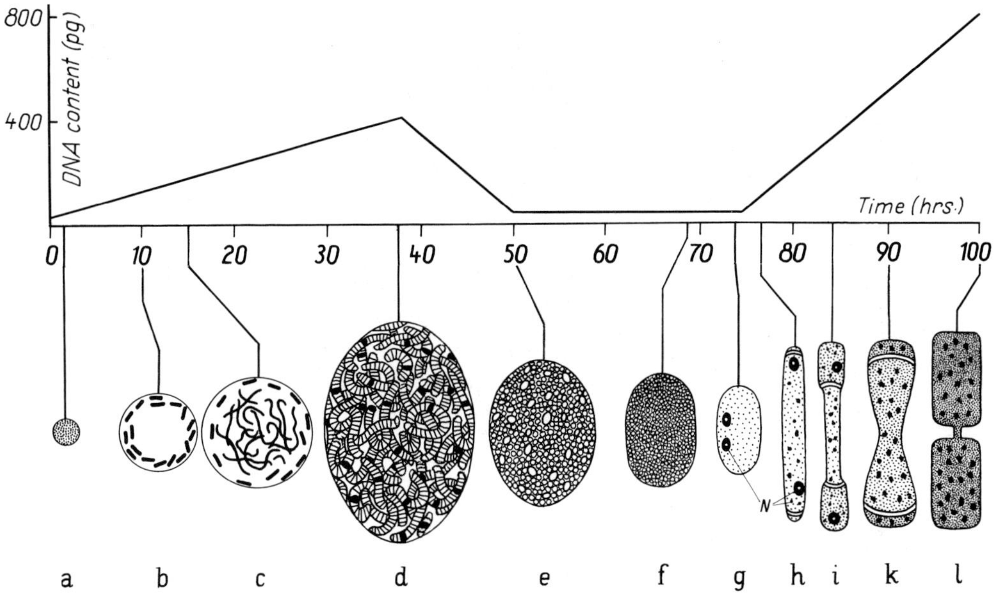

Fig. 5. The development of the macronucleus anlage of *Stylonychia lemnae*. (After Ammermann et al. 1974) *a* After separation of the exconjugant cells (0 h) the young macronucleus anlage has the DNA content and the size of a diploid micronucleus. The increase of the DNA content starts; *b* the anlage swells considerably and small condensed chromosomes become visible at the periphery of the anlage; *c* 20% of the chromosomes despiralize and begin to develop into giant chromosomes. The other (80%) condensed chromosomes are resorbed (first elimination step); *d* the DNA content reaches a first peak and the giant chromosomes show their maximum size at this time; *e, f* giant chromosomes are fragmented, their bands are enclosed into independent vesicles. The DNA content drops drastically; more than 90% of the anlagen DNA is eliminated (second elimination step). A DNA-poor stage follows; *g* the end of the DNA-poor stage is marked by the appearance of two nucleoli (*N*) in the anlage; *h–k* replication bands become visible which double the DNA five times (second DNA replication round). The anlage elongates. The number of nucleoli increases; *l* the anlage has reached its final DNA content and appearance

Most details are known about the giant chromosomes of hypotrich ciliates (Fig. 3). They were first discovered in *Stylonychia* (Ammermann 1965), and later found in several other genera (e.g., *Euplotes, Oxytricha;* review: Raikov 1982). Their development is shown in Fig. 5. Giant chromosomes were also found in *Chilodonella cucullulus* exconjugants (Radzikowski 1979; see Fig. 4).

In summarizing what is known about the giant chromosomes of ciliates, it becomes apparent that the basis for conclusions is very small. It is estimated that about 8000 species exist, but the macronucleus development of less than 20 species has been investigated. Furthermore, the species of several groups have never been studied (Table 1). This gives the following overview a very preliminary status.

Table 1. The classification of ciliates (Corliss 1979), the occurrence of giant chromosomes and the molecular weight of macronuclear DNA

	Giant chromosomes	Size of DNA
Class 1: *Kinetofragminophora*		
Subclass 1: Gymnostomata		
Loxophyllum	Yes [g]	– [a]
Didinium	No [c]	–
Subclass 2: Vestibulifera		
Bursaria	(Oligotene) [g]	High [d, b]
Subclass 3: Hypostomata		
Chilodonella	Yes [g]	High [e]
Subclass 4: Suctoria		
Ephelota	(Oligotene) [g]	–
Class 2: *Oligohymenophora*		
Subclass 1: Hymenostomata		
Paramecium	No [c]	High [f]
Tetrahymena	No [c]	High, but sub-chromosomal [f]
Subclass 2: Peritricha	–	–
Class 3: *Polyhymenophora*		
Subclass 1: Spirotricha		
Order 1: Heterotricha		
Nyctotherus	Yes [g]	–
Blepharisma	No [c]	High [d]
Climacostomum	No [c]	High [d]
Order 2: Odontostomatida	–	–
Order 3: Oligotricha		
Strombidium	–	High [d]
Order 4: Hypotrichida		
Keronopsis	No [g]	Gene-sized [d]
Stylonychia	Yes [g]	Gene-sized [f]
Oxytricha	Yes [g]	Gene-sized [f]
Euplotes	Yes [g]	Gene-sized [f]

[a] Nothing is known. [b] High: more than 20 kbp. [c] Review: Raikov 1982. [d] Steinbrück and Ammermann, unpublished. [e] Steinbrück and Radzikowski, unpublished. [f] Review: Steinbrück 1986. [g] Reference see text.

2 Occurrence of Giant Chromosomes

Giant chromosomes are found in some ciliate genera and species (see above), but are apparently lacking in others (e.g., *Tetrahymena, Paramecium,* see Table 1). Those species with giant chromosomes do not belong to one systematic group of the ciliates, but are scattered throughout the entire phylum (Table 1). Some related species differ in the occurrence of giant chromosomes. In *C. cucullulus* and *C. steini* quite large giant chromosomes are present (Fig. 4), but in *C. uncinata* no giant chromosomes can be observed with the light microscope. Only the electron microscope reveals the existence of oligotene chromosomes in the exconjugants of this species (Pyne 1978). *Keronopsis rubra* shows paired filament chromosomes

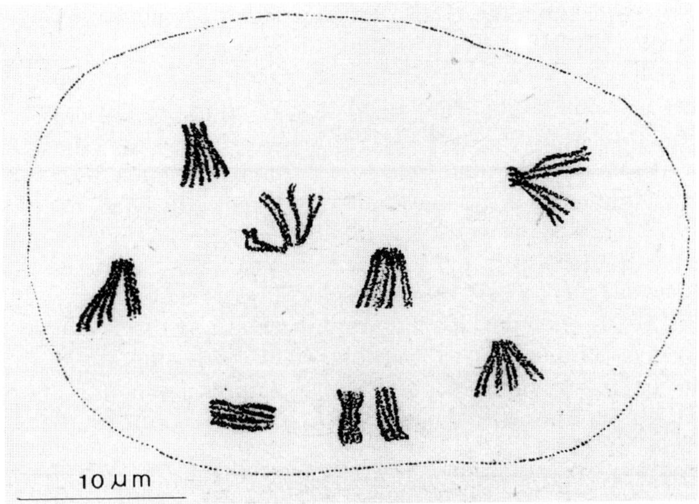

Fig. 6. Oligotene chromosomes in the macronucleus anlage of *Ephelota gemmipara*. (Grell 1949)

in its macronucleus anlagen, but no polytene chromosomes (Ruthmann 1972). At present, this species is the only *Hypotricha* species which lacks giant chromosomes.

The exconjugants of several species show transient stages between chromatids and giant chromosomes, e.g., "tetrads developed by endomitosis" (*Bursaria truncatella,* Poljansky 1934), or oligotene bundles (e.g., *Ephelota gemmipara*, Fig. 6).

It appears that there is no marked difference between species with and species without giant chromosomes. It seems probable that many or all species without visible giant chromosomes have comparable structures (paired chromatin fibrils) in their macronucleus anlage.

3 Chromatin Diminution (Elimination) in Ciliates

The giant chromosome stage in all ciliates is followed by a chromatin elimination stage. This diminution was either demonstrated using spectrophotometric methods (e.g., in several *Hypotricha,* see Fig. 5) or it is very likely in the light of the observed "Feulgen negative stages" and/or chromatin extrusion (several *Heterotricha* species, *N. cordiformis, C. cucullulus,* and *E. gemmipara*). In at least one species lacking giant chromosomes, diminution was also demonstrated spectrophotometrically (*P. bursaria,* Schwartz and Meister 1975). In other species also lacking giant chromosomes, no DNA loss could be demonstrated with spectrophotometric methods despite many efforts (e.g., *Tetrahymena* species, *Paramecium aurelia* species group, review: Raikov 1982). There is, however, no doubt that *T. thermophila* eliminates DNA sequences during macronucleus development (review: Gorovsky 1980). I think there is a convincing explanation for these discrepancies. It has been shown in recent years that the kinetic complexity (about 3×10^{10} Da) of the macronuclei of all ciliates which have been investigated is surprisingly similar (Review: Steinbrück 1986). The DNA content of the micronuclei is, however, quite different, and therefore different portions of the micronuclear genomes are eliminated. If the micronucleus is very large, e.g., in *S. lemnae* (C= 9.2 pg), then the DNA elimination is obvious even if the remaining DNA replicates at the same time (see Fig. 5). If the micronuclei are small, however, e.g., in *P. aurelia* species (C=0.32 pg, Sonneborn 1974) or *T. thermophila* (C=0.25 pg, Gaertig et al. 1986), then the small amount of eliminated DNA escapes detection especially if it is "hidden" by replication of the remaining DNA. In agreement with this, *P. bursaria* (micronuclei: C=3.8 pg, Ammermann, unpublished) clearly shows DNA elimination, while species of the *P. aurelia* group (C=0.32 pg, see above) do not.

There are several differences between the giant chromosomes of different ciliate species. *N. cordiformis* probably has one somatically paired, collective giant chromosome (Fig. 2) while *S. lemnae* has about 70 unpaired giant chromosomes (Ammermann et al. 1974) and *Oxytricha* sp. has about 120 giant chromosomes (Spear and Lauth 1976). In *S. lemnae* and *C. cucullulus* there are a few characteris-

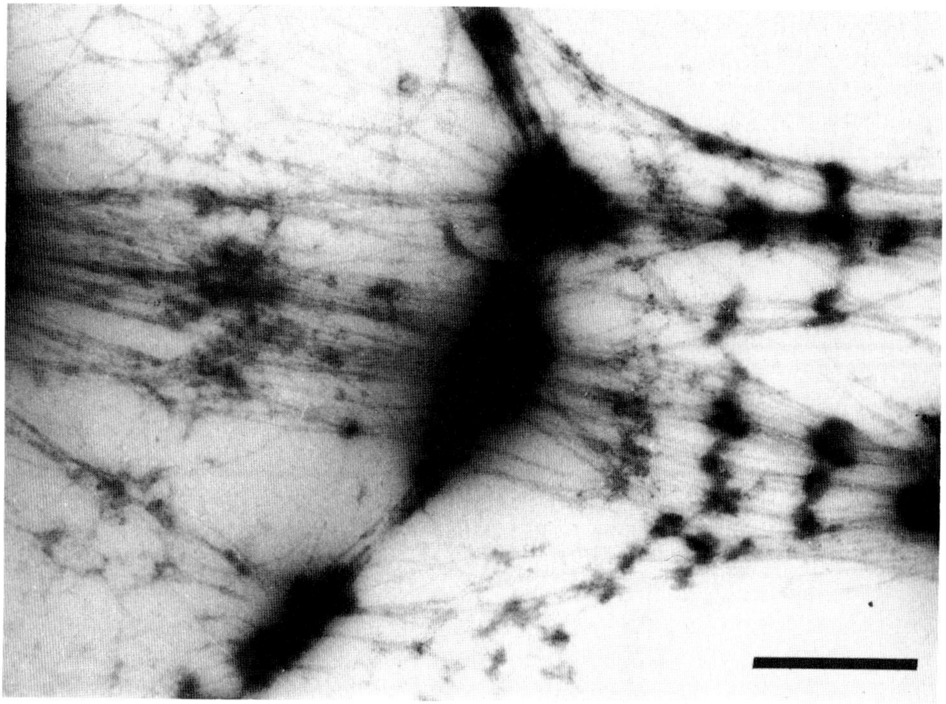

Fig. 7. Spread native polytene chromosomes of *Drosophila*. The organization into dense bands and individual chromatids in the interband regions is visible. Bar = 1 μm. (Meyer and Lipps 1984)

tically large, DNA-rich heterochromatic bands as well as numerous smaller bands. These large bands are lacking in other species. Nevertheless, the overall appearance of giant chromosomes of Ciliata and Diptera is rather similar under the light microscope. The number of bands in *S. lemnae* and *Oxytricha* sp. (about 10,000) is also in accordance with that of several Diptera.

In contrast to these findings are the results of electron microscope observations. Meyer and Lipps (1984) compared surface spread polytene chromosomes of *Drosophila* and *S. lemnae* and they observed clear differences. *Drosophila* giant chromosomes are composed of chromomeres of highly coiled DNA which are interconnected by interband fibers (10 nm) (Fig. 7). However, *S. lemnae* giant chromosomes show chromomeres which appear as aggregates of 30 nm loops laterally attached to 10 nm axial filaments which represent the interband sections of the chromatid (Fig. 8).

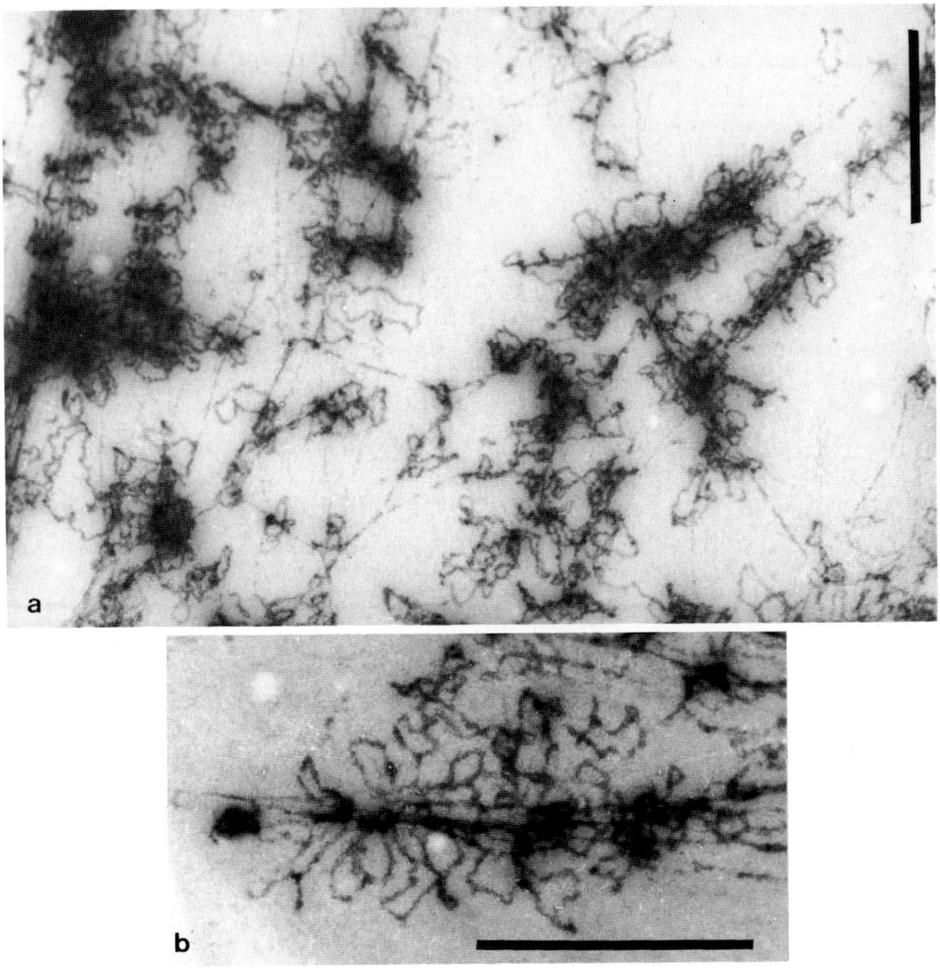

Fig. 8 a, b. Spread native polytene chromosomes of *S. lemnae*. Chromomeres are composed of 30 nm-loops attached to a 10 nm axial fiber representing the interband fiber. Bar = 1 μm. (Meyer and Lipps 1984)

Another important difference between the giant chromosomes of ciliates and diptera is their different activity. The giant chromosomes of all ciliates show no or at most a low RNA synthesis. RNA synthesizing puffs and nucleoli are absent. Transcription is apparently not the main function of the giant chromosomes in ciliates. But what other function could they have?

The investigations of all ciliates over the past years have shown clearly that during the development of the macronucleus anlage the micronuclear genome is more or less subjected to severe changes. Only a part of it (80% in *Tetrahymena thermophila,* less than 2% in *S. lemnae*) is retained in the macronucleus. This part, which derives from micronuclear chromosomal DNA, is cut during development into either still large, but subchromosomal-sized (e.g., *Glaucoma chattoni, T. thermophila*) or into short gene-sized DNA molecules (*Hypotricha*) (review: Steinbrück 1986). In addition, extensive DNA cutting and rejoining occurs. It can be supposed that the function of the giant chromosomes is related to these events.

I think the development of the *Hypotricha* giant chromosomes can give us some hints as to their role. The whole development shows two DNA replication phases (Fig. 5). The first one leads to polytene chromosomes with some remarkable DNA-rich blocks and many smaller bands. Following chromatin diminution, the second phase of DNA replication occurs. The course of this latter replication is the same as that which occurs prior to every cell division. The characteristic structures are the replication bands, which are assumed to duplicate the DNA exactly. I assume that the first replication round (in contrast to the second one) is characterized by over- and/or underreplication of DNA. There are three observations which support this hypothesis:

a) It was demonstrated that there is a shift in buoyant density of the DNA during the development from micronuclei to macronuclear anlagen with giant chromosomes (Ammermann et al. 1974; Spear and Lauth 1976).

b) At the end of giant chromosome development, an unequal DNA replication of certain bands is clearly demonstrable (Ammermann et al. 1974). The same observation was made in the giant chromosomes of *C. cucullulus.* Figure 4a shows a "DNA puff" of this species (Radzikowski 1979).

c) In a recent investigation Helftenbein (in preparation) found that in *S. lemnae* the relative frequencies of gene copies is the same in macronuclei and in macronucleus anlagen with giant chromosomes. If we assume that the diploid micronucleus contains two copies of each gene (this, however, remains to be proved), then the polytenization process seems to overreplicate some genes severely.

The giant chromosomes are (according to this hypothesis) a product of unequal replication. Their structure could enable the cells to sort out the germ line-restricted (= later eliminated) and the somatic (= macronuclear) DNA sequences. The fact that all copies of a gene are paired in a band could be important for this sorting out mechanism. Meyer and Lipps (1981) found that some sections of the giant chromosomes of *S. lemnae* are preserved (they assume: the 10 nm-interband fibrils), and others are degraded (they assume: the 30 nm loops in the bands, see Fig. 6). The giant chromosomes' degradation process is, however, far from being clear and makes further investigation necessary.

Despite their similar structure under the light microscope it seems obvious that the giant chromosomes of Ciliata and Diptera have completely different functions.

References

Ammermann D (1965) Cytologische und genetische Untersuchungen an dem Ciliaten Stylonychia mytilus Ehrenberg. Arch Protistenkd 108:109–152
Ammermann D, Steinbrück G, Berger L von, Hennig W (1974) The development of the macronucleus in the ciliated protozoan Stylonychia mytilus. Chromosoma 45:401–429
Balbiani EG (1890) Sur la structure intime du noyau du Loxophyllum meleagris. Zool Anz 13:110–115, 132–136
Corliss JO (1979) The ciliated protozoa: Characterization, classification and guide to the literature, 2nd edn. Pergamon, Oxford New York, 455 pp
Gaertig J, Seyfert H-M, Kaczanowski A (1986) Post-meiotic DNA synthesis in nocodazole-blocked nuclei during conjugation of Tetrahymena thermophila. Exp Cell Res 164:562–567
Golikova MN (1964) Der Aufbau des Kernapparates und die Verteilung der Nukleinsäuren und Proteine bei Nyctotherus cordiformis Stein. Arch Protistenkd 108:191–216
Gorovsky MA (1980) Genome organization and reorganization in Tetrahymena. Annu Rev Genet 14:203–239
Grassé PP (1952) Embranchement des Protozoaires. Generalités. In: Grassé PP (ed) Traité de Zoologie, vol I. Masson et Cie, Paris, pp 87–88
Grell KG (1949) Die Entwicklung der Makronucleusanlage im Exkonjuganten von Ephelota gemmipara R. Hertwig. Biol Zentralbl 68:289–312
Meyer GF, Lipps HJ (1981) The formation of polytene chromosomes during macronuclear development of the hypotrichous ciliate Stylonychia mytilus. Chromosoma 82:309–314
Meyer GF, Lipps H-J (1984) Electron microscopy of surface spread polytene chromosomes of Drosophila and Stylonychia. Chromosoma 89:107–110
Poljansky G (1934) Geschlechtsprozesse bei Bursaria truncatella O. F. Müll. Arch Protistenkd 81:420–546
Pyne CK (1978) Electron microscopic studies on the macronuclear development in the ciliate Chilodonella uncinata. Cytobiologie 18:145–160
Radzikowski S (1979) Asynchronous replication of polytene chromosome segments of the new macronucleus anlage in Chilodonella cucullulus O. F. Müller. Protistologica 15:521–526
Raikov IB (1982) The protozoan nucleus. Springer, Berlin Heidelberg New York
Ruthmann A (1963) Die Struktur des Chromatins und die Verteilung der Ribonucleinsäure im Makronucleus von Loxophyllum meleagris. Arch Protistenkd 106:422–436
Ruthmann A (1972) Division and formation of the macronuclei of Keronopsis rubra. J Protozool 19:661–666
Schwartz V, Meister H (1975) Die Extinktion der feulgengefärbten Makronucleusanlage von Paramecium bursaria in der DNS-armen Phase. Arch Protistenkd 117:60–64
Sonneborn TM (1974) Paramecium aurelia. In: King RC (ed) Handbook of genetics, vol 2. Plenum, New York, pp 469–594
Spear BB, Lauth MR (1976) Polytene chromosomes of Oxytricha: biochemical and morphological changes during macronuclear development in a ciliated protozoan. Chromosoma 54:1–13
Steinbrück G (1986) Molecular reorganization during nuclear differentiation in ciliates. In: Hennig W (ed) Results and problems in cell differentiation, vol 13. Springer, Berlin Heidelberg New York

The sp-I Genes in the Balbiani Rings of *Chironomus* Salivary Glands

C. Grond [1], H. Saiga [1,2] and J.-E. Edström [1,3]

1 Introduction

The polytene chromosomes of dipteran insects like *Drosophila* or *Chironomus* show a characteristic pattern of bands and interbands. With few exceptions, this pattern is constant, even for cells of different tissues as for instance salivary glands, Malpighian tubules or rectum epithelium (Beermann 1952). However, the pattern of diffuse regions, puffs and Balbiani rings (BRs) along the polytene chromosomes changes according to tissue and developmental stage, which indicates differentiated functions for these structures (Beermann 1962).

The large BRs of the *Chironomus* salivary gland cells are suitable for the study of functional structures in polytene chromosomes. Early electron-microscopical studies (Beermann and Bahr 1954) demonstrated the high amount of putative RNP granules in the BRs. Pelling (1964) showed that BRs are sites of high transcriptional activity. The tissue-specific development of BRs suggested a role in the production of the gland secretion. Indeed, Beermann (1961) demonstrated that granule-like secretion in so-called special cells in *C. pallidivittatus* × *C. tentans* hybrids was inherited as a consequence of the presence of a short chromosome region from *C. pallidivittatus* containing a BR4 that is not present in *C. tentans*. Grossbach (1969, 1973), using a similar design, could subsequently identify a secretory protein component whose presence is correlated to the presence of BR4. Grossbach (1973) also showed that the secretion contains a prominent, very high molecular weight secretory protein, later designated sp-I, a likely product of the 75S RNA originating from BR2 (and BR1) (Daneholt 1972; Daneholt and Hosick 1973). BRs, like the secretory process, occur throughout the larval existence. The secretion proteins form a tube (Fig. 1) in which the larva usually stays, well hidden through the incorporation of bottom material into the tube walls. Probably this tube has also other functions for the larva, but these still remain a matter of speculation.

Case and Daneholt (1978) showed that RNA of BR2 (and BR1) has an extraordinarily large size of approximately 37 kb (75S RNA). This RNA fraction is not measurably smaller as mRNA in the cytoplasm. The sp-I fraction in the sal-

[1] Department of Genetics, University of Lund, Sölvegatan 29, 22362 Lund, Sweden.
[2] Present address: Department of Molecular Biology, University of Occupational and Environmental Health, 1-1 Iseigaoka, Kitakyushu, 807 Japan.
[3] Correspondence to: J.-E. Edström, Department of Genetics, Sölvegatan 29, 22362 Lund, Sweden.

Results and Problems in Cell Differentiation 14
Structure and Function of Eukaryotic Chromosomes
Edited by W. Hennig

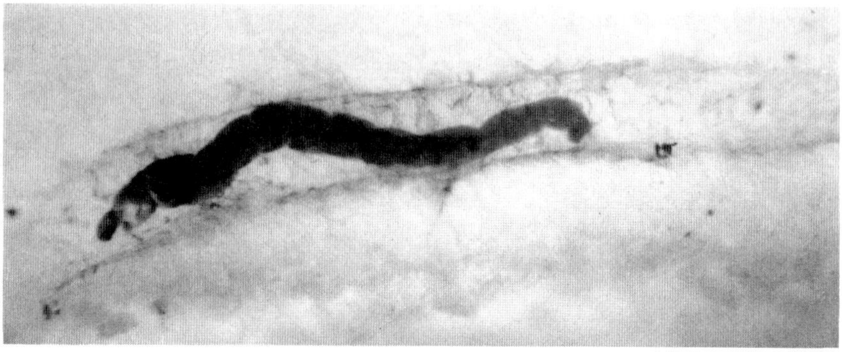

Fig. 1. Fourth instar *Chironomus* larva in its protective tube. The visibility of the tube was improved by allowing the larva to produce it in agarose

ivary gland secretion originally described by Grossbach (1973) was determined to have a size of about 1 000 000 daltons and to be complex (Rydlander and Edström 1980). It is similar to the in vitro translation products of the 75S BR RNAs (Rydlander et al. 1980; Weber et al. 1983). The relationship between BR genes, 75S RNA and sp-I was confirmed after culturing larvae in medium with galactose, glycerol or ethanol: In *C. pallidivittatus* and *C. tentans* a new BR is induced under these conditions (BR6: Beermann 1973), a new 75S RNA, slightly different in migration appears, and a new sp-I component (sp-Ic) is synthesized (Edström et al. 1980, 1982). Improved techniques of sample preparation and electrophoretic separation now permit the visualization of at least three different fractions of 75S RNA. Also several fractions are resolved among the giant secretory proteins: sp-Ia and sp-Ib in animals under standard culture conditions, sp-Ia and sp-Ic after BR6 induction (Galler et al. 1984). Recently, Kao and Case (1985) detected a fourth fraction (sp-Id) using a new method for protein sample preparation.

During the last years BR DNA has been cloned and the molecular structure of the BR genes studied. This review deals with the molecular structure of the sp-I genes. For other aspects of the BR system see Pustell et al. (1984) and reviews by Beermann (1962), Grossbach (1977) and Daneholt (1975, 1982).

2 The Structure of the sp-I Genes in the Large BRs

2.1 Number of sp-I Genes

The sp-I genes are members of a gene family and encode giant secretory proteins. Sequence analysis of cloned DNA fragments of various BRs confirmed that the sp-I genes are evolutionarily related, as will be discussed (see also Pustell et al. 1984 and Wieslander et al. 1984).

In *C. tentans* and *C. pallidivittatus*, both members of the subgenus *Camptochironomus*, the genes for the giant secretory proteins are located in the large BRs

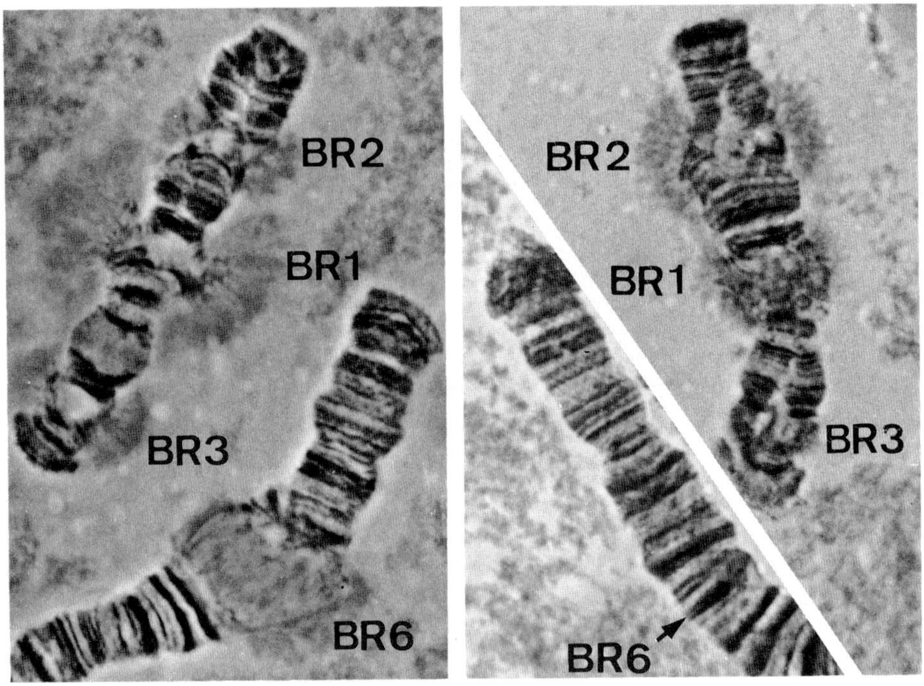

Fig. 2. Squashed salivary gland chromosomes from (*right*) control *C. pallidivittatus* larvae showing chromosome 4 with BR1, BR2 and BR3, and the left arm of chromosome 3 that carries the band for BR6; and (*left*) experimentals (6-day exposure to glycerol medium for the induction of BR6) showing all four BRs. (From Edström et al. 1980)

1,2 and 6 (Fig. 2). The large BRs might contain more than a single gene, and Wieslander (1979) estimated the number of 75S RNA producing genes per BR to be between one and four. The first direct piece of evidence for two genes in a BR came from a *C. pallidivittatus* mutant where two BR1 genes separated by a translocation each formed a separate BR. In fact, BR1 appears to originate from two (closely adjacent) chromosome regions even in wild-type stocks (Beermann 1962). For BR1 in *C. tentans* two genes were identified by molecular techniques: an sp-I gene (Wieslander et al. 1982; Case and Byers 1983) and a second gene producing a 6.5 kb messenger in the salivary glands (Dreesen et al. 1985). For BR2 of *C. pallidivittatus* two sp-I genes with different sequences are reported; these have messengers of slightly different size and with different half-lives after glycerol-induced BR2 regression (Galler et al. 1985). Also on BR2 in *C. tentans* two sp-I genes are likely to occur (Kao and Case 1985). So far, there is no evidence for more than a single gene in BR6.

Hence, the number of sp-I genes appears to be rather low: each of the three large BRs probably contains one or two sp-I genes.

2.2 Structure of the sp-I Polypeptides

Enzyme digestion or partial hydrolysis of sp-I polypeptides reveals tandem arrangements of repeat units with unit sizes in the range of 15 kD (Hertner et al. 1983; Rydlander 1984; Serfling et al. 1983). According to fingerprint analyses of tryptic peptides sp-Ia, sp-Ib and sp-Id appear to be rather similar in structure, in spite of significant differences (Rydlander 1984; Kao and Case 1985). Amino acid analyses by Grossbach (1969) of total secretory protein, dominated by sp-I, and direct analyses of sp-I components by Rydlander (1984) indicated an unusual amino acid composition. The sp-I fractions contain high amounts of basic amino acids (*Lys* + *Arg* in amounts of about 30%), low amounts of acidic amino acids and a high content of *Pro* (Table 1). Nevertheless, these normally produced sp-I fractions (sp-Ia, Ib and Id) are not basic, but instead, like the glycerol-induced sp-Ic, have isoelectric points of 7.2 (Kao and Case, unpublished). This result is likely to be due to the extensive phosphorylation of the normal sp-I components (Galler et al. 1984).

Table 1. Coded composition of common amino acids from repeat units of BR1α (pCp 90), BR1β (pCp · c14), BR2α (pCp · c12-8), BR2β (pCp 41), BR6 (pCp 16) and directly determined composition for sp-Ia and sp-Ib (Rydlander 1984)

	BR1α	BR1β	BR2α	BR2β	BR6	sp-Ia	sp-Ib
Arg	16.9	12.0	5.2	23.1	22.4	11.2	13.5
Lys	14.3	17.3	26.0	9.2	7.5	21.1	13.8
Sum basic	31.2	29.3	31.2[a]	32.3	29.9	32.3	27.3
Asp	1.3	0.0	0.0	0.0	11.9	3.5[b]	2.4[b]
Glu	3.9	8.0	3.9	7.7	26.9	8.2[b]	8.5[b]
Sum acidic	5.2	8.0	3.9	7.7	38.8	11.7[b]	10.9[b]
Ser	16.9	14.7	22.1	13.8	0.0	14.1	15.1
Thr	5.2	2.7	3.9	3.1	0.0	2.1	2.3
Sum Ser + Thr	22.1	17.4	26.0	16.9	0.0	16.2	17.4
Pro	19.5	17.3	10.4	12.3	13.4	14.9	8.6

[a] In addition, 9.1% His.
[b] Includes corresponding amines.

2.3 The Inducible BR6

The observation by Beermann (1973) that culturing of *Chironomus* larvae in media with monosaccharides or some other metabolites like ethanol or glycerol induces a new BR, BR6, and leads to regression of BR2 (Fig. 2) offers a model for the study of BR gene regulation. In *C. pallidivittatus* BR6-inducing agents decrease the concentration of inorganic phosphate in the larval haemolymph, whereas gland phosphate levels remain unchanged. The role of the phosphate level in induction is underlined by the observation that administration of inorganic phosphate together with BR6 inducers eliminates their inductive action (Edström et al. 1982).

As the sequence of BR6 repeat units was analyzed and compared with sequences of BR1 and BR2 genes (Galler et al. 1984; Lendahl and Wieslander 1984), a lead to the role of phosphate was obtained. In BR6 *Glu* and *Asp* code words are present in BR6 to an extent that is largely sufficient to neutralize the high content of *Lys* and *Arg*. This is not the case for BR1 and BR2 sequences that contain codons for *Ser* and *Thr*, instead of the acidic residues. This suggested that the phosphate effect in BR6 induction could be related to protein phosphorylation (at *Ser* and *Thr* positions). Indeed, it was found that BR1 and BR2 products (sp-Ia and sp-Ib) are heavily phosphorylated, mainly at the *Ser* residues (Galler et al. 1984) in contrast to the non-phosphorylated BR6 product, sp-Ic. *Glu* (and *Asp*) residues mimic the phosphoserines also in steric properties as shown by 3D computer predictions applied to the constant parts of BR1, BR2 and BR6 repeat units (Hamodrakas and Kafatos 1984). The switch to sp-Ic during phosphate starvation consequently would allow the production of secretory proteins that, although non-phosphorylated, have properties comparable with the proteins secreted under normal culture conditions. Galler and Edström (1984) also showed that the nascent sp-I polypeptide chains become phosphorylated already in the BR1 and BR2 encoded polysomes.

2.4 Sequence Organization of the sp-I Genes

All sp-I genes identified so far contain tandemly repeated units of 180–300 bp. About half of the sequence of these repeat units is subrepeated. The non-subrepeated part has usually 32 codons (Fig. 3). It is called the constant part, since certain positions are invariant for all repeat units (four *Cys*, one *Met*, one *Phe*), indicating the common origin of these sequences (Fig. 3). Repeat units with these features may be called standard repeat units. The constant parts of all standard repeat units are highly similar, but the situation for the subrepeated regions is more complex (Fig. 4), since number, length and sequence of the subrepeats varies for different repeat units (Höög and Wieslander 1984; Lendahl et al. in preparation). The standard repeat units can be classified on the basis of the subrepeat arrangements into *alpha* units (three or six codon subrepeat), *bèta* units (ten codon subrepeat) and *gamma* units (seven or eleven codon subrepeat) (Fig. 4). The *alpha* subrepeats conserve a high degree of homology within a repeat unit, as well as between repeats of the same gene. The *bèta* subrepeats, however, show great variation in individual base positions within and between repeat units, including amino acid differences (Jäckle et al. 1982; Wieslander and Lendahl 1983). The two known *gamma* units have almost identical constant parts, but their subrepeats

Constant part of standard repeat unit
consensus features

–C– – –M– – – – – – – –C– – – – – – –F– – – –C–C– – – –

Fig. 3. Codon consensus sequence for the constant part of Balbiani ring sp-I genes

```
type of
repeat unit                          constant regions

α BR1p   K C G S A M R R V E G E R C A A R K G R F N D K N C R C T S V G
  BR2p   - - - - - - K - T - A - K - - R K N - - - - S - R - T - - - - -
  BR2t   - - - - - - K - T - A A K - - R K N - - - - S - R - T - - - - -

  BR6p   - - D D E - - E R V K R - - D N E N R - - D A R R - E - G E K K
  BR6t   - - D D E - - E E V K R - - D N E N R - - D A R R - E - G E K K


β BR1p   V - - - - - - - T K A - E - - K - N - - - - G - - - - - - - A N
  BR1t   V - - - - - - - T K A - E - - R - N - - - - G - - - - - - - A N
  BR2p   - - - - K - - - - F A - K - - - - - - - - - S A S K - - - F -
  BR2t   - - - - K - - - - F A - K - - - - - - - - - S A S K - - - F -


γ BR1p   - - - - - - - K A - A - K - - R - - - - - - A N K - - - - - A -
  BR1t   G - - - - - - K A - A - K - - R - N - - - - A S K - - - - - A -
```

α-subrepeats

BR1p	Br2p	BR2t	BR6p	BR6t
K P S	K P S	K P S	R P E	R P D
K P S	K H S	K P S	D N D	D N D
R P S	K P S	K H S	D E D	D N D
K P S	K H S	K P S	R P E	D E D
R P S	K P S	K H S	R P E	R P E
R P S	K H S	K P S	R P E	R P E
R P S	K P S	K H S	R P E	R P E
K P T	K H S	K P S	R P E	E P E
R P S	K P S	K H S	E P E	R P E
K P T	K H S	K P S	R E E	R P E
R P S	K P S	K H S	P E R	R P E
R P S	K H S	K P S	E P	R P E
K P S	K P S	K H S		E P E
K P S	K H S	K P E		R E E
K P E	K P E			P E R
				E P

β-subrepeats

```
BR1p
K P S|K S G P|K P E
R - -|- - - -|- - -
R - -|- - - -|- - -
R - -|- - R -|- - S
R - E|       |

BR1t
K P S|K S G S|K P E
R - -|- - R P|- - -
R - -|- - R P|- - -
R - -|- - - P|- - -
R - -|- - R P|- - S
R - E|       |

BR2p
R P S|R S G L|R P E
- - T|- - - -|- - -
- - -+-------|- - -
- - T|K - - S|- - -

BR2t
R P S|W S G I|R P E
- R -|R - - P|- - K
G - -|K - - S|- - -
```

γ-subrepeats

```
BR1p                     BR1t
K P S|K S G P            ------|K P S|---------
  :  |                   R K S|E - -|K G S K P
R - -|- - - -            - P E|- - -|- E - - -
R - -|- - - -            - - E|- - -|- - - - -
R - -|- - - -            - P E|- - -|- - - - -
R - E|                   - P E|     |
```

Fig. 4. Conceptually translated sequences of sp-I genes from *Camptochironomus* in the one-letter code. *p* and *t* describe *C. pallidivittatus* and *C. tentans* respectively. The first complete repeat units according to the first published sequence from each species and Balbiani ring have been listed: for *alpha* repeats: BR1p (Galler et al. 1984), BR2p (Galler et al. 1985), BR2t (Sümegi et al. 1982), BR6p (Galler et al. 1984), BR6t (Lendahl and Wieslander 1984); for *bèta* repeats: BR1p (Saiga et al., unpublished data), BR1t (Höög and Wieslander, unpublished data), BR2p (Galler et al. 1984), BR2t (Case et al. 1983); for *gamma* repeats: BR1p (Saiga et al. 1987), BR1t (Wieslander et al. 1982). It can be seen that with the present classification the constant parts become well de-have diverged considerably between *C. pallidivittatus* and *C. tentans* (Lendahl et al. in preparation) and contain different numbers of amino acids (eleven for *C. tentans* and seven for *C. pallidivittatus*). The subrepeat structure of the different repeat units will be discussed in more detail in the next section.

Each repeat unit is present in numbers of 60–120 per BR (the highest values recorded in BR6). For BR1 and BR2 this accounts for 15–20 kb. However, Galler et al. (1985) have shown that much larger blocks of DNA, in the range of 30–35 kb can be isolated from BR2 by digestion of high molecular weight DNA with restriction enzymes that do not have recognition sites within the repeat unit. Also the size of the mRNA suggests the sp-I genes to be about 35 kb. Moreover, the *alpha* and *bèta* repeats (the only units known for BR2) reside in different sp-I genes (Galler et al. 1985). The discrepancy between these data cannot be explained.

It is not known whether an sp-I gene contains more than one type of standard repeat unit. So far, there is no clone isolated, in which different types of standard repeat units adjoin. The *gamma* units might either be present in separate genes, or they might be restricted to the 3′ part of genes with *alpha* and *bèta* units in the remaining part.

2.5 Basic Elements of the Subrepeats

The *alpha* repeat shows a 3 (or 2 × 3) codon subrepeat structure. By and large this subrepeat can be formalized as K/R-P-S/E in the one-letter code, or in other words a *Pro* (occasionally a *His*) preceeded by a positively charged and followed by a negatively charged amino acid (Fig. 4). The subrepeated parts of *alpha*-type repeat units are entirely composed of such three codon elements. This element is not only present in *alpha* units. It is also found in the *bèta* and *gamma* repeat units in *C. pallidivittatus* and *C. tentans* and usually surrounds the constant parts of these repeat units. (However, the 5′-end of the BR6 constant part is flanked by an incomplete triplet: E–P.) For the purpose of discussion we here designate this three codon subrepeat structure *s-element*. The *alpha* type subrepeated parts can thus be formalized as $(s)_n$. In *C. pallidivittatus* BR genes of the *alpha* type are found in BR1, BR2 and BR6. In *C. tentans alpha* type genes are found only in BR2 and BR6 (Lendahl and Wieslander 1985). The subrepeat arrangement in BR6 fits only partially the typical *alpha* arrangement. The sequences bordering the constant region are highly degenerate according to this classification.

The subrepeated part of the *bèta* type repeats displays a ten codon periodicity. Within the ten amino acid stretch two *s-elements* are found, surrounding a K-S-G-P sequence (Fig. 4), here called the *t-element*. The *bèta* subrepeat region can thus be formalized as $(s.t.s.)_n$ (Table 2). The *bèta*-type genes are found in BR1 and BR2 of both *C. pallidivittatus* and *C. tentans*.

---

fined to 32 codons with the exception of BR2 *bèta* repeats with only 30 codons. For BR6 the subrepeated region bordering the constant region is highly degenerate according to the present classification. The number of subrepeats for the *gamma* repeat unit in BR1p is not yct known

Table 2. Classification of Balbiani ring (BR) repeat units (C=constant part; s=K/R·P·S/E; t=K·S·G·P; u=K·G·S·K·P, in the one-letter amino acid code)

α Repeat units:	$[C \cdot s_n]$
β Repeat units	
BR1:	$[C \cdot (s \cdot t \cdot s)_n \cdot s]$
BR2:	$[C \cdot (s \cdot t \cdot s)_n]$
γ Repeat units	
C. pallidivittatus:	$[C \cdot s \cdot (t \cdot s)_n \cdot s]$
C. tentans:	$[C \cdot s_2 \cdot (s \cdot u \cdot s)_n]$

In both species repeat units have been described that belong neither to the *alpha* nor to the *bèta* type: they are called *gamma* repeats (Lendahl et al. in preparation). They are similar in their constant parts, but their subrepeated parts differ considerably. The *C. pallidivittatus* subrepeat is relatively simple: an *s-element* followed by a *t-element* (Fig. 4). The *C. tentans* sequence is considerably more complex and contains a new element, K-G-S-K-P, a *u-element,* surrounded by *s-elements*. *Gamma* repeats have been found at the 3′-end of BR1 genes in both *C. pallidivittatus* and *C. tentans*. The composition of the different repeat units can be described as in Table 2.

2.6 Sequence Organization of the 3′-End

Recently, 3′-end regions of the BR1 and BR2 mRNA have been isolated in *C. pallidivittatus* (Saiga et al. 1987). Genomic clones of 3′-ends of BR genes have been obtained from *C. tentans* (Höög et al. 1986) and *C. thummi* (Bäumlein et al. 1986), the results being similar for the three species. Gamma repeats are followed by a new repeat unit at the 3′-end of BR1. This new unit is as a rule only 18 codons long and without subrepeats. It has been designated the Cys-1 unit, and is interspersed with constant parts of the standard type repeat unit without the usual excess of positive charges (Fig. 5). The structure of the Cys-1 unit differs from that of the other repeat units, although Bäumlein et al. (1986) consider it to be distantly related to the other BR repeats. This shorter type of repeat unit has an excess of acidic amino acid codons.

The last Cys-1 repeat is followed by a unique sequence of 111 amino acids, with physical properties different from those of the giant secretory proteins. The sequence is non-repetitive and encodes two short blocks of hydrophobic residues with partial homology. Genomic DNA sequencing, comparison with the cDNA clones and S1 protection mapping have shown that a small intron of 55 bp length separates the 3′-end unique sequence from the repetitive BR gene sequences upstream in *C. tentans*. An intron of similar length and location is present in *C. thummi* sp-I genes. Comparison of three clones from the 3′-end region (*C. pallidivittatus* BR1 and BR2 and *C. tentans* BR1) revealed that these sequences downstream from the last constant region have been more strongly conserved

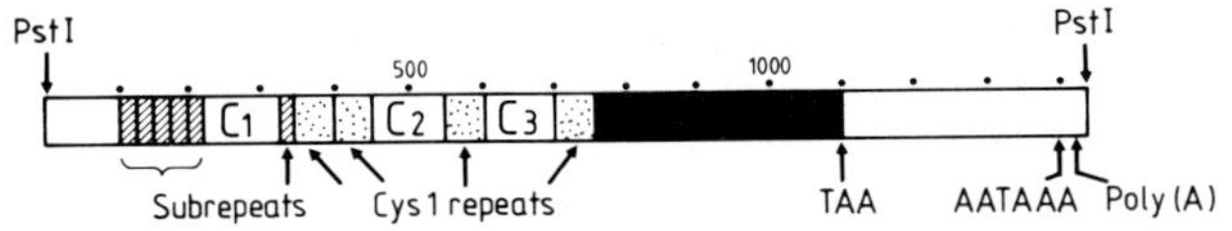

Fig. 5. Schematic arrangement of a cDNA clone of the 3′-end of a BR1 sp-I gene in *C. pallidivittatus*. C_1, C_2 and C_3 are the constant regions, the *black area* is the unique, translated region, and the *white area* to the *right side* is the untranslated region. (From Saiga et al. 1987)

than sequences further upstream. There is, furthermore, a decrease in rate of corrected silent mutations in the 3′ direction, the lowest mutation rate being found in the non-translated region (Saiga et al. 1987). The presence of a homogenizing mechanism is also apparent in the *C. thummi* results. Here, two BRc clones show a high degree of homology up to the point where polyadenylation occurs, after which all homology disappears. This suggests that the transcript has been used for gene homogenization.

3 Conclusion and Perspectives

In this review we have focussed on the molecular structure of the sp-I genes in the BRs and their translational products. The sequences are usually repeated and have diverged at various hierarchical levels (genes, repeat units, subrepeat components). It has been proposed (Pustell et al. 1984, for discussion) that the repeated BR sequences arose from a single founder sequence by duplication, translocation and diversification. Sümegi et al. (1982) suggested a 9-bp unit bordering the constant region as a founder sequence for the entire subrepeat region. Sequence analyses of the 3′-ends of BR transcription units have led to the identification of a new type of repeat unit (Cys-1 repeat), an intron, and an exon with a unique sequence that differs markedly from the rest of the sp-I gene. The Cys-1 repeat may have originated from the founder sequence of the repeat units (Bäumlein et al. 1986). Finally, there is some evidence that also secretory protein genes with translational products in the 200 kD range may share sequence features with the sp-I genes (Dreesen et al. 1985). The 3′-exon, however, is different from the repetitive sequences both in origin and in properties. Its evolutionary constancy is one striking feature to which unconventional sequence correction mechanisms may have contributed.

Differences in sequence of the kind found between 3′-ends and the cores of the sp-I genes, or between constant and subrepeated parts of the BR repeat units also reflect the selective advantage of certain sequence features and thus ask for a functional interpretation. A clear example is given by Galler et al. (1984) for the sequence of the BR6 encoded protein that lacks phosphorylated residues and is produced as a response to phosphate starvation. Hamodrakas and Kafatos (1984) predict the 3D structure of several conceptual BR amino acid sequences and interprete some of the characteristic aspects of the repeat units (conserved *Cys* residues in the constant parts, high amount of *Pro* residues in the subrepeated parts) as important factors for the configuration of the large secretory protein. The deviant amino acid composition of the BR6 encoded protein could also be interpreted in conformational terms by the same approach: the *Glu* residues in the BR6 product mimic the conformational effect of the phosphoserine residues in the protein products of BR1 and BR2 (Hamodrakas and Kafatos 1984). The unique 3′-end exon of the BR genes might also encode a part of the secretory proteins with a unique conformation and function. A distinct function of this domain would be likely since its sequence is strongly conserved among the different genes and also in comparisons of *C. thummi* with *C. pallidivittatus* or *C. tentans,* between which other known sp-I sequences diverge considerably.

BRs have outstanding features for cytological investigation and the extraordinary dimensions of BRs (Beermann 1962), their nascent and finished transcripts (for reviews, see Daneholt 1975, 1982), their RNP granules (Andersson et al. 1980; Skoglund et al. 1986) and their polysomes (Francke et al. 1982; Olins et al. 1980) permit many detailed morphological observations.

In eukaryotes, cytological studies were in the past a major tool for the study of the structure of interphase chromosomes, of the process of transcription and of the regulation of gene activity. The role of cytology has changed since techniques developed that permit a more direct molecular analysis of eukaryotic gene regulation. For the BR system the accumulation of molecular descriptions might offer new opportunities to interprete the unique wealth of cytological observations. Such an integration of cytological observations with genetic and molecular data justifies also at present the choice of BRs as a model for the study of gene expression.

Acknowledgements. We are indebted to Drs. L. Wieslander, C. Höög, U. Lendahl, S. Case, H. Bäumlein, J. Pustell, U. Wobus and F. Kafatos for communicating unpublished results, and to L. Botella for critical reading of the manuscript. C. G. was recipient of an EMBO long-term fellowship. H. S. was recipient of a fellowship from the Carl Thesdorp Foundation. Work from the authors' laboratory described here was supported by a grant from the Swedish Cancer Society.

References

Andersson K, Björkroth B, Daneholt B (1980) The in situ structure of the active 75S RNA genes in Balbiani rings of *Chironomus tentans*. Exp Cell Res 130:313–327

Bäumlein H, Pustell J, Wobus H, Case ST, Kafatos FC (1986) The 3'ends of two genes in the Balbiani ring c locus of *Chironomus thummi*. J Mol Evol 24:72–82

Beermann W (1952) Chromomerenkonstanz und spezifische Modifikation der Chromosomenstruktur in der Entwicklung und Organdifferenzierung von *Chironomus tentans*. Chromosoma 5:139–198

Beermann W (1961) Ein Balbiani-Ring als Locus einer Speicheldrüsenmutation. Chromosoma 12:1–15

Beermann W (1962) Riesenchromosomen. Protoplasmologia VI C. Springer, Wien, pp 1–165

Beermann W (1973) Directed changes in the pattern of Balbiani ring puffing in *Chironomus:* Effects of a sugar treatment. Chromosoma 41:297–326

Beermann W, Bahr GF (1954) The submicroscopic structure of the Balbiani ring. Exp Cell Res 6:195–201

Case ST, Daneholt B (1978) The size of the transcription unit in Balbiani ring 2 of *Chironomus tentans* as derived from analysis of the primary transcript and 75S RNA. J Mol Biol 124:223–241

Case ST, Byers MR (1983) Repeated nucleotide sequence arrays in Balbiani ring 1 of *Chironomus tentans* contains internally nonrepeating and subrepeating elements. J Biol Chem 258:7793–7799

Case ST, Summers RL, Jones AG (1983) A variant tandemly repeated nucleotide sequence in Balbiani ring 2 of *Chironomus tentans*. Cell 33:555–562

Daneholt B (1972) Giant RNA transcript in a Balbiani ring. Nature New Biol 240:229–232

Daneholt B (1975) Transcription in polytene chromosomes. Cell 4:1–9

Daneholt B (1982) Structural and functional analysis of Balbiani ring genes in the salivary glands of *Chironomus tentans*. In: King R, Akai H (eds) Insect ultrastructure, vol 1. Plenum, New York, pp 382–401

Daneholt B, Hosick H (1973) Evidence for transport of 75S RNA from a discrete chromosome region via nuclear sap to cytoplasm in *Chironomus tentans*. Proc Natl Acad Sci USA 70:442–446

Dreesen TD, Bower JR, Case ST (1985) A second gene in a Balbiani ring. J Biol Chem 260:11824–11830

Edström J-E, Rydlander L, Francke C (1980) Concomitant induction of a Balbiani ring and a giant secretory protein in *Chironomus* salivary glands. Chromosoma 81:115–124

Edström J-E, Sierakowska H, Burvall K (1982) Dependence of Balbiani ring induction in *Chironomus* salivary glands on inorganic phosphate. Dev Biol 91:131–137

Francke C, Edström J-E, McDowall AW, Miller OL Jr (1982) Electron microscopic visualization of a discrete class of giant translation units in salivary gland cells of *Chironomus tentans*. EMBO J 1:59–62

Galler R, Edström J-E (1984) Phosphate incorporation into secretory protein of *Chironomus* salivary glands occurs during translation. EMBO J 3:2851–2855

Galler R, Rydlander L, Riedel N, Kluding H, Edström J-E (1984) Balbiani ring induction in phosphate metabolism. Proc Natl Acad Sci USA 81:1448–1452

Galler R, Saiga H, Widmer RM, Lezzi M, Edström J-E (1985) Two genes in Balbiani ring 2 with metabolically different 75S transcripts. EMBO J 4:2977–2982

Grossbach U (1969) Chromosomen-Aktivität und biochemische Zelldifferenzierung in den Speicheldrüsen von *Camptochironomus*. Chromosoma 28:136–187

Grossbach U (1973) Chromosome puffs and gene expression in polytene cells. Cold Spring Harbor Symp Quant Biol 38:619–627

Grossbach U (1977) The salivary gland of *Chironomus* (Diptera): a model system for the study of cell differentiation. In: Beermann W (ed) Results and problems in cell differentiation, vol 8. Springer, Berlin Heidelberg New York, pp 147–196

Hamodrakas SJ, Kafatos FC (1984) Structural implications of primary sequences from a family of Balbiani ring-encoded proteins in *Chironomus*. J Mol Evol 20:296–303

Hertner T, Eppenberger HM, Lezzi M (1983) The giant secretory proteins of *Chironomus tentans* salivary glands: the organization of their primary structure, their amino acid and carbohydrate composition. Chromosoma 88:194–200

Höög C, Wieslander L (1984) Different evolutionary behaviour of structurally related repetitive sequences occurring in the same Balbiani ring gene in *Chironomus tentans*. Proc Natl Acad Sci USA 81:5165–5169

Höög C, Engberg C, Wieslander L (1986) A BR1 gene in *Chironomus tentans* has a composite structure: a larger repetitive core block is separated from a short unrelated 3′ terminal domain by a small intron. Nucl Acids Res 14:703–719

Jäckle H, de Almeida JC, Galler R, Kluding H, Lehrach H, Edström J-E (1982) Constant and variable parts in the Balbiani ring 2 repeat unit and the translation termination region. EMBO J 1:883–888

Kao W-Y, Case ST (1985) A novel giant secretion polypeptide in *Chironomus* salivary glands: Implications for another Balbiani ring gene. J Cell Biol 101:1044–1051

Lendahl U, Wieslander L (1984) Balbiani ring 6 gene in *Chironomus tentans:* A diverged member of the Balbiani ring gene family. Cell 36:1027–1034

Lendahl U, Wieslander L (1985) Abrupt evolutionary change in the Balbiani ring gene family in two sibling species of *Chironomus*. J Mol Biol 22:63–68

Olins AL, Olins DE, Francke WW (1980) Stereo-electron microscopy of nucleoli, Balbiani rings and endoplasmatic reticulum in *Chironomus* salivary gland cells. Eur J Cell Biol 22:714–723

Pelling C (1964) Ribonukleinsäuresynthese der Riesenchromosomen: autoradiographische Untersuchungen an *Chironomus tentans*. Chromosoma 15:71–122

Pustell J, Kafatos FC, Wobus U, Bäumlein H (1984) Balbiani ring DNA: sequence comparisons and evolutionary history of a family of hierarchically repetitive protein-coding genes. J Mol Evol 20:281–295

Rydlander L (1984) Isolation and characterization of the two giant secretory proteins in salivary glands of *Chironomus tentans*. Biochem J 220:423–431

Rydlander L, Edström J-E (1980) Large-sized nascent protein as dominating component during protein synthesis in *Chironomus* salivary glands. Chromosoma 81:85–99

Rydlander L, Pigon A, Edström J-E (1980) Sequences translated by Balbiani ring 75S RNA in vitro are present in giant secretory protein from *Chironomus tentans*. Chromosoma 81:101–113

Saiga H, Grond C, Schmidt ER, Edström J-E (1987) Evolutionary conservation of the 3′-ends of members of a family of giant secretory proteins in *Chironomus*. J Mol Evol 25:20–28

Serfling E, Meyer L, Rudolph A, Steiner K (1983) Secretory proteins and Balbiani ring gene activities in salivary glands of *Chironomus thummi* larvae. Chromosoma 88:16–23

Skoglund U, Andersson K, Strandberg B, Daneholt B (1986) Three-dimensional structure of a specific pre-messenger RNP particle established by electron microscope tomography. Nature 319:560–564

Sümegi J, Wieslander L, Daneholt B (1982) A hierarchic arrangement of the repetitive sequences in the Balbiani ring 2 gene of *Chironomus tentans*. Cell 30:579–587

Weber F, Mähr R, Meyer B, Eppenberger HM, Lezzi M (1983) Cell-free translation of Balbiani ring RNA (75S) of *Chironomus tentans* salivary glands into high molecular weight products. Wilhelm Roux's Arch Dev Biol 192:200–203

Wieslander L (1979) Number and structure of the Balbiani ring 75S RNA transcription units in *Chironomus tentans*. J Mol Biol 134:347–367

Wieslander L, Lendahl U (1983) The Balbiani ring 2 gene in *Chironomus tentans* is built from two types of tandemly arranged major repeat units with a common evolutionary origin. EMBO J 2:1169–1175

Wieslander L, Sümegi J, Daneholt B (1982) Evidence for a common ancestor sequence for the Balbiani ring 1 and Balbiani ring 2 genes in *Chironomus tentans*. Proc Natl Acad Sci USA 97:6956–6960

Wieslander L, Höög C, Höög J-O, Jörnvall H, Lendahl U, Daneholt B (1984) Conserved and non-conserved structures in the secretory proteins encoded in the Balbiani ring genes of *Chironomus tentans*. J Mol Evol 20:304–312

The *white* Locus of *Drosophila melanogaster*

B. H. JUDD [1]

1 Introduction

In an article entitled *Sex Limited Inheritance in Drosophila* which appeared in the July 22, 1910 issue of *Science,* T. H. Morgan wrote:

> "In a pedigree culture of *Drosophila* which had been running for nearly a year through a considerable number of generations, a male appeared with white eyes. The normal flies have brilliant red eyes."

Morgan went on to describe the inheritance of the new character. A sentence from a later paper (Bridges and Morgan 1923) reads:

> "The first eye-color, and the character first clearly recognized as a sharp "mutation" in *Drosophila* was 'white', found in April 1910, by Morgan."

Compare the following paragraph from Lutz (1941), who in 1910, at the Station for Experimental Evolution in Cold Spring Harbor, New York, was carrying inbred strains of *Drosophila* and studying the inheritance of abnormal wing-veination patterns:

> "Meanwhile something far more worthwhile happened, ... Professor T. H. Morgan visited the Station and I told him that a white-eyed *Drosophila* had appeared in one of the pedigreed strains but that I was too busy with abnormal veins to attend to it. He took live descendants of this white-eyed 'sport' and bred from them. Eventually he got the white eye back. On working out the inheritance of this character he discovered an important addition to the Mendelian Laws. Not only that but he and his students and his students' students have been breeding descendants of those *Drosophila* ever since and have built up a truly magnificent structure of knowledge concerning the way in which characters are inherited. This knowledge can be applied not only to the breeding of domestic animals and plants but to the breeding of Man himself. This little story is no particular credit to me. If I had realized how valuable that white-eyed mutant was destined to be, I would not have been happy to give it away. However, it fell into good hands ..."

[1] National Institute of Environmental Health Sciences, P.O. Box 12233, Research Triangle Park, North Carolina, 27709, USA.

Results and Problems in Cell Differentiation 14
Structure and Function of Eukaryotic Chromosomes
Edited by W. Hennig

Studies of *white* locus mutants have indeed played prominent roles in almost every phase of advancement in 20th century genetics. By virtue of its mode of inheritance, linked to sex determination, it was the first gene to be located to a specific chromosome. As mutations at other X chromosome loci were discovered by the Morgan group, the concept of linkage and crossingover was developed. Sturtevant (1913), with brilliant insight, used the frequency of crossingover between *white* and other X-linked genes to deduce relative positions of the marker loci in the X chromosome and thus constructed the first genetic map.

By 1925, Morgan et al. (1925) reported that approximately 25 separate occurrences of mutations at the *white* locus had been documented and that the mutants comprised 11 distinct types. This series of "allelomorphs", as they were then called, figured prominently in attempts to define the abstract Mendelian unit in operational terms, i.e., as units of mutation (Muller 1920), recombination (Lewis 1952; MacKendrick and Pontecorvo 1952) and function. I cannot here recount all the details of the role the *white* locus has played in the development of modern genetic concepts. It is, however, among the loci most thoroughly studied genetically, cytologically, biochemically, and molecularly. Those studies have often yielded pivotal information, as will become evident as we examine the structure and function of the locus as we understand it today.

2 Genetic and Cytological Position of *white*

The *white* locus is located in the X chromosome at map position 1.5 (Lindsley and Grell 1968). The cytological position is still not settled precisely, although a large number of deficiency and rearrangement break points in the *white* region have been examined cytologically. Lefevre and Wilkins (1966) did a comprehensive study of rearrangements involving the *white* locus and concluded that it lies in band 3C2 of Bridges (1938) polytene chromosome map.

Sorsa et al. (1973) examined with the electron microscope the cytogenetic fine structure of $In(1)z^{64b9}$, which has a break point in or near the *white* locus. Their results are compatible with the interpretation that *white* resides in band 3C2 if it occupies only a small part of the distal edge of the band, because 3C2 or a major part of it is transferred by this inversion to a new position at 12C but *white* function remains in the original 3C position. An alternative explanation they proposed is that another faint band exists between 3C1 and 3C2 in which *white* resides. This latter view is supported by observations of Lefevre and Green (1972) and Judd (1976). Goldberg et al. (1982) showed that the break point of $In(1)z^{64b9}$ is indeed proximal to the *white* locus on the molecular map.

In an effort to answer questions about gene size and number in *Drosophila* and the relationship between the band/interband units of the polytene chromosomes relative to units of genetic functions, Judd et al. (1972) saturated the region around *white* with mutations and positioned them cytologically. Their cytogenetic map indicated that genes in *Drosophila* have very large sizes averaging about 25 kb of DNA. Further, their data were consistent with the concept that a band/interband unit corresponds to a unit of genetic function, a concept developed largely through the work of Beermann (1972). Further work shows that this relationship is not precise, however (Young and Judd 1978). The molecular cloning of the entire *white* locus region (Bingham et al. 1981; Goldberg et al. 1982; Pirrotta et al. 1983) and the determination of the nucleotide sequence (O'Hare et al. 1984) has not given any further evidence about the relationship of the gene to the chromomeres in that region of the polytene chromosomes. The molecular analysis has, however, shed light on the organization of the locus and the nature of many of the mutations that have occurred there.

3 Molecular Cloning of *white*

The cloning of the locus was in itself a significant advance in genetic analysis. A strategy called "transposon tagging" was employed by Bingham et al. (1981) to recover parts of the gene flanking the site of the insertion of the transposon *copia*. It had been demonstrated by Gehring and Paro (1980) that a copy of *copia* is found in the distal part of 3C in chromosomes carrying the mutation *white-apricot*, w^a, but not in chromosomes carrying other *white* alleles or in an X-ray induced partial revertant of w^a (Rasmuson et al. 1960). Bingham and Judd (1981) by fine scale genetic mapping demonstrated that *copia* in the w^a chromosome is located within the locus between the two alleles *white-carrot* (w^{crr}) and *white-cherry* (w^{ch}) and inseparable from the w^a site. With that knowledge, Bingham et al. (1981) constructed a gene library of w^a and screened for those clones containing homology to *copia*. By in situ hybridization of those clones to chromosomes devoid of *copia* in 3C, a clone containing unique sequences of *white* locus was identified.

The original clone was extended on both sides by the molecular cloning of successive overlapping fragments. The limits of the *white* gene were roughly established as chromosomal rearrangement break points flanking the locus were crossed by that molecular walk. Goldberg et al. (1982) also cloned the locus, taking advantage of a large transposable element containing *white* and *roughest* that had inserted into the third chromosome near the previously cloned heat-shock puff sequences at 87A7. Pirrotta et al. (1983) employed a remarkably precise dissection of the 3C region of polytene chromosomes to isolate and clone the *white* locus region.

A physical map of the *white* locus produced by cleaving DNA with various restriction endonucleases (Levis et al. 1982; Goldberg et al. 1982) set the stage for extending to the molecular level the information about the gene structure gathered over many years of genetic and cytological studies. The numerous alleles and rearrangements at the *white* locus exhibit a wide range of phenotypes that make the locus an excellent model for examining mechanisms of gene mutation, transmission and regulation.

4 Phenotypic Expression of the *white* Locus

Null alleles of *white,* equivalent to a deletion of the locus, result in the absence of all but a trace of pigment in the compound eyes, ocelli, testis sheath and larval Malphigian tubules. Viability and fertility are not significantly affected. White-eye flies are positively phototaxic, but they fail to perceive or respond to movement around them. In normal individuals pigments of two groups, the drosopterins, red in color, and the ommochromes, which are brown, are found in varying amounts in the pigmented tissues. Intermediate alleles of *white* modify amounts of one or both groups of pigments in one or more of the tissues. One of the still unanswered questions concerns the function of the *white* locus protein and how it can affect the synthesis or deposition of two unrelated groups of compounds. From the nucleotide sequence, O'Hare et al. (1984) determined that the mRNA upon translation would encode a polypeptide of 78,000 Da. The protein would be rather hydrophilic, particularly so near the C terminus. This suggests that it might be membrane-associated and therefore could be involved in pigment deposition or pigment granule structure.

5 Genetic and Molecular Fine Structure Map

Bringing together information from the genetic and molecular analyses of the gene provides a look at the organization of the locus and the nature of the mutational changes characterizing some of the alleles. Figure 1 superimposes the ge-

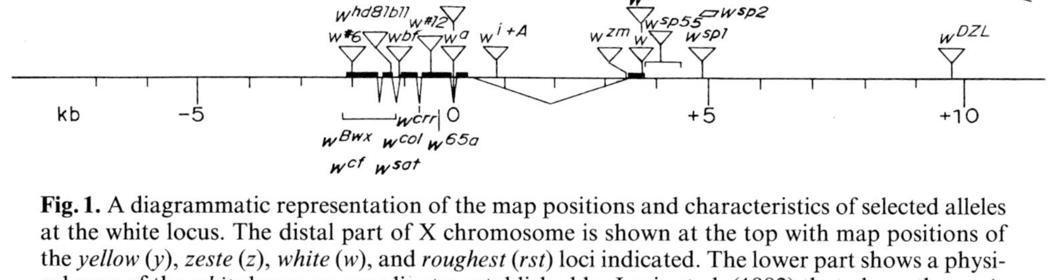

Fig. 1. A diagrammatic representation of the map positions and characteristics of selected alleles at the white locus. The distal part of X chromosome is shown at the top with map positions of the *yellow* (*y*), *zeste* (*z*), *white* (*w*), and *roughest* (*rst*) loci indicated. The lower part shows a physical map of the *white* locus on coordinates established by Levis et al. (1982) that places the *copia* insertion of w^a at coordinate 0.0. Sequences are measured in kilobases (kb), negative on the distal and positive on the proximal sides of w^a. Mutants above the line designated with triangles are associated with insertions of tansposons. See Table 1 for explanations. *Open parallelograms* designate deficiencies. Mutants listed below the line are in approximate map positions, but because they are not associated with major sequence changes have not been accurately positioned on the physical map. The transcript exons are shown by *solid bars* connected by lines indicating introns. The 5'end of the transcript is centromere proximal

netic map on a molecular map of the *white* locus to show the position and the nature of several alleles.

The work of Zachar and Bingham (1982), Collins and Rubin (1982) and Levis and Rubin (1982) showed that insertional mutagenesis by transposable elements is a major cause of spontaneous mutation at the *white* locus. As reviewed by Rubin (1983), this can be generalized to essentially all of the loci in *Drosophila* that have been studied at the molecular level. Transposons interfere with gene expression in a variety of ways, particularly at the level of transcription, and they also perturb the process of recombination and contribute significantly to chromosomal rearrangements. Analysis of some of the transposon-mediated mutations at *white* have provided considerable information about the function and regulation of the locus. Consider first some examples of the effects of transposable elements on gene expression at the level of transcription.

6 RNA Transcripts of *white* Alleles

In wild type, the major transcript of the *white* locus is a poly-$(A)^+$ RNA about 2.6 kb in length (O'Hare et al. 1983; Pirrotta et al. 1983), which is present in very low quantity at all stages of development. The total accumulation of this transcript peaks at third larval instar at a level of about 0.003% of total poly-$(A)^+$ RNA (Fjose et al. 1984). Smaller, even less abundant transcripts can be detected at some stages (Pirrotta and Bröckl 1984; Fjose et al. 1984). By determining the nucleotide sequence of more than 14 kb of DNA from the *white* locus region, and that of a cDNA clone complementary to *white* locus, O'Hare et al. (1984) identified the 3′ end of the transcript and the probable exon-intron structure of the gene. Further refinement of the gene structure by S1 nuclease mapping (Steller and Pirrotta 1985) shows six exons with interspersed introns transcribed from coordinates +3737 (5′) to −2233 (3′) as diagrammed in Fig. 1.

Table 1 summarizes some of the mutant characteristics and effects of the mutational lesions on transcription. Mutant alleles created by insertion of transposable elements are shown in the figure as triangles positioned at the sites of insertion. Those inserted into exons, such as $w^{hd81b11}$, w^{12}, w^{6} and the original *w*, produce the null, bleached white phenotype. Exceptions appear to be w^{e} and w^{h}, which are partial revertant derivatives of *w* and show some pigment deposition. The supposition is that the insertion of the original element at *w* was into a transcribed region 5′ to the start of translation (Levis et al. 1984), and the w^{e} and w^{h} alleles are modifications of the original transposon. In w^{e}, Pirrotta and Bröckl (1984) observed only a 2.6 kb RNA much reduced in abundance compared to w^{+}, while Levis et al. (1984) saw a 5 kb w^{e} RNA in addition to the 2.6 kb species. It is possible that both the 2.5 kb and 5 kb species have the same 5′ end with 3′ termini at different positions within the insertion. The termination of transcription within the insertion, particularly in the case of retrovirus-like transposons having long terminal repeats (LTR's), is a typical pattern for the effects of transposons on transcription.

Table 1. Selected *white* locus alleles

Allele		Phenotype	Mutation	Position	RNAs(kb)	Reference
w^{+}	wild type	bright red	normal		2.6	a, b
w	white	white	6kb F	+3696	–	a, c
w^{a}	apricot	yellow orange	copia	−1	5.7; 1.25	a, b, c, d, e, f, g
w^{a4}	apricot 4	yellow orange	BEL	−1	1.0	c, e, h
w^{bf}	buff	pale yellow	roo	−1134	2.1	a, c, e, g, i
w^{Bwx}	Brownex	dark brown	–	distal to w^{bf}	–	j
w^{c}	crimson	dark red	9kbFB in w^{i}	+188	–	i, k, l, m
w^{cf}	coffee	dark brown	–	distal to w^{bf}	–	j
w^{col}	colored	dark red-brown	–	distal to w^{a}	–	j
w^{crr}	carrot	orange	–	distal to w^{a}	–	d, j
w^{DZL}	dominant zeste-like	yellow orange	13kb FB	+9772	3.9	l, m, n, o
w^{h}	honey	light yellow	5.7kb F	+3696	–	c, i
$w^{hd81b11}$		white	copia	−1288	–	g, i
w^{i}	ivory	light yellow	2.96kb dup	+2795 to −173		i, l, p, q
w^{i+A}		red	3kb F	+833	2.6	e, i
w^{sat}	satsuma	red-brown	–	distal to w^{a}	–	j
w^{sp}	spotted	red spots/orange	roo	+4922	2.6; 5.5	c, e, f, g, i, r
w^{sp2}	spotted 2	red spots/orange	def	+4552 to +4668	–	g, m
w^{sp4}	spotted 4	red spots/orange	def	+4781 to +5850	–	g, i, m
w^{sp55}	spotted 55	red spots/orange	5.8kb insert	+3500 to +4700	none	e
w^{zm}	zeste mottle	brown-red mottle	BEL	+3428	2.6	c, e, g, s
$w^{\#6}$	white-6	white	1.1kb P	−2025	–	i
$w^{\#12}$	white-12	white	1.6kb P	−511	–	i

References
[a] O'Hare et al. (1983)
[b] Pirrotta et al. (1983)
[c] Zachar and Bingham (1982)
[d] Bingham and Judd (1981)
[e] Levis et al. (1984)
[f] Pirrotta and Bröckl (1984)
[g] Zachar et al. (1985)
[h] Goldberg et al. (1983)
[i] O'Hare et al. (1984)
[j] Lindsley and Grell (1968)
[k] Green (1967a)
[l] Collins and Rubin (1982)
[m] Davison et al. (1985)
[n] Bingham (1980)
[o] Bingham and Zachar (1985)
[p] Karess and Rubin (1982)
[q] Muller (1920)
[r] Lewis (1956)
[s] Judd (1963)

In those cases where transposons have inserted into introns such as w^{a}, w^{a4} and w^{bf} shown in Fig. 1, the mutation is leaky, giving partial pigmentation. In the case of w^{a} (Pirrotta and Bröckl 1984; Levis et al. 1984; Zachar et al. 1985), it appears that transcripts initiated at the appropriate 5′ site frequently terminate at one or the other, but primarily the 3′, LTR's of the transposon. The leaky nature of the mutation can be explained if some transcription continues through the transposon with termination at the proper 3′ site of *white*. Normal splicing would then remove the transposon transcript to provide a small amount of w^{+} mRNA.

The w^{i} mutation is a tandem duplication of the major part of the first intron and extending into the third exon (Karess and Rubin 1982; O'Hare et al. 1984). This mutant has very lightly pigmented eyes. A partial revertant *white-crimson* (w^{c}) arose by insertion of an FB element into the proximal copy of the duplication. This proved to be a highly mutable allele (Green 1967a) with precise excisions of FB giving reversions to w^{i} and excision of the insertion and one copy of the w^{i} duplication returning the gene to wild type function (Collins and Rubin 1982). One such wild type revertant is w^{i+A} (Fig. 1) where one copy of the duplication has been lost and a transposon has been inserted into the first intron of *white* (Karess and Rubin 1982).

The insertion of the retrovirus-like transposon *roo* 5′ to the initiation of transcription in w^{sp} has an interesting effect on *white* locus phenotype (Lewis 1956), reducing the overall pigmentation of the eye and causing a variegated pattern of pigmented spots on a light background. Despite the reduction in pigmentation, Pirrotta and Bröckl (1984) and Levis et al. (1984) find the w^{sp} mRNA to be like that of w^{+} in size, structure, and abundance. This variegated phenotype is also produced by w^{sp55}, which is due to a transposon inserted in the region of the 5′ exon of *white*. No *white* locus RNA is detected in w^{sp55} flies. The mutants w^{sp2}, w^{sp3} and w^{sp4} have similar variegated phenotypes and all are due to deletions with break points in the same 0.9 kb region (Zachar and Bingham 1982) just 1.0–1.9 kb upstream from the 5′ end of *white* locus transcription initiation (O'Hare et al. 1983). Davison et al. (1985) note that this region has some characteristics similar to the enhancer regions of SV40 and bovine papilloma virus and thus may exert positive effects on *white* locus transcription from outside the transcribed region.

Two other insertions shown in Fig. 1, w^{zm} and w^{DZL}, will be discussed in reference to the interaction of the *white* and *zeste* loci.

The alleles noted in Fig. 1, that are not associated with detectable insertions or sequence rearrangements are w^{Bwx}, w^{cf}, w^{col}, w^{crr} and w^{sat} all of which have reduced amount of pigments, and w^{65a} which is bleached-white in color. There is no available information about transcripts from these mutant strains.

7 Asymmetrical Crossingover at *white* Locus

Construction of a fine structure map of the *white* locus by intralocus recombination experiments gave results consistent with the map in Fig. 1 (see Judd 1976 for review). The picture was often complicated, however, when recombinant classes other than those expected were recovered from particular crosses (MacKendrick 1953; Green 1959; Judd 1959, 1961; Rasmuson 1962). These unusual recombinants had characteristics of duplications and deficiencies for all or part of the *white* locus and were interpreted as arising from asymmetrical pairing and crossingover within the locus. As it became possible to analyze these unusual products at the molecular level, transposons inserted in and near the *white* locus were shown to be involved in the asymmetrical exchanges. Goldberg et al. (1983) showed that the unusual products of exchange in w^{a}/w^{a4} heterozygotes arose by pairing of a copy of the transposon BEL that causes the w^{a4} mutation with a copy of BEL located some 60 kb proximal to the *white* locus in the w^{a} chromosome. Crossingover within the paired elements thus creates a duplication and a deficiency of 60 kb. The deficiency has a white-eye phenotype and is viable and fertile, while the duplication is wild-type except for its interaction with the *zeste* (*z*) locus, a topic that is discussed below.

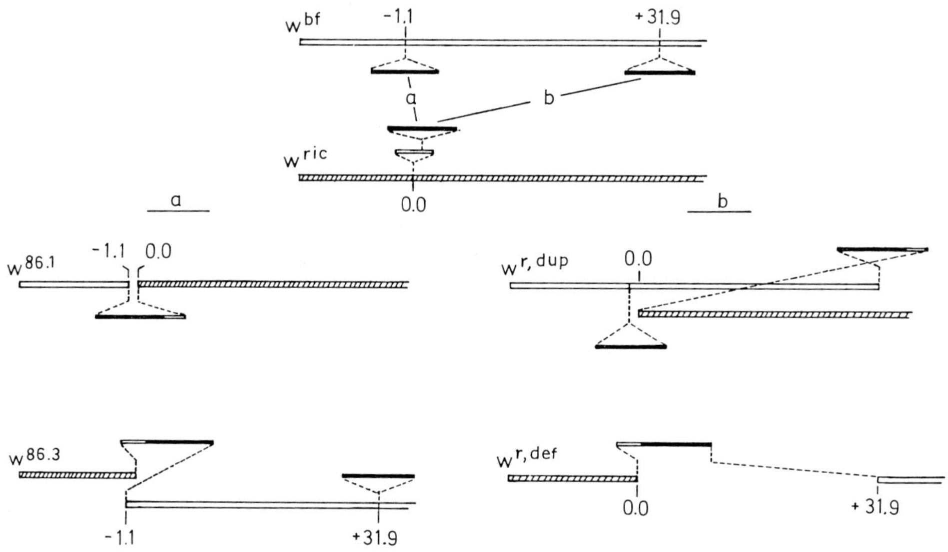

Fig. 2. A diagram of patterns of synapsis between two chromosomes, w^{bf} and w^{ric}. The w^{bf} chromosome contains two copies of the transposon *roo* (*solid bar*). The w^{ric} chromosome contains one copy of *roo* inserted into the transposon *copia* (*open bar*). Pattern (a) pairing and exchange produces reciprocal products shown as the deficiency $w^{86.1}$ and duplication $w^{86.3}$. Pairing and exchange of pattern (b) produces reciprocal products $w^{r,def}$, deficient for sequences between coordinates 0.0 and +31.9, and $w^{r,dup}$, the duplication for the same region. All products give a white-eye phenotype except $w^{r,dup}$, which has a white-buff phenotype. All products are viable and fertile

Another case of asymmetrical exchange involving transposons has been examined at the molecular level by Judd and his associates (Davis et al. 1987). From w^{bf}/w^{a} heterozygotes, Judd (1959, 1961) reported that none of the expected classes of recombinants appeared but four classes that were unexpected were recovered. The molecular analysis showed that the w^{bf} chromosome contains a transposon, *roo* in the *white* locus that accounts for the w^{bf} mutations and another copy of *roo* some 33 kb proximal to the w^{bf} site. The w^{a} chromosome contains a copy of *copia* at the w^{a} site but in the strain used by Judd for these experiments a copy of *roo* had invaded *copia* without changing the apricot phenotype perceptibly. Asymmetrical pairing and exchange between the copies of *roo* in these chromosomes then occurs in two patterns, producing two sets of duplication and deficiency products as shown in Fig. 2. The molecular analyses of chromosomes such as these have provided explanations for unusual recombination events that have been puzzles for decades. Further, the role of transposons in the processes of mutation and gene transmission has been brought into sharper focus.

8 Expression and Regulation of *white*

Several aspects of *white* locus expression and regulation can be addressed through mutant phenotypes and the molecular characterization of the normal and mutant alleles. It should also be noted, however, that a most powerful and direct approach to these questions has been through the construction of P element transposon vectors (Spradling and Rubin 1982; Rubin and Spradling 1982) carrying different *white* locus DNA sequences used to transform flies that were then tested for various aspects of *white* locus expression (Hazelrigg et al. 1984; Gehring et al. 1984; Levis et al. 1985 a and b; Pirrotta et al. 1985). It has been established, most definitively by Levis et al. (1985 b) and Pirrotta et al. (1985), that multiple *cis*-acting regulatory elements in the 5′ flanking region of the gene are responsible for the control of expression. These studies show that the DNA sequences required in *cis* for normal *white* locus expression are contained within a 9.9 kb segment extending from coordinates +5.6 to −4.3 on the map in Fig. 1.

9 Dosage Compensation

As is the case for most X chromosome genes, *white* shows a dosage compensation such that males with one copy express the gene as strongly as females with two copies. Green (1959) noted that mutants unable to compensate for dosage map to the proximal part of the gene. The transposon constructs of Levis et al. (1985 b) and Pirrotta et al. (1985) identify the region responsible for dosage compensation to be within a few hundred base pairs of the start of transcription, or possibly within the 5′ transcribed region. The alleles that fail to compensate, the primary one being w^{e}, are located precisely in that region (Fig. 1). The mechanistic basis for dosage compensation continues to go unrecognized, however.

10 Expression in Testes and Malphigian Tubules

Other aspects of regulation relate to expression of the gene in the testes and Malphigian tubules. The transformation experiments identify a region necessary for Malphigian tubule pigmentation that is not necessary for expression in the eye or in the testes. It is located within a region between 0.74 and 0.22 kb 5′ to the transcribed region. Testes pigmentation is produced by transposons containing 1.9 kb 5′ to the start of transcription but is not carried out by those carrying only 1.1 kb 5′ (Levis et al. 1985 b; Pirrotta et al. 1985). The w^{sp3} mutation, which is a deletion beginning about 0.9 kb 5′ to the start of transcription and removing sequences upstream from there (Davison et al. 1985), gives unpigmented testes, while w^{sp}, a *roo* insertion at about 1.2 kb upstream from the transcribed region, and w^{sp2}, a deletion of about 0.1 kb in that region, appear to enhance testes pigmentation (Pirrotta et al. 1985).

11 *zeste-white* Interaction

An aspect of *white* expression that has drawn much attention is its interaction with the *zeste* (*z*) locus. First described by Gans (1953), *zeste,* only 0.5 map units distal to *white,* also affects eye pigmentation, with z^1/z^1 females having lemon-yellow eye color. The hemizygous z^1 male on the other hand has wild-type eye color. Gans determined that this is because the male has only one copy of w^+. Males with tandemly duplicated *white* loci express a zeste phenotype while homozygous z^1 females with one *white* gene deleted are wild-type. Gans also discovered that some *white* locus alleles act essentially as dominant suppressors of the *zeste* phenotype in homozygous z^1 females. Green (1959) established that such suppressor alleles map in the proximal portion of the gene, and this was followed (Judd 1961; Green 1963) with the demonstration that direct tandem duplication of only the proximal part of w^+ promotes the *zeste* phenotype in z^1 males while the complementary deletion results in wild-type phenotype in homozygous z^1 females.

In her original paper, Gans (1953) had noted that translocations of w^+ to new positions produced wild-type phenotype in homozygous z^1 flies even though two copies of w^+ were present and Gelbart (1971) confirmed those results. These interactions were indeed puzzling and became even more so as experiments such as those of Green (1967 b) were added. Green X-rayed males carrying tandem duplication or triplication of the proximal part of w^+ and looked for changes in the capacity of such a duplication to enhance the expression of z^1 in males. A total of 19 cases were recovered where males showed either wild-type or zeste-variegated phenotypes. Cytological analysis of 13 of them showed that all had a chromosome break in or near the *white* locus. The curious result was that females homozygous for the rearranged X chromosome were *zeste* in phenotype while the males were not.

All these observations neatly fit an interpretation proposed by Jack and Judd (1979), who presented data that z^1 and z^+ are antagonistic in expression and proposed that z^1 represses w^+ activity but that the repression is effective only if there are two or more w^+ genes in close proximity, i.e., as in chromosomes that can pair somatically or in tandem array in the same chromosome. Jack and Judd interpreted the *zeste-white* interaction as a version of the transvection phenomenon described for certain alleles of the *bithorax* complex by Lewis (1954). It is of interest to note here that the z^a allele, which is the null state of the locus, acts as an enhancer of certain *bithorax* alleles (Kaufman et al. 1973). Gelbart and Wu (1982) added further information supporting the transvection model for the *zeste-white* interactions and noted that *zeste* mutations can also interact with the *decapentaplegic* gene complex (*dpp*) in much the same manner as with *bithorax*. Lifschytz and Green (1984) and Green (1984) used the *zeste-white* interactions to search for and characterize additional classes of mutations at the *zeste* locus in an effort to extend the understanding of the regulating mechanisms involved.

The *zeste-white* interaction has enjoyed considerable attention at the molecular level. Because the interaction could be perturbed by changing the spatial relationships between w^+ genes, and because there were specific mutations that do not respond to the action of *zeste,* some of which show some allelic complementation that also is proximity-dependent (Jack and Judd, unpublished observations), Jack and Judd (1979) were prompted to speculate that the interaction might be modulated by RNA produced in the proximal portion of the *white* locus. However, studies of the *white* locus transcripts in z^+ and z^1 individuals do not support this model. O'Hare et al. (1983) showed that the size and abundance of *white* locus poly(A)$^+$ RNA from adult flies is the same for both genotypes. Essentially the same result was reported by Pirrotta and Bröckl (1984). However, ingham and Zachar (1985) examined transcripts in adult head and body tissues and reported that in z^1 flies *white* transcript is reduced in quantity in heads. The same effect was noted for the mutant w^{DZL}, which is due to the insertion of a transposon at coordinate +9.77 (Levis and Rubin 1982; Zachar and Bingham 1982; O'Hare et al. 1983; Pirrotta and Bröckl 1984). The w^{DZL} allele produces a *zeste*-like phenotype in w^{DZL}/w^+ and w^{DZL}/w^{DZL} individuals. Its ability to suppress w^+ activity in heterozygotes is, like the z^1 effect, proximity- or pairing-dependent (Bingham 1980).

The region of the *white* gene that is required for *zeste* interaction has been identified through the transformation experiments described above by Hazelrigg et al. (1984), Levis et al. (1985 b) and Pirrotta et al. (1985). Those sequences are positioned in the region between 1.1 and 1.9 kb upstream of the transcribed region. It is important to note that the w^{sp} alleles that show partial complementation with other *white* alleles that is proximity- or pairing-dependent (Jack and Judd, unpublished observations) are located in this region also.

The mutant w^{zm}, due to the insertion of a transposon near the 5′ junction of the first intron, has an interesting *zeste* interaction. In z^+ flies, w^{zm} produces a wild-type eye color and the RNA transcript is like that in wild-type. In z^1w^{zm} males there is a definite reduction in pigmentation and a mottled pattern of pigment deposition (Becker 1960; Judd 1963); the transcript remains normal, however (Levis et al. 1984). A spontaneous derivative of w^{zm} called w^{zl} (Becker 1960) has a definite clonal pattern of wild-type pigmentation on a *zeste* background in z^1w^{zl} males. In z^+w^{zl} flies, however, the phenotype is completely normal. Another derivative of this w^{zm} family involved the transposition of the *white* and *roughest* loci to the left arm of the third chromosome at 61D (Judd 1975). The transposition called $Tp(1 \rightarrow 3)w^{zh}$ created a deficiency, $Df(1)w^{zh}$, in the X chromosome. Function of *white* is normal in flies carrying one copy of the transposition. However, when the transposition is made homozygous in z^1 flies, pigment is deposited around the periphery of the eyes grading to *zeste* color at the center. Various spontaneous derivatives of the transposition modify the position of the pigmented areas around the edge of the eye and the amount of pigment deposited. Most interesting is that in the z^1; $Tp(1 \rightarrow 3)w^{zh}$ homozygotes, *white* function is not autonomous and the pigmented regions are not clonal in origin. Such shifts from autonomous to nonautonomous expression are seen in some cases of *white* locus transformants using transposon vectors (Hazelrigg et al. 1984; Levis et al. 1985). This type of expression appears as a part of the *zeste-white* interaction and seems dependent on the genomic position where *white* has been inserted. This neighborhood effect may be a way of exploring the mechanistic basis for the *cis* and *trans* regulation of *white* locus expression.

Clearly the last chapter on understanding *white* locus regulation, particularly the *cis*-acting elements that in close proximity to other *white* genes can have an effect in *trans*, has not yet been written. Despite having a complete nucleotide sequence of the gene, the ability to design modified versions of the locus and use them in transformation experiments, and despite all the information about the transcripts and where and when they are found, we have still a lot to learn about how the *white* locus functions. To put it in perspective, however, this gene has served 20th century geneticists very well indeed. I believe Frank Lutz and Thomas Hunt Morgan would be pleased.

References

Becker HJ (1960) Variegation in the zeste eye color alleles and its bearing on gene action during the development of the eye of *Drosophila melanogaster*. Genetics 45:519–534
Beermann W (1972) Chromomeres and genes. In: Beermann W (ed) Results and problems in cell differentiation, vol 4. Springer, Berlin Heidelberg New York, pp 1–33
Bingham PM (1980) The regulation of white locus expression: A dominant allele at the white locus of *Drosophila melanogaster*. Genetics 95:341–353
Bingham PM, Judd BH (1981) A copy of the *copia* transposable element is very tightly linked to the w^{a} allele at the *white* locus of *D. melanogaster*. Cell 25:705–711
Bingham PM, Zachar Z (1985) Evidence that two mutations, w^{DZL} and z^{1}, affecting synapsis-dependent genetic behavior of *white* are transcriptional regulatory mutations. Cell 40:819–825
Bingham PM, Levis R, Rubin GM (1981) Cloning of DNA sequences from the white locus of *D. melanogaster* by a novel and general method. Cell 25:693–704
Bridges CB (1938) A revised map of the salivary gland X chromosome of *Drosophila melanogaster*. J Hered 29:11–13
Bridges CB, Morgan TH (1923) The third-chromosome group of mutant characters of *Drosophila melanogaster*. Carnegie Inst Wash 327
Collins M, Rubin GM (1982) Structure of the Drosophila mutable allele *white-crimson*, and its *white-ivory* and wild-type derivatives. Cell 30:71–79
Davis PS, Shen MW, Judd BH (1987) Asymmetrical pairing of transposons in and proximal to the white locus of *Drosophila* account for four classes of regularly-occurring exchange products. Proc Natl Acad Sci USA 84:174–178
Davison D, Chapman CH, Wedeen C, Bingham PM (1985) Genetic and physical studies of a portion of the white locus participating in transcriptional regulation and in synapsis-dependent interactions in Drosophila adult tissues. Genetics 110:479–494
Fjose A, Polito LC, Weber U, Gehring WJ (1984) Developmental expression of the *white* locus of *Drosophila melanogaster*. EMBO J 3:2087–2094
Gans M (1953) Etude génétique et physiologique du mutant *z* de *Drosophila melanogaster*. Bull Biol Fr Belg (Suppl) 38:1–90
Gehring WJ, Paro R (1980) Isolation of a hybrid plasmid with homologous sequences to a transposing element of Drosophila melanogaster. Cell 19:897–904
Gehring WJ, Klemenz R, Weber U, Klotter U (1984) Functional analysis of the $white^{+}$ gene of *Drosophila* by P-factor-mediated transformation. EMBO J 3:2077–2085
Gelbart WM (1971) Cytogenetics of *zeste* expression in *Drosophila melanogaster*. Thesis, Univ Wisconsin-Madison, WI
Gelbart WM, Wu C-T (1982) Interactions of *zeste* mutations with loci exhibiting transvection effects in *Drosophila melanogaster*. Genetics 102:179–189
Goldberg ML, Paro R, Gehring WJ (1982) Molecular cloning of the *white* locus region of *Drosophila melanogaster* using a large transposable element. EMBO J 1:93–98

Goldberg ML, Sheen J-Y, Gehring WJ, Green MM (1983) Unequal crossing-over associated with asymmetrical synapsis between nomadic elements in the *Drosophila melanogaster* genome. Proc Natl Acad Sci USA 80:5017–5021

Green MM (1959) Spatial and functional properties of pseudoalleles at the white locus in *Drosophila* melanogaster. Heredity 13:302–315

Green MM (1963) Unequal crossing-over and the genetical organization of the white locus of *Drosophila melanogaster*. Z Vererbungsl 94:200–214

Green MM (1967a) The genetics of a mutable gene at the *white* locus of *Drosophila melanogaster*. Genetics 56:467–482

Green MM (1967b) Variegation of the eye color mutant *zeste* as a function of rearrangements at the *white* locus in *Drosophila melanogaster*. Biol Zentralbl 86 (Suppl):211–220

Green MM (1984) Genetic instability in *Drosophila melanogaster:* Transpositions of the white gene and their role in the phenotypic expression of the zeste gene. Mol Gen Genet 194:275–278

Hazelrigg T, Levis R, Rubin GM (1984) Transformation of *white* DNA in Drosophila: Dosage compensation, *zeste* interaction, and position effects. Cell 36:469–481

Jack JW, Judd BH (1979) Allelic pairing and gene regulation: A model for the zeste-white interaction in *Drosophila melanogaster*. Proc Natl Acad Sci USA 76:1368–1372

Judd BH (1959) Studies on some position pseudoalleles at the white region in *Drosophila melanogaster*. Genetics 44:34–42

Judd BH (1961) Formation of duplication-deficiency products by asymmetrical exchange within a complex locus of *Drosophila melanogaster*. Proc Natl Acad Sci USA 47:545–550

Judd BH (1963) The genetic fine structure of the mutants z^{m} and z^{1} in *Drosophila melanogaster*. Genetics Today. Proc XI Int Cong Genet 1:3–4

Judd BH (1975) Genes and chromomeres of Drosophila. In: Peacock WJ, Brock RD (eds) The eukaryote chromosome. Aust Natl Univ Press, Canberra, pp 169–184

Judd BH (1976) Complex loci. In: Asburner M, Novitski E (eds) The genetics and biology of *Drosophila* 1b:767–799

Judd BH, Shen MW, Kaufman TC (1972) The anatomy and function of a segment of the X chromosome of *Drosophila melanogaster*. Genetics 71:139–156

Karess RE, Rubin GM (1982) A small tandem duplication is responsible for the unstable *white-ivory* mutation in Drosophila. Cell 30:63–69

Kaufman TC, Tasaka SE, Suzuki DT (1973) The interaction of two complex loci, zeste and bithorax in *Drosophila melanogaster*. Genetics 75:299–321

Lefevre G Jr, Green MM (1972) Genetic duplication in the white-split interval of the X chromosome in *Drosophila melanogaster*. Chromosoma 36:391–412

Lefevre G Jr, Wilkins MO (1966) Cytogenetic studies on the white locus in *Drosophila melanogaster*. Genetics 53:175–187

Levis R, Rubin GM (1982) The unstable w^{DZL} mutation of Drosophila is caused by a 13 kilobase insertion that is imprecisely excised in phenotypic revertants. Cell 30:543–550

Levis R, Bingham PM, Rubin GM (1982) Physical map of the *white* locus of *Drosophila melanogaster*. Proc Natl Acad Sci USA 79:564–568

Levis R, O'Hare K, Rubin GM (1984) Effects of transposable element insertions on RNA encoded by the *white* gene of Drosophila. Cell 38:471–481

Levis R, Hazelrigg T, Rubin GM (1985a) Effects of genomic position on the expression of transduced copies of the *white* gene of Drosophila. Science 229:558–561

Levis R, Hazelrigg T, Rubin GM (1985b) Separable *cis*-acting control elements for expression of the *white* gene of Drosophila. EMBO J 4:3489–3499

Lewis EB (1952) The pseudoallelism of white and apricot in *Drosophila melanogaster*. Proc Natl Acad Sci USA 38:953–961

Lewis EB (1954) The theory and application of a new method of detecting chromosomal rearrangements in *Drosophila melanogaster*. Am Nat 88:225–239

Lewis EB (1956) An unstable gene in *Drosophila melanogaster*. Genetics 41:651

Lifschytz E, Green MM (1984) The *zeste-white* interaction: induction and genetic analysis of a novel class of *zeste* alleles. EMBO J 3:999–1002

Lindsley DL, Grell EH (1968) Genetic variations of *Drosophila melanogaster*. Carnegie Inst Wash 627

Lutz FE (1941) A lot of insects. Putnam, New York
MacKendrick ME (1953) Further examples of crossing over between alleles of the *w* series. Dros Inform Serv 27:100
MacKendrick ME, Pontecorvo G (1952) Crossingover between alleles at the *w* locus in *Drosophila melanogaster*. Experientia 8:390–391
Morgan TH (1910) Sex limited inheritance in *Drosophila*. Science 32:120–122
Morgan TH, Bridges CB, Sturtevant AH (1925) The genetics of *Drosophila*. Bibliogr Genet II
Muller HJ (1920) Further changes in the white-eye series of *Drosophila* and their bearing on the manner of occurrence of mutation. J Exp Zool 31:443–473
O'Hare K, Levis R, Rubin GM (1983) Transcription in the *white* locus in *Drosophila melanogaster*. Proc Natl Acad Sci USA 80:6917–6921
O'Hare K, Murphy C, Levis R, Rubin GM (1984) DNA sequence of the *white* locus of *Drosophila melanogaster*. J Mol Biol 180:437–455
Pirrotta V, Bröckl C (1984) Transcription of the *Drosophila white* locus and some of its mutants. EMBO J 3:563–568
Pirrotta V, Hadfield C, Pretorius GHJ (1983) Microdissection and cloning of the *white* and 3B1-3C2 region of the *Drosophila* X chromosome. EMBL J 2:927–934
Pirrotta V, Steller H, Bozzetti MP (1985) Multiple upstream regulatory elements control the expression of the *Drosophila white* gene. EMBO J 4:3501–3508
Rasmuson B (1962) An intragenic duplication in *Drosophila melanogaster* and its significance for gene function. Hereditas 48:587–611
Rasmuson B, Green MM, Ewertson G (1960) Qualitative and quantitative analyses of eye pigments and pteridines in back-mutations of the mutant w^a in *Drosophila melanogaster*. Hereditas 46:635–650
Rubin GM (1983) Dispersed repetitive DNAs in *Drosophila*. In: Shapiro JA (ed) Mobile genetic elements. Academic Press, New York
Rubin GM, Spradling AC (1982) Genetic transformation of Drosophila with transposable element vectors. Science 218:348–353
Sorsa V, Green MM, Beermann W (1973) Cytogenetic fine structure and chromosomal localization of the white gene in *Drosophila melanogaster*. Nature New Biol 245:34–37
Spradling AC, Rubin GM (1982) Transposition of cloned P elements into *Drosophila* germ line chromosomes. Science 218:341–347
Steller H, Pirrotta V (1985) Expression of the *Drosophila white* gene under the control of the *hsp70* heat shock promoter. EMBO J 4:3765–3772
Sturtevant AH (1913) The linear arrangement of six sex-linked factors in Drosophila, as shown by their mode of association. J Exp Zool 14:43–59
Young MW, Judd BH (1978) Nonessential sequences, genes and the polytene chromosome bands of *Drosophila melanogaster*. Genetics 88:723–742
Zachar Z, Bingham PM (1982) Regulation of *white* locus expression: the structure of mutant alleles at the *white* locus of *Drosophila melanogaster*. Cell 30:529–541
Zachar Z, Davison D, Garza D, Bingham PM (1985) A detailed developmental and structural study of the transcriptional effects of insertion of the *copia* transposon into the white locus of *Drosophila melanogaster*. Genetics 111:495–515

The Genetic and Molecular Organization of the Dense Cluster of Functionally Related, Vital Genes in the DOPA Decarboxylase Region of the *Drosophila melanogaster* Genome

T. R. F. WRIGHT[1]

1 Introduction

The presence in the genome of clusters of functionally related genes exhibiting some sequence homology with each other has generally been interpreted to be due to the previous occurrence of the appropriate number of gene duplication events along with varying amounts of subsequent, divergent evolution of the duplicated genes. Just how far the genes within a cluster have diverged probably has been dependent on a number of different factors such as how early in the phylogeny the duplication event occurred, dosage considerations, i.e., the need for large amounts of more or less identical proteins or RNA molecules, and selective pressures to maintain similar, but new functions. Furthermore, the existence of clusters of duplicated genes, some of which may include pseudogenes, may be due to random chance, with the duplications providing no selective advantage or disadvantage to the organism (Loomis and Gilpin 1986). On the other hand, clusters of homologous, but diverged genes with related functions may be maintained in the genome to facilitate the coordinate or sequential regulation of genes in time and space.

Gene clusters in *Drosophila melanogaster* can be classified into four categories; reiterated genes, gene complexes, functionally but not evolutionarily (structurally) related clusters, and functionally and evolutionarily (structurally) related clusters. Examples of reiterated genes are the rRNA genes (Long and Dawid 1980) and the histone gene cluster (Lifton et al. 1977). A gene complex is defined by the presence of two or more adjacent transcription units that interact in *cis* with the primary example being the Bithorax Complex (Lewis 1978). There are very few examples of functionally related but not evolutionarily (structurally) related clusters that are well-documented. An example is found in the 88F region, where the indirect flight muscle specific actin gene, *act88F,* is located within 140 kb of two transcription units which code for tropomyosin isoforms and tropomyosinlike proteins, which are also found in indirect flight muscle (Karlik et al. 1984). On the other hand, there are numerous examples of gene clusters which include both functionally and evolutionarily related genes. Some of these are characterized briefly below.

In *Drosophila melanogaster* most of the clustered genes which exhibit sequence homologies code for nonenzymatic proteins. These include the two chorion protein gene clusters at 7F1–2 and 66D11–15 (Spradling et al. 1980; Spradling 1981),

[1] Department of Biology, University of Virginia, Charlottesville, Virginia 22109, USA.

Results and Problems in Cell Differentiation 14
Structure and Function of Eukaryotic Chromosomes
Edited by W. Hennig

the two yolk protein genes separated from each other by 1 kb at 8F–9A (Barnett et al. 1980) and the three salivary gland glue protein genes within 5 kb of DNA at 68C (Meyerowitz and Hogness 1982; Crowley et al. 1983). In addition, there are seven small heat shock protein genes clustered within 15 kb of DNA at 67B that crosshybridize with each other by means of two conserved regions, one of which is an 83 amino acid sequence showing strong homology to the mammalian α-crystallin B2 chain (Ayme and Tissières 1985). Another heat shock protein, *hsp70,* is coded for by several duplicated genes at 87A and 87C (Moran et al. 1979). Furthermore, there are two clusters of genes within 11 kb of each other at 44D on the right arm of the second chromosome. One cluster includes the four homologous larval cuticle protein genes plus one pseudogene within 7.9 kb of DNA (Snyder et al. 1981; Snyder et al. 1982a). The other cluster consists of three genes that are 55 to 60% homologous in their DNA sequences (Snyder and Davidson 1983). These genes, designated H, D, and L, are located within an 8 kb region of DNA and show no homology with the larval cuticle protein genes 11 kb away. Sequence analysis shows that a 12 amino acid sequence in all three genes is "very similar" to the calcium-binding region of myosin light chains, troponin C, parvalbumin, and calmodulin. The larval cuticle protein genes are expressed coordinately primarily throughout the third larval instar whereas the H, D, and L genes are expressed during the first, second, and early third larval instars and in adults. Thus, although the genes within a cluster are coordinately expressed, the two clusters located in a 27 kb region of DNA are not (Snyder and Davidson 1983). Although most of the genes in the clusters listed above are coordinately expressed in the same tissue and at the same time in development or respond to the same stimulus, i.e., heat shock and/or ecdysone, some individual genes within the clusters may also be expressed separately at other times and in other tissues, e.g., *hsp26* and *hsp28* in the 67B cluster (Zimmerman et al. 1983), indicating that genes even within a cluster are subject to differential and multiple regulation (Cohen and Meselson 1985).

A striking example of differentially expressed genes in a cluster is found in the tightly linked cluster of seven genes located within approximately 10 kb of DNA at the 71E ecdysone puff site (Restifo and Guild 1986). All the genes are expressed in larval or prepupal salivary glands, but six of them are expressed as "late" ecdysone genes in prepupae, and in contrast one is expressed as an "intermolt" gene in late third instar larvae (for a review of ecdysone-induced puffing see Ashburner et al. 1975). The six "late" genes are arranged as three divergently transcribed gene pairs with the seventh "intermolt" gene interspersed between two of the gene pairs. Restifo and Guild (1986) have yet to detect any cross-homology among the seven genes, using stringent hybridization conditions. It will be most interesting, but unlikely, if the six "late" genes continue to show no cross-homology after lower stringency hybridization and sequencing. Restifo and Guild (1986) suggest the "intermolt" gene codes for a salivary gland glue protein and that six "late" gene code for a group of small molecular weight proteins which accumulate during the prepupal period in the pupation fluid (Sarmiento and Mitchell 1982). Thus with the one exception, this cluster of genes is expressed coordinately at one time in development and in one tissue and is probably producing related low molecular weight proteins with similar, unknown functions.

The *serendipity* (*sry*) gene cluster at 99D is an interesting mixture of sequence-related and unrelated genes (Vincent et al. 1985). Five genes are located within 8.4 kb of DNA with the two flanking genes structurally unrelated to each other or to the three interior genes. One of the flanking genes codes for the ribosomal protein rp49, and it and the other flanking gene are transcribed in the opposite direction from the three internal genes. The three internal genes, *beta, alpha,* and *delta,* in that order 5′ to 3′, produce five different, partially overlapping, developmentally regulated mRNA's which are all present maximally at the blastoderm stage of embryogenesis. Two of the mRNA's, D and E, are maternal, being synthesized during oogenesis, while the other three, A, C, and B, appear between 0 and 2 h of embryogenesis. All five are degraded at different times and at different rates subsequent to the blastoderm stage. mRNA's E, C, and D are coded for by *beta, alpha,* and *delta,* respectively. mRNA B contains all the *beta* and *alpha* coding sequences plus the spacer sequences between *beta* and *alpha,* and mRNA A similarly contains all the coding sequences for *alpha* and *delta* plus the spacer sequences between *alpha* and *delta.* Vincent et al. (1985) suggest that B and A are read-through transcripts that fail to undergo 3′ cleavage and polyadenylation. The conceptualized *beta* and *delta* polypeptides are partially homologous, suggesting an evolutionary relationship, but *alpha* located between *beta* and *delta* shows no homology with them. The functions of the proteins coded for by these developmentally regulated mRNA's are unknown.

It should be apparent that the consequences of inactivating by mutation a single member of a functionally and structurally related gene cluster would have different phenotypic effects, depending on how the individual genes had diverged from one another in evolution and whether they had acquired related, but substantially distinct, functions not provided by any other gene in the genome. If the different, individual functions are vital, inactivating mutations will be lethal. Considering the time in development when the *beta, alpha,* and *delta sry* genes are expressed, one would predict that mutations in these genes would be lethal (see also the *Ddc* cluster below). If the individual functions are required for normal female or male fertility, the inactivating lesions will be female and/or male sterile mutations, e.g., the female sterile mutation *fs(1)1163* affecting the production of yolk protein YP1, but not YP2, (Bownes and Hodson 1980; Giorgi and Postlethwait 1985) may be just such a mutation (see also the *Ddc* cluster below). If the individual genes of a cluster have not diverged significantly, inactivity of members should have little consequence unless quantitative considerations are critical. For example, Synder et al. (1982 b) report the inactivation of one of the larval cuticle protein genes in the 44D cluster by the insertion of a transposable element with no apparent phenotypic effects. Furthermore, all three clustered salivary gland glue protein genes in 68C5,6 are dispensable, with no phenotypic effects even when all three are simultaneously deleted in overlapping double deficiency heterozygotes (Crosby and Meyerowitz 1986) probably because other glue protein genes are active.

The DOPA decarboxylase (*Ddc*) gene cluster, subject of this chapter, is not strictly comparable to any of the gene clusters mentioned above. Almost all the genes in this region are functionally related, being involved in catecholamine metabolism, in the formation, sclerotization, and pigmentation of cuticle, and in female fertility. Similar to other densely clustered genes, recent evidence on sequence homologies indicates that the genes are evolutionarily related. However, unlike most other clustered genes investigated to date in *Drosophila,* sufficient genetic and biochemical information has been accumulated to conclude that most of the genes in this region have evolved far enough from each other to have acquired separate, vital functions and that the proteins specified are not all structural proteins, but that at least some of the genes in the region specify related enzymes in the same pathway.

2 Genetic, Cytological, and Molecular Organization of the *Ddc* Cluster

The *Ddc* region which encompasses the *Ddc* gene cluster is arbitrarily defined by *Df(2L)TW130,* 37B9-C1,2;37D1–2, an 8- to 12-band deletion near the proximal end of the left arm of the second chromosome (Wright et al. 1981 a). The uncertain band count is due to the difficulty of determining the precise site of the distal breakpoint of *Df(2L)TW130* in relation to the four very fine bands 37B10, 11, 12, and 13 (Wright et al. 1981 a). Wright et al. (1981 a, 1982; T. R. F. Wright, unpublished) have screened more than 23,000 mutagenized chromosomes over *Df(2L)TW130* for recessive lethals and approximately 12,000 mutagenized chromosomes for female sterile mutations. To date approximately 315 "point" mutations have been isolated in this 8–12 band region including 43 mutations isolated by other investigators. Of these 315 mutations, 282 are lethals assigned to 16 vital genes (lethal complementation groups), 5 are in the nonvital, visible gene *hook* (*hk*), and 28 are female steriles with 13 alleles assigned to a single female sterile locus, *fs(2)TW1*. Thus 18 mutant complementation groups (genes) have been established in the *Ddc* region. Since mutations recovered from among the last 5000 chromosomes screened (mostly mutagenized with both EMS and γ-rays) permitted the identification of four of the genes in the region for the first time, one is reluctant to conclude that the region has been saturated (see also the molecular data below). Using overlapping deficiencies to localize the genes cytologically, it is clear that they are not randomly distributed (see Table 1). At the two extremes seven genes have been localized in the one- to three-band region 37C1,2–C4, no mutations have been isolated in the one- to two-band region 37C7–D1 and only two genes have been assigned to the three- to six-band region 37C3–C7. Although additional deficiencies with breakpoints in the *Df(2L)TW130* region have made it possible to assign the 18 genes to 10 genetic regions (Fig. 1), the cytological localization of these breakpoints has not further refined the gene-to-band assignments beyond those presented in Table 1.

Approximately 160 kb of DNA in the *Ddc* region have been cloned (Hirsh and Davidson 1981; Gilbert et al. 1984; Steward et al. 1984; E. S. Pentz, G. R. Hankins and T. R. F. Wright, unpublished). The physical location in the cloned DNA of 17 different deficiency breakpoints (11 depicted in Fig. 1) make it possible to assign all 18 genes to defined segments of the cloned DNA (Gilbert et al. 1984; Pentz and Wright 1986; E. S. Pentz, G. R. Hankins and T. R. F. Wright, unpublished).

Table 1. Nonrandom gene to band distributions in the *Ddc*-region

No. of genes	No. of bands	Genes	Bands	Distal limit	Proximal limit
18	8–12	All genes listed below	37B10–D1	Df(2L)TW130^{D} = 37B9–C1,2	Df(2L)TW130^{P} = 37D1–2
6	1–4	*hk, l(2)37Ba, Bc Be, Bb, Dox-A2*	37B10–13	Df(2L)TW130^{D} = 37B9–C1,2	Df(2L)hk18^{P} = 37B12–C1,2
3	1–2	*l(2)37Bg, amd*	37B13–C1,2	Df(2L)hk18^{P} = 37B12–C1,2	Df(2L)VA17^{D} = 37C1,2
7	1–3	*Ddc, l(2)37Cc, Cb Cd, Ca, Cg, Ce*	37C1,2–C4	Df(2L)VA17^{D} = 37C1,2	Df(2L)VA12^{D} = 37C1,2–C5
2	3–6	*l(2)37Cf, fs(2)TW1*	37C3–C7	Df(2L)VA12^{D} = 37C1,2–C5	Df(2L)E71^{P} = 37C6–D1
0	1–2		37C1–D1	Df(2L)E71^{P} = 37C6–D1	Df(2L)TW130^{P} = 37D1–2

Data from Wright et al. 1981a; Gilbert et al. 1984; Wright, unpublished.
Band counts from electron micrographs of Saura and Sorsa (1979) including a previously unidentified band designated 37C5.5 (Wright et al. 1981a) and counting 37C1,2 as a single band (Saura and Sorsa, 1979).
Superscripts D and P designate distal and proximal deficiency breakpoints respectively.

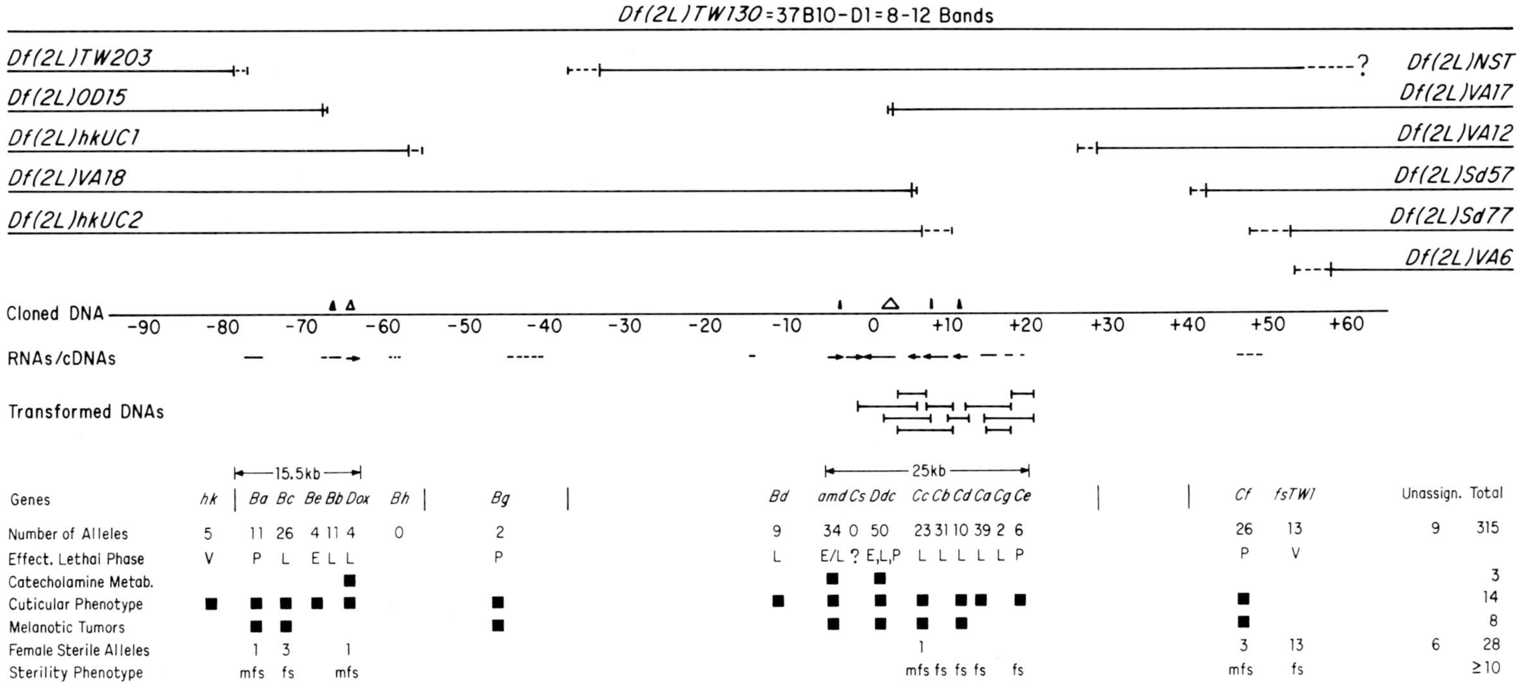

Fig. 1. The genetic and molecular organization of the *Ddc*-region. Deficiencies: *Solid lines* represent deleted DNA with *dashed lines* indicating uncertainty of the position of the breakpoint. Cloned DNA coordinates in kb from Gilbert et al. (1984). *Small triangles* above the cloned DNA line physically locate small deletion mutations and *short lines underneath* designate regions which hybridize to mRNA's or cDNA's with *arrowheads* representing direction of transcription. Transformed DNA lines indicate the segments of DNA that have been transformed by P elements. All the gene symbols except *hk*, *Dox*=*Dox-A2*, *Bh*, *amd*=*l(2)amd*, *Cs*, *Ddc*, and *fsTW1*=*fs(2)TW1* should be preceded by "*l(2)37*", e.g., *l(2)37Ba*. Effective lethal phase designations: *E* embryonic; *L* larval; *P* pupal; *V* viable. *Solid squares* underneath a gene symbol indicate that mutant alleles of that gene alter catecholamine metabolism, express a mutant cuticular phenotype, or produce melanotic tumors. Sterility phenotype: Individuals hemizygous for female sterile, ts, or hypomorphic alleles or heterozygous for complementing heteroalleles are female sterile=*fs* or both male and female sterile=*mfs*. See text for the sources of the information included in this figure

In addition, 12 mutations have been physically located in the cloned DNA by screening Southern blots for differences in restriction digests of genomic DNA from a total of 149 of the 315 mutations in the *Ddc* region. Except for one putative insertion mutation of *hk,* all of the physically detected mutations appear to be small deletions (Gilbert et al. 1984; Pentz and Wright 1986; Black et al. 1987; G.R. Hankins, J. Kullman, E.S. Pentz and T.R.F. Wright, unpublished). The sites of six of these small mutations are designated in Fig. 1 by triangles immediately above the cloned DNA line.

The most striking feature of the *Ddc* region as shown in Fig. 1 is the location of at least 13 of the 18 genes in two dense subclusters. The proximal subcluster includes eight mutually exclusive lethal complementation groups, *amd, Ddc, l(2)37Cc, Cb, Cd, Ca, Cg,* and *Ce,* plus one unmutated transcription unit *Cs* in a maximum of 25 kb of DNA of which at least 70% is transcribed. The distal subcluster consists of five mutually exclusive complementation groups, *l(2)37Ba, Bc, Be, Bb,* and *Dox-A2* in a maximum of 15.5 kb of DNA. In addition, *hook* may belong to this subcluster, being grossly located by the insertion mutation in the 7-kb fragment immediately distal to the *Df(2L)TW203* breakpoint. Of the four remaining "scattered" genes in the overall *Ddc* region cluster, two, *l(2)37Cf* and *fs(2)TW1,* have been physically localized fairly accurately and two, *l(2)37Bg* and *l(2)37Bd,* have not (Fig. 1).

Of the 17 segments of DNA in the *Ddc* region that have been shown to hybridize to RNA's or cDNA's (Hirsh and Davidson 1981; Gilbert and Hirsh 1981; Gilbert 1984; Pentz and Wright 1986; Spencer et al. 1986a; Marsh et al. 1986; M.E. Freeman, G.R. Hankins, J. Kullman, E.S. Pentz and T.R.F. Wright, unpublished), 11 have been assigned to mutant complementation groups either by the localization in the DNA of alleles that are small deletions or by the rescue of mutant alleles by P-element-mediated transformed segments of DNA or both. All of these assignments except the original localization of *Ddc* by Hirsh and Davidson (1981) were made with the use of mutations, and the physical site of *Ddc* was confirmed by the intragenic location in *Ddc* of both Ddc^{n27}, a 2.3 kb deletion, and the distal breakpoint of Df(2L)VA17 (Gilbert et al. 1984). Transcriptional units that have been identified by small deletion mutations are: *l(2)37Bb* and *Dox-A2* in the distal subcluster, *amd, Ddc, l(2)37Cc, Cb,* and *Cd* in the proximal subcluster and *l(2)37Cf,* one of the "scattered" genes (Pentz and Wright 1986; Black et al. 1987; G.R. Hankins, J. Kullman, E.S. Pentz and T.R.F. Wright, unpublished). Furthermore, rescue of mutations by P-element-mediated transformants has identified transcription units for *l(2)37Ca, Cg,* and *Ce,* and in addition has confirmed unequivocally the identification of four transcription units, *Ddc, l(2)37Cc, Cb,* and *Cd* (Scholnick et al. 1983; J. Kullman and T.R.F. Wright, unpublished). Of the six unassigned transcriptional units, the 3.5 kb and 0.7 kb units in the distal subcluster region should eventually be assigned to *l(2)37Ba, Be,* or *Bc,* and perhaps the one in the −45 to −41 region to *l(2)37Bg.* No small deletions were found in the attempt (Black et al. 1987) to assign *l(2)37Bd* to either the transcription unit at −15.4 to −14.6 (Hirsh and Davidson 1981) or the transcription unit designated *Cs* from −3.7 to −1.5. This latter transcript characterized by Spencer et al. (1986a) is transcribed off the strand opposite from *Ddc* and actually overlaps *Ddc* by 88 bp at the 3′ terminus (Spencer et al. 1986b). Since its expression is barely detectable in the third larval instar and is maximally expressed in male testes, and since *l(2)37Bd* is a larval lethal, it is highly unlikely that it is the *Bd* transcript, and therefore we have designated it as *Cs* with no mutant lesions in it. The *Bd* transcript might still be the −15.4 to −14.6 transcript (Black et al. 1987). The last transcription unit, *Bh,* is also devoid of mutant lesions and is defined by two cDNA clones which hybridize to the −59.4 to −58.6 region just proximal to *Dox-A2* (E. S. Pentz and T. R. F. Wright, unpublished). Thus there are at least two transcriptional units for which we have no mutations, *Cs* and *Bh,* and perhaps also the units at −45 to −41 and −15.4 to −14.6. Furthermore, both Gilbert (1984) and D. D. Eveleth (personal communication) report the recovery of cDNA clones that hybridize to the sequence between the 3′ end of *Cc* and the 5′ end of *Ddc;* another gene without mutations. In addition, *l(2)37Bg* and *Bd* may be specified by as yet undetected transcription units. Transcription mapping of the entire *Ddc* region has not yet been completed, and five of the 18 genes remain unordered relative to each other; *l(2)37Ba, Bc,* and *Be* in the distal subcluster and *l(2)37Ca* and *Cg* in the proximal subcluster.

3 The Functional Relatedness of the Genes in the *Ddc* Region

Evidence on the specific function of the 18 genes in the *Ddc* region is still far from complete. However, it is known that at least three of the genes function in catecholamine metabolism, and on the basis of mutant phenotypes, the functions of 14 genes, including the catecholamine metabolizing genes, have been inferred to be required for the formation, sclerotization, and pigmentation of cuticle (see Fig. 2). Mutant phenotypes of many of these genes are surprisingly similar (see below), suggesting that they are involved in the same physiological process, but since most of the mutations are lethal, the function of each is indispensable. In addition, melanotic tumors have been observed in mutant larvae and/or pupae of eight of the genes, including *Ddc* and *amd,* and are, perhaps, symptomatic of altered catecholamine pools. Furthermore, it has been determined that the activity of at least ten of the genes in the *Ddc* region is required for adequate female fertility. This is a very high concentration of female fertility genes in the *Drosophila* genome. Whether or not the female sterility observed derives directly or indirectly from defects in catecholamine metabolism has not been determined. More detailed information on the function of the genes in the *Ddc* region is summarized in the sections below.

3.1 Catecholamine Metabolism

In addition to *Ddc* two other genes in the *Ddc* region have been shown to function in catecholamine metabolism. These are *Dox-A2* (Pentz et al. 1986; Pentz and Wright 1986) and the alpha methyl DOPA hypersensitive gene, *amd* (Black et al. 1987) (Fig. 2). Information on these three genes and the enzymes they specify is presented below.

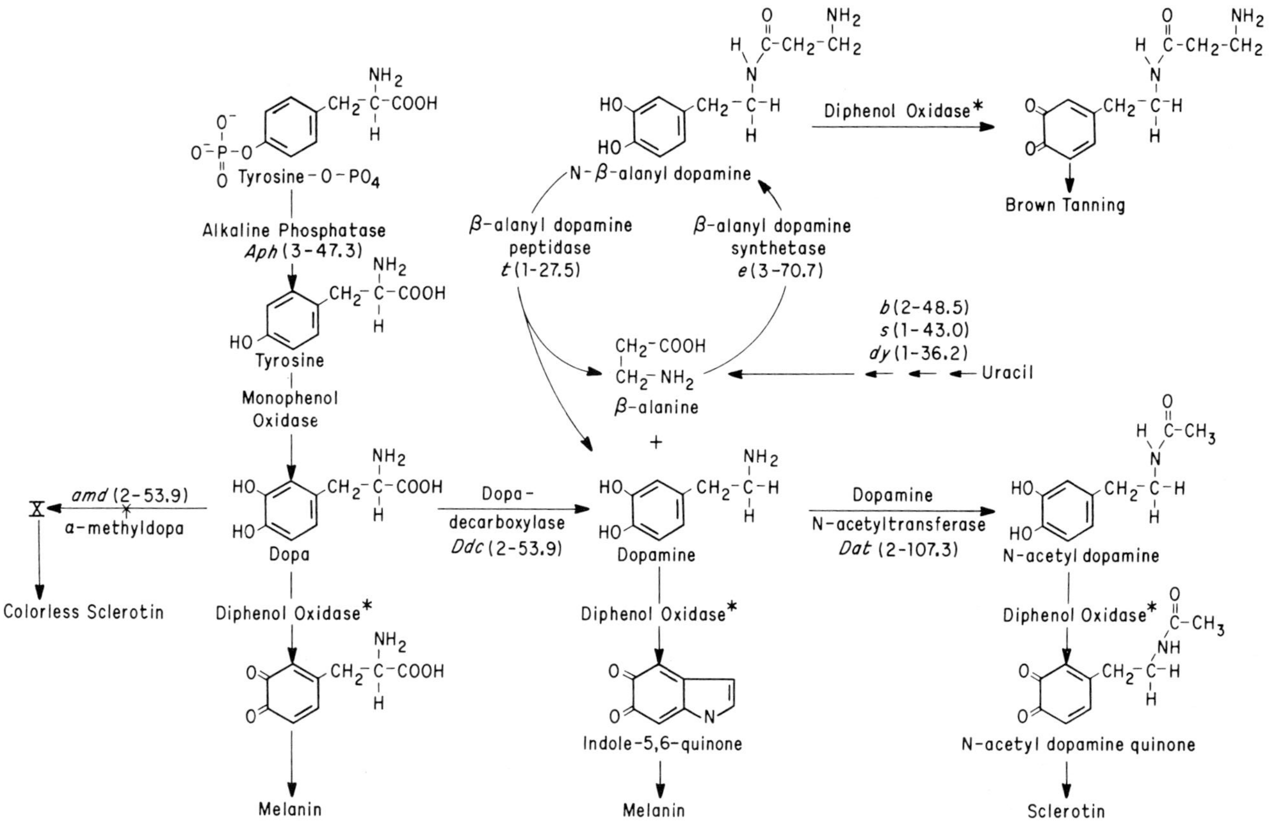

*Diphenol Oxidase: A_2-*Dox-A2* (2-53.9); A_3-*Dox-3* (2-53.1+); *qs* (1-39.5); *tyr-1* (2-54.5)

Fig. 2. Proposed pathway in *Drosophila* for catecholamine metabolism involved in sclerotization and melanization of cuticle and neurotransmitter synthesis. Mutations affecting diphenol oxidase activities are listed at the bottom *Figures in brackets* after genes symbols specific chromosome and locus. References for gene assignments to reactions in the pathway are: *Aph* (Harper and Armstrong 1972); *amd* (Black et al. 1987); *Ddc* (Wright et al. 1976a); *Dat* (Huntley 1978); *t* and *e* (B.C. Black and T.R.F. Wright, unpublished); *b, s,* and *dy* (B.C. Black, G.R. Hankins and T.R.F. Wright, unpublished); *Dox-A2* (Pentz et al. 1986); *Dox-3* (Rizki et al. 1985); *qs* (B.C. Black, E.S. Pentz and T.R.F. Wright, unpublished); *tyr-1* (Lewis and Lewis 1963)

The DOPA Decarboxylase, Ddc, Locus. Ddc (*DOPA decarboxylase*, 2-53.9+) located in salivary gland chromosome band 37C1,2 (Gilbert et al. 1984) is the sole structural gene for the enzyme DOPA decarboxylase (DDC) in the *Drosophila melanogaster* genome. DDC catalyzes the decarboxylation of DOPA to dopamine (Lunan and Mitchell 1969) and 5-hydroxytryptophan to serotonin (5-hydroxytryptamine) (Fig. 2) but not tryosine to tyramine (Livingstone and Tempel 1983). *Drosophila* DDC is a homodimer with a subunit molecular weight of 54,000 Da (Clark et al. 1978). The dopamine produced by DDC is necessary to effect sclerotization of the cuticle by being further metabolized both to N-acetyldopamine and N-β-alanyldopamine, which, after oxidation to their respective quinones, crosslink cuticular proteins (Fig. 2). Thus in adults and white prepupae more than 90% of the DDC activity is located in the epidermis (Lunan and Mitchell 1969; Scholnick et al. 1983). Some DDC activity ($\sim$5%) is found in the central nervous system of white prepupae and adults, where it produces the neurotransmitters dopamine and serotonin (Wright 1977; Livingstone and Tempel 1983). Limited amounts are also found in the ovaries (Wright et al. 1981 b) and in the proventriculus (Wright and Wright 1978). By assaying whole organisms throughout development, five peaks of DDC activity are evident. These occur at the end of embryogenesis, the two larval molts, puparation, and imaginal eclosion (Marsh and Wright 1980; Kraminsky et al. 1980). The largest peak, which occurs at pupariation, is induced by a coincident peak of the molting hormone, ecdysone (Marsh and Wright 1980), and has been shown to be attributable to a rapid increase in translatable DDC mRNA following administration of 20-OH-ecdysone (Kraminsky et al. 1980). Administration of a pulse of ecdysone also induces DDC in isolated imaginal discs (Clark et al. 1986). As yet no causal relationship has been established between the other DDC peaks and the peaks of ecdysone titer which precede them.

Most mutations in *Ddc* are homozygous or hemizygous lethal. The effective lethal phases (ELP) of the first eight lethal alleles, Ddc^{n1}–Ddc^{n8}, were almost identical. As hemizygotes over *Df(2L)TW130* almost all mortality was exclusively late embryonic with actively moving larvae, exhibiting unpigmented cephalopharyngeal apparatuses and denticle belts, unable to hatch. When homozygous there is a fairly uniform shift in ELP with mean mortalities for all eight alleles in the cross of $Ddc^{n}/CyO \times Ddc^{n}/cn\ bw$ being 13.6% embryonic, 14.1% larval, and 4.8% pupal (Wright and Wright 1978). Many larvae hemizygous for null alleles, e.g., the 2.3 kb deletion Ddc^{n27} over *Df(2L)TW130,* when released from the egg membranes, will continue development to the 3rd larval instar and even to the pharate adult stage.

Genotypes which produce individuals with drastically reduced DDC activities (0.5–5% of wild type) exhibit an "escaper" phenotype characterized by incomplete pigmentation and sclerotization of the cuticle (see colored illustration in Pentz et al. 1986). Developmental time can be prolonged for as many as 4 or 5 days and mutant puparia are easily scored by the melanization at each end of the greenish-gray pupa case. Adults often die or get stuck in the food within 24 h of eclosion, have macrochaetae that may be very thin, long, and straw-colored or colorless. The whole adult body remains light, i.e., does not take on its normal pigmentation: (1) abdominal markings are apparent but do not darken, (2) upon aging a few hours wing axillae become melanized similar to the phenotype of the mutant *speck* (*sp*), and leg joints also become melanized, perhaps due to the phenol oxidase wound reaction brought on by the rupture of weakened cuticle, and (3) flies walk on their tibiae rather than on their tarsi, but leg movements appear to be coordinated (Wright et al. 1976a). Genotypes which produce flies that exhibit the "escaper" phenotype include heteroallelic intragenic complementing heterozygotes with less then 5% of the expected number of survivors (e.g., Ddc^{n8}/Ddc^{n6}, Ddc^{n5}/Ddc^{n6}, Ddc^{n5}/Ddc^{n4}, Ddc^{n5}/Ddc^{n3}, and Ddc^{n5}/Ddc^{n2}) (Wright et al. 1976a), hemizygotes of the ts allele Ddc^{ts2} raised at 22 °C or 25 °C continuously, or homozygotes of the ts alleles Ddc^{ts1} or Ddc^{ts2} exposed to the restrictive temperature 30 °C for 24- or 48-h pulses at the end of the pupal stage (T. R. F. Wright, unpublished). To date, 50 mutagen-induced alleles have been isolated, including five temperature-sensitive alleles, a low activity allele, Ddc^{lo1}, (Wright et al. 1982) and two alleles that affect the developmental expression of *Ddc*, Ddc^{+4} (Estelle and Hodgetts 1984a, b) and Ddc^{DE1} (Bishop and Wright 1987).

Genotypes with reduced levels of DDC activity, e.g., Ddc^{n5}/Ddc^{n8} and Ddc^{n1}/Ddc^{n8} with less than 4% DDC activity, are not more sensitive to dietary alpha methyl DOPA, nor are genotypes with increased levels of DDC activity more resistant (Marsh and Wright 1986). In fact, the reverse may be true: reduced DDC, more resistant; increased DDC more sensitive.

Ddc mutations reduce learning acquisition approximately in proportion to their effect on enzymatic activity (Tempel et al. 1984). Ddc^{ts2} homozygotes and $Ddc^{ts1}/Df(2L)TW130$ hemizygotes were raised at 20 °C and kept at 20 °C for 3 days after eclosion before shifting to 29 °C for 3 more days. After shifting down to 25 °C, behavioral assays were run which showed that the associative learning was reduced significantly. Experience-dependent male courtship depression was absent in $Ddc^{ts1}/Df(2L)TW130$ (29 °C) males. Although electroretinograms of $Ddc^{ts1}/Df(2L)TW130$ (29 °C) flies were normal, ~5% of the population showed little positive phototaxis, and negative geotaxis, strength, coordination, walking ability, and olfactory acuity were normal. The threshold for proboscis extension in response to sucrose is significantly raised from 0.004 M to 0.025 M sucrose in $Ddc^{ts1}/Df(2L)TW130$ (29 °C) flies. Memory retention is unaltered (Tempel et al. 1984). Valles and White (1986) have shown that the central nervous systems of DDC-deficient larvae ($Ddc^{n27}/Df(2L)TW130$) are completely lacking in serotonin immunoreactivity. However, in these mutant larvae the same CNS cells, which if wild type would presumably show serotonin immunoreactivity, are capable of serotonin uptake from the medium. These results suggest that serotonin synthesis is not required for the differentiation of these particular specialized nerve cells.

Ddc was cloned by Hirsh and Davidson (1981). The gene has a 0.1 to 0.3 kb 5′ exon, a 0.98 ± 0.03 kb first intron, a 0.1 ± 0.05 kb middle exon, a 1.15 ± 0.02 kb second intron, and a 1.46 ± 0.04 kb 3′ large exon as established by R-loop mapping of RNA:DNA hybrids using single-stranded DNA (Beall and Hirsh 1984). 5′ is centromere proximal (Gilbert and Hirsh 1981). Beall and Hirsh (1984) find five species of poly-$(A)^+$ RNA in 16–20-h embryos: 4 kb, 3 kb, 2.7 kb, 2.3 kb, and 2.0 kb. All except the 2.0 kb RNA contain first intron sequences. No *Ddc* RNA's were found prior to 16 h of embryogenesis. At pupariation all RNA species except the 2.3 kb RNA are found, and pre-eclosion adults contain primarily 2.0 kb and 3.0 kb RNA's with no 2.3 kb RNA. Presumably the 2.0 kb RNA is a mature mRNA (see also Gietz and Hodgetts 1985; Beall and Hirsh 1986; Spencer et al. 1986a). Recently the genomic nucleotide sequence of DDC has been completely determined along with the partial sequence of two cDNA's (Eveleth et al. 1986). Evidence from these sequences suggests the existence of two RNA splicing alternatives which would encode two DDC protein isoforms. On the basis of different thermolability, Bishop and Wright (1987) have provided evidence for the presence of a different DDC isoform at pupariation in comparison to adult eclosion.

Using the same 7.5 kb *Pst*I restriction enzyme fragment that straddles the *Ddc* gene, but using different P element vector constructs, both Scholnick et al. (1983) and Marsh et al. (1985) have effected the germline transformation of *Ddc*$^+$ DNA which rescues *Ddc* mutant homo- and hemizygotes. All except two of the total of 16 transformed strains examined showed approximately normal levels of DDC activity along with normal tissue and temporal expression of the transposed *Ddc* genes. One strain had the expected level of DDC activity at pupariation but unexpectedly low levels in both sexes of newly emerged adults, and the other strain gave elevated DDC activities at all stages (Marsh et al. 1985). Of the two X-linked transformants, one was dosage-compensated (Scholnick et al. 1983) and the other was not (Marsh et al. 1985).

Hirsh and co-workers (Hirsh 1986) have begun to define the 5′ flanking sequences necessary for the developmentally regulated expression of *Ddc* by deleting cloned *Ddc* in vitro and reintroducing different deletion constructs back into the genome by P element-mediated transformation. The in vivo developmental expression of these constructs is then monitored. All *Ddc* genes containing 208 or more base pairs of 5′ flanking DNA appear to be normally regulated. *Ddc* genes with just 22 bp of 5′ flanking sequences are not normally regulated but are expressed using the usual wild-type *Ddc* RNA start site even though the "*TATA*" box sequences are deleted. Hirsh (1986) speculates that use of the normal start site is regulated by the adjacent *Adh* gene also present in the P element constructs. Further reports on this work from Hirsh's laboratory should be most interesting.

The Diphenol Oxidase-A2, Dox-A2, Locus. Dox-A2 (*Diphenol oxidase-A2,* 2-53.9; 37B10–13) is located in the distal subcluster approximately 63 kb from *Ddc*. Pentz et al. (1986) report that *Dox-A2* function is necessary for the production of the A2 component of the complex phenol oxidase enzyme and is probably its structural gene. Mitchell and co-workers (Geiger and Mitchell 1966; Seybold et al. 1975) have characterized this complex enzyme system as being made up of at least three protein components, A1, A2, and A3, all of which are activated by a reaction cascade involving at least three additional proteins. Pre-S interacts with S-activator to yield S, and then S acts on P to produce P′, which interacts with the A components to yield active phenol oxidase. The activated enzyme complex utilizes both monophenol and diphenol substrates. The three A components can be separated from one another by gel electrophoresis of homogenates prior to activation. After electrophoresis the gels are first incubated in an activator solution and then in a suitable substrate solution. If the substrate is oxidized, pigment is deposited in the gel at the A component sites. Monophenol substrates are oxidized primarily by the A1 component and diphenol substrates including DOPA, dopamine, and N-acetyldopamine by the A2 and A3 components (for methods see Warner et al. 1975; Rizki et al. 1985; Pentz et al. 1986).

The three lethal alleles of *Dox-A2* as heterozygotes over the *CyO* balancer chromosome reduce diphenoloxidase activities to 47–79% of wild type and have no effect on monophenol oxidase activity. Pool sizes of DOPA, dopamine, and N-acetyldopamine are elevated in a mixture of 1 *Dox-A2*^{1}/*Dox-A2*^{1}: 2 *Dox-A2*^{1}/*CyO*: 1 CyO/CyO 20–24-h embryos in comparison to suitable controls. This incidates that *Dox-A2* function is necessary to oxidize these substrates to their respective quinones (Fig. 2) (Pentz et al. 1986). *Dox-A2* mutants reduce only the A2 component activity after separation of the A components in polyacrylamide gels. Thus Dox-A2 may be the structural locus for the A2 component of the diphenol oxidase enzyme system.

Hemizygotes, *Dox-A2*^{n}/*Df(2L)TW130*, of all three Dox-A2 lethal alleles die during the first larval instar, having normally pigmented mouth parts and denticle belts and showing no other abnormal cuticular phenotypes. However, the dead larvae never turn black. A rare *Dox-A2*^{1} homozygous mutant individual survived to the pharate adult stage and was released alive from the pupa case by dissection. The mutant was completely unpigmented, with bristles and cuticle being totally colorless (see colored illustration in Pentz et al. 1986). This mutant never developed melanization in the joints of the legs, axillae of the wings or, even when the very weak cuticle of the abdomen eventually ruptured. The fact that this *Dox-A2*^{1} homozygote never tanned or melanized suggests that normal A2 component must be present at this stage to have in vivo any functional phenol oxidase activity at all, i.e., for the A1 and A3 components to be active in vivo also (Fig. 2). The pigment deposited during embryogenesis in the mouth parts and denticle belts of *Dox-A2* homozygotes may be due to the presence of a maternal component, protein or mRNA, or be due to independent A1 or A3 component activity. No experiments have been done to verify these possibilities.

The *Dox-A2*^{1} and *Dox-A2*^{2} alleles are 0.1 kb and 1.1 kb deletions respectively located within 3.5 to 4.8 kb of the proximal breakpoint of Df(2L)OD15 (Fig. 1) (Pentz and Wright 1986). A 1.7 kb *Dox-A2* mRNA has been identified in 15–17-h embryos, crawling third instar larvae, and 1–4-day-old adults. cDNA clones indicate that the 3′ end is centromere proximal and that the coding region contains at least one small intron (Pentz and Wright 1986).

The alpha methyl dopa, l(2)amd, Locus. amd (*l(2)amd; alpha methyldopa* 2-53.9+) is located 0.002 centimorgans distal ot *Ddc* (Wright et al. 1981). As heterozygotes, amorphic mutations of *amd* (*amd*/+) are hypersensitive to the dietary administration of the DDC analog inhibitors α-methyl DOPA (αMD) (Sparrow and Wright 1974) and N^1(DL-seryl)-N^2-(2,3,4-trihydroxybenzyl)hydrazine (Hoffman-LaRoche No. 4-4602/1) (Wright et al. 1976a). It has been demonstrated that resistance to dietary αMD is directly correlated with *amd*$^+$ gene dosage; the more doses of the *amd*$^+$ gene the more resistant and vice versa (Wright et al. 1976b; Marsh and Wright 1986). This relationship suggests that the in vivo function of the *amd*$^+$ protein product is inhibited by the binding of the modified catecholamine, αMD. Amorphic *amd* homo- and hemizygotes die as normally pigmented larvae both prior to and just after larval hatching. In addition to having necrotic, extruded anal organs, these larvae burst very easily when manipulated, suggesting incomplete sclerotization of the colorless body wall cuticle (Wright 1977). Electron micrographs indicate that the anal organ defect arises from the incomplete sclerotization of the cuticular suture between anal organ cells and the normal cells of the epidermis (J. C. Sparrow, personal communication). Some hypomorphic *amd* alleles as hemizygotes permit larvae to hatch and complete larval development, forming normally pigmented pseudopupae that are abnormally flexible (T. R. F. Wright, unpublished). *amd* heterozygotes (*amd*/+) and *amd* intragenic complementing heteroallelic heterozygotes (amd^{H1}/amd^{H89}) do not affect DDC or dopamine acetyl transferase activity in any way whatsoever; neither as adults nor as white prepupae, and neither in the epidermis nor the central nervous system of white prepupae (Wright et al. 1976a; Wright 1977; Huntley 1978; Wright et al. 1981a; E. Y. Wright and T. R. F. Wright, unpublished).

Black et al. (1987) infer that *amd* activity is necessary for colorless sclerotization. They have demonstrated that in catecholamine pools a prominent electroactive compound is missing in intragenic complementary heteroallelic heterozygous adults (amd^{H1}/amd^{H89}). This compound has been identified as a catecholamine by low resolution mass spectroscopy and other criteria and has been designated as Catecholamine X. Its complete structure has yet to be solved. Catecholamine X pools are found in embryos at the time of initial colorless sclerotization, which is prior to the appearance of DDC activity and dopamine. In Ddc^{ts2} homozygotes at 22 °C which have markedly reduced levels of dopamine, N-acetyldopamine, and N-β-alanyldopamine, levels of Catecholamine X are slightly elevated. From these facts, the inference is made that the *amd* gene codes for an enzyme which converts DOPA to Catecholamine X in a separate branch of the pathway (see Fig. 2).

A transcription unit between DNA coordinates −5.7 and −3.75 characterized and sequenced by Marsh et al. (1986) has been identified as *amd* by the determination that the two alleles, amd^{37} and amd^{40}, are small deletions of DNA located between coordinates −4.5 and −3.5 within this transcriptional region (Black et al. 1987). A 2 kb *amd* transcript is most abundant at about 12 h of embryogenesis and lower levels are detected throughout most of embryogenesis and in adult females, but not in males. Unique stage-specific transcripts of 1 kb and 0.6 kb are produced in mid 3rd instar and late 3rd instar, respectively (Marsh et al. 1986).

3.2 The Formation, Sclerotization, and Pigmentation of Cuticle

Inspection of the gross morphology of the mutant phenotypes at various stages of development, but particularly at the pharate adult stage, has led to the conclusion that the function of at least 14 of the genes in the *Ddc* region are required for the normal formation, sclerotization, or pigmentation of the cuticle (see Fig. 1) (T. R. F. Wright, unpublished). In order to inspect the effects of inadequate gene activity on the cuticle at the pharate adult stage, hypomorphic and ts alleles and partially complementing heteroallelic heterozygotes were used for those genes with effective lethal phases prior to this late stage (Fig. 1).

Hemizygotes (*l/DfTW130*) of hypomorphic alleles of *l(2)37Be* and *Bd*, and amorphic alleles of *l(2)37Ba* and *Ce* exhibit a very similar phenotype in pharate adults incapable of hatching. Hemizygotes of hypomorphic alleles of *l(2)37Bg* and *Bc* exhibit what appear to be more extreme manifestations of the same mutant phenotype. For most of these genes, more than one allele will produce the same phenotype. In the less extreme manifestation, cuticle is incompletely formed over much or all of the abdomen and in these areas no bristles form and there are no indications of either segmentation or the normal pattern of melanization. The epidermis is complete and has probably laid down a thin procuticle without the layered epicuticle being laid down under it. This is an inference from gross external observation; no sectioned material has been examined. The expression can be quite variable and often small or large regions of the abdomen will be covered by normal cuticle which can include bristles, manifestations of segmentation, and some patterns of melanization even in the *l(2)37Ba* and *Ce* amorphic pupal lethal alleles. In some cases the phenotype is more extreme. For example, in some *l(2)37Bd* alleles the incomplete formation of cuticle extends anteriorly, involving much of the thorax and even parts of the head and eyes. Hemizygotes of $l(2)37Bg^{1}$ consistently produce pupa cases that are larger and darker than normal and which contain pupae with this incomplete cuticle formation phenotype over the entire head, thorax, and abdomen with the curious exception of the genital disc area, where the cuticle formed is more complete and is tanned. In these *Bg* pupae, head eversion is complete but there is no evidence of legs or wings nor of differentiation of ommatidia. Hemizygotes of $l(2)37Bc^{11}$ produce distinctly darker pupa cases and usually show evidence of regions of internal melanization resulting from the rupture of adult abdominal epidermis. Occasional individuals have only minor rupturing and when dissected exhibit the above incomlete cuticle formation phenotype described above. Perhaps Bc^{11} usually lays down no cuticle at all. For all six of the above mutations the thoracic, leg, and head cuticle appears to be formed, but the thoracic cuticle in particular does not appear normal, i.e., appears incompletely sclerotized (T.R.F. Wright, unpublished).

Eclosed adults homozygous for ts alleles of *l(2)37Ca* and *Cc* at permissive temperatures have more or less normal abdominal but abnormal thoracic cuticle. Although the thoracic cuticle appears to be normally formed, it is not sclerotized normally. Contractions of the indirect flight muscles deform the thorax so that the thoracic sutures are particularly prominent and other irregular indentations are apparent and perhaps due to the collapse of the thorax, wings in Cc^{ts} individuals are never expanded and are only occasionally expanded normally in Ca^{ts} individuals. *l(2)37Cd* hemizygotes that pupate show a similar but less extreme phenotype than *Cc* and *Ca*. Somatic mosaic patches homozygous for a *Ca* amorphic allele exhibit abnormally tiny bristles (C.P. Bishop, unpublished), and Ca^{ts} and Cc^{ts} hemizygotes have moderate *Minute*-sized scutellar bristles. Intragenic complementary heteroallelic heterozygotes of alleles of *l(2)37Ca* and *Cc* and *Cf* "escapers" eclose after an extended development time, often are small, usually are a darker tan color than wild type and have moderate *Minute* bristles also. The cuticle of pharate adult hemizygotes of amorphic alleles of *1(2)37Cf* appears essentially normal except that the melanized stripes on the abdomen become extremely dark (T.R.F. Wright, unpublished).

The mutant cuticular phenotypes of *Ddc, Dox-A2,* and *amd* are described in Section 3.1 above and the phenotype of *hook* (hooked bristles) is described in Lindsley and Grell (1968) and Mitchell and Lipps (1978). Except for *l(2)37Ba, Bg, Ce,* and *Cf,* which are pupal lethals, the mutant cuticular phenotypes outlined above are derived from hypomorphic or ts alleles. For those amorphic mutations with effective lethal phases prior to pupation no obvious morphological differences are apparent in whole and dissected homo- or hemizygous individuals except for some with melanotic pseudotumors (see below) and except for *Ddc* and *amd* (see above). Many have extended development times and a few are abnormally small or have undersized imaginal discs or other organs (T. R. F. Wright, unpublished).

3.3 Melanotic Pseudotumors

In *Drosophila melanogaster,* melanotic tumors are formed by aggregations of hemolymph cells around foreign substances such as bacteria or parasitic wasp larvae and in numerous mutant strains around various different endogenous tissues (Rizki and Rizki 1984). These aggregations then melanize. Most of them are not true tumors (see Sparrow 1978). At least one allele, usually many more, of eight genes in the *Ddc* region produces melanotic tumors in hemizygous or homozygous larvae and/or pupae (see Fig. 1). This is the highest concentration of known melanotic tumor-producing genes in the genome. Why these genes produce melanotic tumors is unknown, but it might be related to unusual catecholamine pools resulting from blocks in catecholamine metabolism. For example, *Ddc* mutants may be prone to produce melanotic tumors because of very high pools of DOPA which could be metabolized to melanin very easily. This then may be an indication that the genes *l(2)37Ba, Bc, Bg, Cc, Cd,* and *Cf* also function in catecholamine metabolism in some way.

3.4 Female Sterility

Using hypomorphic alleles that produce a few adult progeny, ts alleles, complementary heteroallelic heterozygotes, and female and male sterile alleles, it has now been established that the activity of at least ten of the genes in the *Ddc* region is required for female fertility and activity of at least four of these is also required for male fertility (see Fig. 1) (T. R. F. Wright, unpublished). This is an unusually high concentration of female fertility genes in the *Drosophila* genome, and their coordinate activity in the ovary might provide a reason for these genes remaining clustered. Of the remaining eight genes, three, *hk, amd,* and *Ddc,* have been tested using some of the variety of genotypes listed above and have not been shown to be required for female fertility, even though DDC activity (Wright et al. 1981) and *amd* transcripts (K. Konrad and J. L. Marsh, personal communication) have been demonstrated to be present in ovaries. It has not been possible to obtain adult mutant females to test the other five genes, *l(2)37Be, Bb, Bg, Bd,* and *Cg,* for female fertility (T. R. F. Wright, unpublished).

Approximately 12,000 mutagenized second chromosomes have been screened for female sterile mutations over *Df(2L)TW130*. Eighteen female sterile (fs) mutations and one male and female sterile (mfs) mutation were recovered along with 11 dominant female sterile (Fs) mutations (T. R. F. Wright, unpublished). An additional 9 *Df(2L)TW130* female sterile mutations have been isolated by other laboratories; 2 by D. L. Lindsley (personal communication) and 7 by T. Schupbach (personal communication) in her global screen of the genome for female sterile mutations. Of the total 28 fs and mfs mutations in the *Ddc* region, 9 have been determined to be alleles of five lethal loci, *l(2)37Ba, Bc, Dox-A2, l(2)37Cc,* and *Cf* (Fig. 1) and 13 are alleles of the only nonvital, female sterile locus, *fs(2)TW1,* established to date in the *Ddc* region.

The first *fs(2)TW1* allele was identified when the female sterile lesion in the Ddc^{ts1} chromosome (Wright et al. 1981) was subsequently separated from Ddc^{ts1} by recombination. This means that the ovary transplants reported in that paper established the ovary autonomous sterile phenotype of *fs(2)TW1* and not Ddc^{ts1}, and the now inexplicable interactions affecting female fertility reported were between *fs(2)TW1* and the series of *Ddc* alleles tested as heterozygotes over Ddc^{ts1}, i.e., Ddc^{ts1} *fs(2)TW1*/Ddc^{x}+. Homo- or hemizygous *fs(2)TW1* females lay many eggs which remain white and do not develop at all. The gene is the most proximal one in the *Ddc* region located distal to the *Df(2L)VA6* breakpoint and within *Df(2L)Sd77* in the vicinity of DNA coordinates +49 to +58. No transcript has been as yet identified (G. R. Hankins and T. R. F. Wright, unpublished).

As hemizygotes, the three mutations designated as fs alleles of *Cf* and the one *Cc* fs allele show reduced viability particularly at elevated temperatures, and, therefore, probably are not ovary specific lesions. The fs mutation in *l(2)37Ba,* the three in *Bc,* and the one in *Dox-A2* do not reduce viability significantly (T. R. F. Wright, unpublished) and are good candidates for lesions in ovary-specific control regions.

It has not been possible to assign the remaining six female sterile mutations to known loci in the *Ddc* region because they produce high levels of female sterility when heterozygous over nonoverlapping deficiencies and lethal alleles from noncontiguous vital genes (T. R. F. Wright, unpublished). Although not complete yet, cleaning the tester chromosome has not eliminated this problem, which continues to be investigated because this behavior is what might be expected from lesions in regional ovary-specific enhancers.

3.5 Mutations of *l(2)37Cf* Produce Malignant Brain Tumors

Third instar larvae homozygous for 13 of the 14 *l(2)37Cf* alleles examined have enlarged brains (G. R. Hankins and T. R. F. Wright, unpublished). This is also true for three of the alleles as hemizygotes over *Df(2L)TW130*. In some cases the brains are over eight times the normal volume. In each case the enlargement appears greatest in the presumptive optic center. Although the ventral nerve cord is unaffected, the eye-antennal imaginal disc may be enlarged in some cases. Transplantation of pieces of mutant (*l(2)37*Cf^{14}) third instar brains into the abdomens of wild type adults by E. Gateff (personal communication) results in a tremendous proliferation of the transplanted tissue, which fills and bloats the entire abdominal and thoracic cavities with neurons. These cells are invasive with, for example, metastases in the indirect flight muscle of the thorax. Thus the loss of *l(2)37Cf*$^{+}$ function results in the excessive (or precocious?) growth of the brain hemispheres in vivo and in the malignant behavior of this tissue when transplanted into the adult hemocoel.

Most homozygous *l(2)37Cf* larvae with enlarged brains are able to pupate and subsequently die as pharate adults. Some carrying hypomorphic or ts alleles eclose and have very much reduced female and male fertility (G. R. Hankins, unpublished). The latter is a behavioral problem, since the lethargic males will mate successfully if they accidentally bump into a female (A. C. Kenyon and G. R. Hankins, unpublished). It has not yet been determined if female infertility is also a behavioral problem. Catecholamine pools have yet to be assayed for this mutation. It will be most interesting if this mutant phenotype results from a defect in catecholamine metabolism. This phenotype appears to be unique to *l(2)37Cf*, for so far no indication of neural overgrowth has been seen in any other *Ddc* region mutation (T. R. F. Wright, unpublished).

4 Sequence Homologies Within the *Ddc* Region

The structures and sequences of *Ddc* and *amd* have been reported by Eveleth et al. (1986) and Marsh et al. (1986), respectively. The two sequences have been examined by Eveleth and Marsh (1986) for homologies using computerized dot matrix analysis. Although *amd* has only one intron whereas *Ddc* has two, striking homology was found between exon II and exon III of *Ddc* and the two exons of *amd*. "Overall approximately 55% of the bases match between the two sequences (excluding intron regions), however two areas of more extensive homology are apparent. One area beginning near the second exon of *amd* is over 80% homologous over 100 bp and a second run of 124 bp (700 bp from the 3′ end) is approximately 90% homologous" (Eveleth and Marsh 1986). Comparison of the deduced amino acid sequences from *Ddc* and *amd* show two regions of considerable amino acid sequence conservation, and analysis of deduced secondary structure indicates the conservation of considerable structural similarity between the two proteins. Eveleth and Marsh (1986) conclude that *Ddc* and *amd* are products of a gene duplication event with subsequent divergence to related subspecialties in catecholamine metabolism. Since the transcripts of the two genes are not found at the same time in development, Eveleth and Marsh (1986) concluded that they are not coordinately regulated.

Gilbert et al. (1984) reported that "the region flanking *Ddc*" (−3 to +38 not including *amd*) "differs from other clusters in that under *normal hybridization* conditions there is no detectable intragenic sequence homology". Recent preliminary results of J. Kullman and T. R. F. Wright (unpublished) using low (55%) stringency hybridization conditions demonstrate sequence homology in this region between *l(2)37Cb* and *amd* but not *Ddc*. In addition, J. Kullman finds low stringency homologies between sequences from the proximal subcluster and sequences from the distal subcluster and from "scattered" gene regions, e.g., the transcribed region of *Cs* crosshybridizes to genomic DNA from the putative location of *hk* and to genomic DNA from the distal subcluster. Furthermore, E.S. Pentz, G.R. Hankins, and T.R.F. Wright (unpublished) have established that under high (>80%) stringency hybridization conditions the 0.84 kb *Bh* cDNA from the −60 to −58.5 region hybridizes with cloned genomic DNA from the +40 to +46.1 region containing the coding region from *l(2)37Cf*. Also a cDNA from the *Cf* region crosshybridizes with genomic sequences in the *Bb Dox-A2 Bh* region, and a cDNA from *Bb* crosshybridizes with genomic DNA from the *Cf fs(2)TW1* region at low (~50%) stringency. Precisely which sequences are involved in the crosshybridizations is still being determined. In conclusion, it is important to note that there are both intra- and intersubcluster homologies and also homologies to the "scattered" genes, suggesting that many of the genes in the *Ddc* region are evolutionarily related.

5 The Function of Densely Clustered Genes Individually Transposed to Other Chromosomal Sites

It is possible that dense gene clusters are maintained in the genome because they provide some selective advantage which is lost when the genes are dispersed. It might be that one or more densely clustered gene may not function at all if individually transposed out of cluster, or more likely does function, but suboptimally. This problem has been approached by using P element-mediated germ line transformation. Unfortunately, negative results, i.e., reduced function of the transposed gene, may be due to factors irrelevant to the question, such as insertion site chromosomal position effects, effects due to other genes or DNA cotransformed in the P element vector, and changes in the DNA of the gene caused by the experimental manipulations. Thus the analysis of negative results must perforce become statistical.

In at least two cases genes which have been transposed out of dense clusters appear to be normally expressed both quantitatively and developmentally at the proper times in development and in the correct tissues. These two genes are *Ddc* (Scholnick et al. 1983; Marsh et al. 1985; Hirsh 1986) and the salivary gland glue protein gene *sgs3* in 68C (Richards et al. 1983; Bourouis and Richards 1985).

Since most of the genes in the *Ddc* region are vital genes, transposed segments of DNA can be assayed for their ability to rescue lethal alleles and thereby determine if the transposed genes function at all, and if they do, make it possible to assign genes to specific segments of the DNA. However, since as little as 10% DDC activity permits complete viability and effects a wild-type cuticular phenotype, the ability of a transposed segment to rescue lethality does not mean that the transposed gene is functioning optimally. To determine whether or not transposed genes function optimally, levels of mRNA or protein products must be measured at different times in development and in the appropriate tissues. The initial part of this procedure, i.e., transposition and rescue of lethal alleles, has been completed by J. Kullman and T.R.F. Wright (unpublished) for the "C" genes, *l(2)37Cc, Cb, Cd, Ca, Cg,* and *Ce,* in the proximal subcluster of the *Ddc* region.

These six genes are densely packed in approximately 15 kb. *l(2)37Cc, Cb,* and *Cd* are transcribed in the same direction as *Ddc*. The exact limits of the transcriptional units have yet to be determined. Gilbert (1984) described the isolation of a collection of densely packed cDNA's homologous to sequences from +3.7 to +16.2 which could have been derived from six separate transcription units. In addition, her Northern analysis of poly-$(A)^+$ RNA's suggested that some RNA's may be hybridizing to probes that span at least half the above distance and beyond. J. Kullman (unpublished) has effected the P element-mediated transposition of the series of overlapping DNA segments in this region depicted in Fig. 1 and tested their ability to rescue lethal alleles of the six vital genes located in this region. With a few very important exceptions, the results are straightforward; normally densely clustered genes can function well enough in ectopic positions to rescue lethal alleles and thereby permit the genes to be ordered. The exceptional results include two different cases when after one segment of DNA capable of rescuing mutations from two different vital genes is divided approximately in half, neither of the resulting "half" segments of DNA is capable of rescuing lethal alleles of either of the two genes. Logically, unless the genes overlap or share one or more common regulatory elements, at least one of the "half" segments should have rescued lethal alleles of one or the other of the two genes. Thus the proximal subcluster may include two sets of genes in addition to *Ddc* and *Cs* which either overlap or share regulatory elements. These are, however, negative results and trivial explanations must still be eliminated, but these data of J. Kullman, although preliminary, indicate that the results obtained with *Ddc* and *sgs3* may not be the final word on the function of individually transposed, normally clustered genes.

6 Concluding Remarks

Almost by definition, the distribution of most clustered genes violates the one gene-one band hypothesis (Beermann 1972; Judd et al. 1972; Lefevre 1974; Young and Judd 1978). In the *Ddc* region and in numerous examples cited in the Introduction several genes (transcription units) have been shown to be clustered in segments of DNA less than 10 kb long. Since many of the fine salivary gland chromosome bands in *Drosophila* contain approximately 10 kb of DNA or less (Beermann 1972; Rudkin 1972), there can not be a unitary gene-to-band ratio for these clustered genes. Yet there are regions of the *Drosophila* genome which appear to approximate, within a factor of 2, a one-to-one gene-to-band correspondence, e.g., the *zeste-white* region (Young and Judd 1978), the *Adh* region (Woodruff and Ashburner 1979) and the fourth chromosome (Hochman 1976). However, in addition to the data from the *Ddc* region (Table 1), Zhimulev et al. (1981 b) and Crosby and Meyerowitz (1986) find nonrandom cytological distributions of genes in the 10A1,2 and 68A3 to 68C11 regions, respectively, which seriously violate the one gene-one band hypothesis. The results in all these studies are based primarily, but not exclusively, on the distribution of lethal genes (see reviews by Zhimulev et al. 1981a and Lefevre and Watkins 1986). On the other hand, Spierer and co-workers (Spierer and Spierer 1983; Hall et al. 1983; Bossy et al. 1984; Gausz et al. 1986) have carefully mapped the distribution of transcripts in 315 kb of DNA in the 87D5 to 87E5,6 region which includes the *rosy* and *Ace* loci. Although this is a 14-band region and apparent saturation mutagenesis (Hilliker et al. 1980) has established 12 genes in this region, Spierer and co-workers find that in different tissues and times in development 43 distinct transcripts are produced by apparently 37 different transcription units (genes) located in this 14-band region. These transcription units are not uniformly distributed in the DNA with, for example, 18 transcription units in the proximal 63 kb of DNA and only seven units in the next 153 kb. Also at the cytological level 15 transcription units are represented as being in the 10-band region 87D5–D14, nine regions of transcription in the huge 87E1,2-band, and 13 transcription units in the three bands 87E3, E4, and E5,6 (Gausz et al. 1986). Curiously Bossy et al. (1984) find a 1:1 correspondence between the number of transcripts isolated from salivary glands and the number of bands present in the salivary gland chromosomes in this 315 kb region. Whether this is a significant observation has not yet been determined. Mirkovitch et al. (1986) have demonstrated that the loop organization as determined by the position of nuclear scaffold attachment regions in this 315 kb of DNA in Kc cells does not correspond to the banding pattern seen in the salivary gland polytene chromosomes.

These data from the 87D,E region, the data from the *Ddc* region with, for example, ten transcription units in 25 kb of DNA in two to three bands along with the density of transcription units found in gene clusters in general suggest that a consideration of the ratio of genes to bands will not be useful in determining the functional significance of bands (chromomeres) (see the extensive discussion by Richards 1986).

The cluster of genes in the *Ddc* region is unique for a number of reasons. First, except for the histone, rRNA, 5sRNA, and tRNA reiterated gene clusters, it is the largest cluster described to date with at least 18 genes, more than twice as many as the seven genes in the 67B heat shock gene cluster. Second, although sequence homologies do exist among genes in the *Ddc* region, all 18 genes appear to have separate functions and mutations in any one of 16 of the genes are lethal. Third, it is the only cluster in which it has been demonstrated that at least some of the genes code for proteins with enzymatic functions. Since genes in this region are evolutionarily related, it is not unexpected that the genes with known enzymatic functions, *Ddc, amd,* and *Dox-A2,* are involved in the same metabolic pathway, i.e., catecholamine metabolism, but it is remarkable that the three enzymes appear to be capable of binding the same substrate, DOPA. In this regard, it is interesting that in vertebrates a number of catecholamine metabolizing enzymes also show sequence homologies (Joh et al. 1984). Fourth, the *Ddc* region has a unique organization made up of two dense subclusters along with a number of "scattered" genes separate from the subclusters, but functionally and evolutionarily part of the *Ddc* cluster. Except for the known and putative overlapping genes in the *Ddc* region, the density of the distribution of genes in the subclusters is not unusual in comparison to other clusters, e.g., the *serendipity* and 67B heat shock gene clusters. Whether there is any functional or evolutionary significance to the 88 bp overlap of the 3′ termini of *Ddc* and *Cs* remains to be determined, and the existence of additional examples of overlapping genes in the *Ddc* region has yet to be established.

With some exceptions, most genes within a cluster are coordinately expressed in development, often responding to the same external stimulus. The genes in the *Ddc* cluster again appear to be exceptional, for evidence has accumulated which indicates that the genes in the *Ddc* region are not coordinately expressed in development (Beall and Hirsh 1984; Gilbert 1984; Spencer et al. 1986a, b; Pentz and Wright 1986; Marsh et al. 1986). Therefore, it has been suggested (Eveleth and Marsh 1986) that the genes in the *Ddc* cluster are not regulated on the basis of their being in a chromosomal domain (Weintraub 1985). It may, however, be too soon to draw this conclusion, for it may yet be documented that at one particular time in development in one specific tissue (e.g., the ovaries) all the genes in the *Ddc* region are coordinately expressed as members of a chromosomal domain, and that the genes can, however, also be individually regulated at other times in development and in other tissues (see also the 67B heat shock gene, Zimmerman et al. 1983; Cohen and Meselson 1985). The regulation of most of the genes as members of a chromosomal domain at just one time in development could conceivably be sufficiently advantageous to the organism to provide the selection pressure necessary to maintain the genes in a cluster. Clearly, genes need not be in a cluster to be coordinately regulated, but clustering could under certain circumstances be very advantageous, with the primary example being the clustering of the chorion genes into two domains, which permits their coordinate amplification at precise times in oogenesis (Spradling 1981).

It is obvious that the work on the *Ddc* cluster is incomplete. The biochemical function of most of the genes still needs to be established in order to document precisely the extent of their functional relatedness. Many additional genes need to be sequenced to establish the extent of homologies throughout the region and perhaps permit inferences to be made on how the cluster evolved. Answers to questions on whether or not the region is at some time in development regulated as a chromosomal domain or whether or not any of the genes only function optimally in situ within the cluster may permit one to decide whether the cluster is maintained because of a selective advantage or whether it is just an evolutionary relic. It does, however, already provide a concrete example of how a portion of a specialized metabolic pathway has evolved in *Drosophila*.

Acknowledgments. The contributions, many of them unpublished, of present and past members of my laboratory at the University of Virginia are gratefully acknowledged. They are Clifton P. Bishop, Bruce C. Black, Patricia O. Cecil, Pamela N. Fornili, Mark E. Freeman, Gerald R. Hankins, John Kullman, Lee Litvinas, Ellen S. Pentz and Eileen Y. Wright. The author particularly appreciates the courtesies extended by Dr. J. Lawrence Marsh and his co-workers at the University of California, Irvine and Dr. Ross B. Hodgetts and his co-workers at the University of Alberta by providing pre-prints of papers submitted for publication. The author's research is supported by the National Institutes of Health Research Grant GM 19242.

References

Ashburner M, Chihara C, Meltzer P, Richards G (1974) Temporal control of puffing activity in polytene chromosomes. Cold Spring Harbor Symp Quant Biol 38:655–662

Ayme A, Tissières A (1985) Locus 67B of *Drosophila melanogaster* contains seven, not four, closely related heat shock genes. EMBO J 4:2949–2954

Barnett T, Pachel C, Gergen JP, Wensink PC (1980) The isolation and characterization of Drosophila yolk protein genes. Cell 21:729–738

Beall CJ, Hirsh J (1984) High levels of intron-containing RNAs are associated with expression of the Drosophila DOPA decarboxylase gene. Mol Cell Biol 4:1669–1674

Beall CJ, Hirsh J (1986) *Drosophila melanogaster* gene transcripts are not expressed at high levels during early embryogenesis. Devel Biol 114:258–259

Beermann W (1972) Chromosomes and genes. In: Beermann W (ed) Developmental studies on giant chromosomes. Springer, Berlin Heidelberg New York, pp 1–33

Bishop CP, Wright TRF (1987) Ddc^{DE1}, a mutant differentially affecting both stage and tissue specific expression of DOPA decarboxylase in Drosophila. Genetics 115:477–491

Black BC, Pentz ES, Wright TRF (1987) The alpha methyl dopa hypersensitive gene, *l(2)amd,* and two adjacent genes in *Drosophila melanogaster:* Physical localization and direct effects of *amd* on catecholamine metabolism. Mol Gen Genet (in press)

Bossy B, Hall LMC, Spierer P (1984) Genetic activity along 315 kb of the Drosophila genome. EMBO J 3:2537–2541

Bourouis M, Richards G (1985) Remote regulatory sequences of the *Drosophila* glue gene *sgs3* as revealed by P element transformation. Cell 40:349–357

Bownes M, Hodson BA (1980) Mutant *fs(1)1163* of *Drosophila melanogaster* alters yolk protein secretion from the fat body. Mol Gen Genet 180:411–418

Clark WC, Pass PS, Vankataraman B, Hodgetts RB (1978) Dopa decarboxylase from *Drosophila melanogaster*. Purification, characterization and an analysis of mutants. Mol Gen Genet 162:287–297

Clark WC, Doctor J, Fristrom JW, Hodgetts RB (1986) Differential response of the dopa decarboxylase gene to 20-OH-ecdysone in *Drosophila melanogaster*. Dev Biol 114:141–150

Cohen RS, Meselson M (1985) Separate regulatory elements for the heat-inducible and ovarian expression of the Drosophila *hsp26* gene. Cell 43:737–746

Crosby MA, Meyerowitz EM (1986) Lethal mutations flanking the 68C glue gene cluster on chromosome 3 of *Drosophila melanogaster*. Genetics 112:785–802

Crowley TE, Bond MW, Meyerowitz EM (1983) The structural genes for three Drosophila glue proteins reside at a single polytene chromosome puff locus. Mol Cell Biol 3:623–634

Estelle MA, Hodgetts RB (1984a) Genetic elements near the structural gene modulate the level of dopa decarboxylase during *Drosophila* development. Mol Gen Genet 195:434–441

Estelle MA, Hodgetts RB (1984b) Insertion polymorphisms may cause stage specific variation in mRNA levels for dopa decarboxylase in *Drosophila*. Mol Gen Genet 195:442–451

Eveleth DD, Marsh JL (1986) Evidence for evolutionary duplication of genes in the dopa decarboxylase region of Drosophila. Genetics 114:469–483

Eveleth DD, Gietz RD, Spencer CA, Nargang FE, Hodgetts RB, Marsh JL (1986) Sequence and structure of the dopa decarboxylase gene of Drosophila: Evidence for novel RNA splicing varients. EMBO J 5:2663–2672

Geiger HR, Mitchell HK (1966) Salivary gland function in phenol oxidase production in *Drosophila melanogaster*. J Insect Physiol 12:749–754

Gausz J, Hall LMC, Spierer A, Spierer P (1986) Molecular genetics of the *rosy-Ace* region of *Drosophila melanogaster*. Genetics 112:65–78

Gietz RD, Hodgetts RB (1985) An analysis of dopa decarboxylase expression during embryogenesis in *Drosophila melanogaster*. Dev Biol 107:142–155

Gilbert DM (1984) The isolation and characterization of a gene cluster flanking the Drosophila dopa decarboxylase gene. Ph D Thesis, Harvard Univ, Cambridge

Gilbert D, Hirsh J (1981) The dopa decarboxylase gene locus of *Drosophila melanogaster:* Orientation of the gene and preliminary mapping of genetic markers. In: Brown DD (ed) Developmental biology of purified genes. Academic Press, New York, pp 11–16

Gilbert D, Hirsh J, Wright TRF (1984) Molecular mapping of a gene cluster flanking the *Drosophila* dopa decarboxylase gene. Genetics 106:679–694

Giorgi F, Postlethwait JH (1985) Yolk polypeptide secretion and vitelline membrane deposition in a female sterile Drosophila mutant. Dev Genet 6:133–150

Hall LMC, Mason PJ, Spierer P (1983) Transcripts, genes and bands in 315,000 base-pairs of *Drosophila* DNA. J Mol Biol 169:83–96

Harper RA, Armstrong FB (1972) Alkaline phosphatase of *Drosophila melanogaster*. I. Partial purification and characterization. Biochem Genet 6:75–82

Hilliker AJ, Clark SH, Chovnick A, Gelbart WM (1980) Cytogenetic analysis of the chromosomal region immediately adjacent to the *rosy* locus in *Drosophila melanogaster*. Genetics 95:95–110

Hirsh J (1986) The *Drosophila melanogaster* dopa decarboxylase gene: Progress in understanding the in vivo regulation of a higher eukaryotic gene. In: Bogorad L (ed) Molecular and developmental biology. Liss, New York, pp 103–114

Hirsh J, Davidson N (1981) Isolation and characterization of the dopa decarboxylase gene of *Drosophila melanogaster*. Mol Cell Biol 1:475–485

Hochman B (1976) The fourth chromosome of *Drosophila melanogaster*. In: Ashburner M, Novitski E (eds) The genetics and biology of Drosophila, 1b. Academic Press, London, pp 903–928

Huntley MD (1978) Genetics of catecholamine metabolizing enzymes in *Drosophila melanogaster*. Ph D Thesis, Univ Virginia, Charlottesville

Joh TH, Baetge EE, Reis DJ (1984) Molecular biology of catecholamine neurons: similar gene hypothesis. Hypertension 6:II-1–II-6

Judd BH, Shen MW, Kaufman TC (1972) The anatomy and function of a segment of the X chromosome of *Drosophila melanogaster*. Genetics 71:139–156

Karlik CC, Mahaffey JW, Coutre MD, Fyrberg EA (1984) Organization of contractile protein genes within the 88F subdivision of the *D. melanogaster* third chromosome. Cell 37:469–481

Kraminsky GP, Clark WC, Estelle MA, Gietz RB, Sage BA, O'Conner JD, Hodgetts RB (1980) Induction of translatable mRNA for dopa decarboxylase in Drosophila: an early response to ecdysterone. Proc Natl Acad Sci USA 77:4175–4179

Lefevre G Jr (1974) The relationship between genes and polytene chromosome bands. Annu Rev Genetics 8:51–63

Lefevre G, Watkins W (1986) The question of the total gene number in *Drosophila melanogaster*. Genetics 113:869–895

Lewis EB (1978) A gene complex controlling segmentation in Drosophila. Nature 276:565–570

Lewis HW, Lewis HS (1963) Genetic regulation of dopa oxidase activity in *Drosophila*. Ann NY Acad Sci 100:827–839

Lifton RP, Goldberg ML, Karp RW, Hogness DS (1977) The organization of the histone genes in *Drosophila melanogaster:* Functions and evolutionary implications. Cold Spring Harbor Symp Quant Biol 42:1047–1051

Lindsley DL, Grell EH (1968) Genetic variations of *Drosophila melanogaster*. Carnegie Inst Wash Publ 627

Livingstone MS, Tempel BL (1983) Genetic dissection of monoamine neurotransmitter synthesis in Drosophila. Nature 303:67–70

Long EO, Dawid IB (1980) Repeated genes in eukaryotes. Annu Rev Biochem 49:727–764

Loomis WF, Gilpin ME (1986) Multigene families and vestigial sequences. Proc Natl Acad Sci USA 83:2143–2147

Lunan KD, Mitchell HK (1969) The metabolism of tryosine-o-phosphate in Drosophila. Arch Biochem Biophys 132:450–456

Marsh JL, Wright TRF (1980) Developmental relationship between dopa decarboxylase, dopamine acetyltransferase, and ecdysone in *Drosophila*. Dev Biol 80:379–387

Marsh JL, Wright TRF (1986) Evidence for regulatory variants of the dopa decarboxylase and alpha-methyldopa hypersensitive loci in Drosophila. Genetics 112:249–265

Marsh JL, Gibbs PDL, Timmons PM (1985) Developmental control of transduced dopa decarboxylase genes in *D. melanogaster*. Mol Gen Genet 198:393–403

Marsh JL, Erfle MP, Leeds CA (1986) Molecular localization, developmental expression and nucleotide sequence of the alpha methyldopa hypersensitive gene of Drosophila. Genetics (114:453–467
Meyerowitz EM, Hogness DS (1982) Molecular organization of a Drosophila puff site that responds to ecdysone. Cell 28:165–176
Mirkovitch J, Spierer P, Laemmli UK (1986) Genes and loops in 300,000 base-pairs of the *Drosophila melanogaster* chromosome. J Mol Biol 190:255–258
Mitchell HS, Lipps LS (1978) Heat shock and phenocopy induction in Drosophila. Cell 15:907–918
Moran L, Mirault M-E, Tissières A, Lis J, Schedl P, Artavanis-Tsakonas S, Gehring WJ (1979) Physical map of two *D. melanogaster* DNA segments containing sequences coding for the 70,000 dalton heat shock protein. Cell 17:1–8
Pentz ES, Wright TRF (1986) A diphenol oxidase gene is part of a cluster of genes involved in catecholamine metabolism and sclerotization in *Drosophila*. II. Molecular localization of the *Dox-A2* coding region. Genetics 112:843–859
Pentz ES, Black BC, Wright TRF (1986) A diphenol oxidase gene is part of a cluster of genes involved in catecholamine metabolism and sclerotization in *Drosophila*. I. Identification of the biochemical defect in *Dox-A2* (l(2)37Bf) mutants. Genetics 112:823–841
Restifo LL, Guild GM (1986) An ecdysterone-responsive puff site in *Drosophila* contains a cluster of seven differentially-regulated genes. J Mol Biol 188:517–528
Richards G (1986) Into the lair of the gene. Arch Insect Biochem Physiol Suppl 1:143–155
Richards G, Cassab A, Bourouis M, Jarry B, Dissous C (1983) The normal developmental regulation of a cloned *sgs3* "glue" gene chromosomally integrated in *Drosophila melanogaster* by P element transformation. EMBO J 2:2137–2142
Rizki TM, Rizki RM (1984) The cellular defense system of *Drosophila melanogaster*. In: King RC, Akai M (eds) Insect ultrastructure, vol 2. Plenum, New York, pp 579–604
Rizki TM, Rizki RM, Bellotti RA (1985) Genetics of a *Drosophila* phenoloxidase. Mol Gen Genet 201:7–13
Rudkin GT (1972) Replication in polytene chromosomes. In: Beermann W (ed) Developmental studies on giant chromosomes. Springer, Berlin Heidelberg New York, pp 59–85
Sarmiento LA, Mitchell HK (1982) Drosophila salivary gland proteins and pupation. Dev Genet 3:255–272
Saura AO, Sorsa V (1979) Electron microscopic analysis of the banding pattern in the salivary gland chromosomes of *Drosophila melanogaster:* Divisions 37, 38, and 39 of 2L. Hereditas (Lund) 91:5–18
Scholnick SR, Morgan B, Hirsh J (1983) The cloned dopa decarboxylase gene is developmentally regulated when reintegrated into the Drosophila genome. Cell 34:37–45
Seybold WD, Meltzer PS, Mitchell HK (1975) Phenol oxidase activation in Drosophila: a cascade of reactions. Biochem Genet 13:85–108
Snyder M, Davidson N (1983) Two gene families clustered in a small region of the Drosophila genome. J Mol Biol 166:101–118
Snyder M, Hirsh J, Davidson N (1981) The cuticle genes of Drosophila: A developmentally regulated gene cluster. Cell 25:165–177
Snyder M, Hunkapiller M, Yuen D, Silvert D, Fristrom J, Davidson N (1982a) Cuticle protein genes of Drosophila: Structure, organization and evolution of four clustered genes. Cell 29:1027–1040
Synder MP, Kimbrell D, Hunkapiller M, Hill R, Fristrom J, Davidson N (1982b) A transposable element that splits the promotor region inactivates a Drosophila cuticle protein gene. Proc Natl Acad Sci USA 79:7430–7434
Sparrow JC (1978) Melanotic "tumours". In: Ashburner M, Wright TRF (eds) The genetics and biology of Drosophila, 2b. Academic Press, London, pp 277–313
Sparrow JC, Wright TRF (1974) The selection for mutants in *Drosophila melanogaster* hypersensitive to α-methyl dopa, a dopa decarboxylase inhibitor. Mol Gen Genet 130:127–141
Spencer C, Gietz RD, Hodgetts RB (1986a) Analysis of the transcription unit adjacent to the 3′-end of the DOPA decarboxylase gene in *Drosophila melanogaster*. Dev Biol 114:260–264
Spencer C, Gietz RD, Hodgetts RB (1986b) Overlapping transcription units in the dopa decarboxylase region of Drosophila. Nature 322:279–281

Spierer P, Spierer A (1983) Molecular mapping of genetic and chromomeric units in *Drosophila melanogaster*. J Mol Biol 169:35–50
Spradling AC (1981) The organization and amplification of two chromosomal domains containing Drosophila chorion genes. Cell 27:193–201
Spradling AC, Digan ME, Mahowald AP, Scott M, Craig EA (1980) Two clusters of genes for major chorion proteins of *Drosophila melanogaster*. Cell 19:905–914
Steward R, McNally FJ, Schedl P (1984) Isolation of the *dorsal* locus of Drosophila. Nature 311:262–265
Tempel BL, Livingstone MS, Quinn WG (1984) Mutations in the dopa decarboxylase gene affect learning in *Drosophila*. Proc Natl Acad Sci USA 81:3577–3581
Valles AM, White K (1986) Development of serotonin-containing neurons in *Drosophila* mutants unable to synthesize serotonin. J Neuroscience 6:1482–1491
Vincent A, Colot HV, Rosbash M (1985) Sequence and structure of the *serendipity* locus of *Drosophila melanogaster*. A densely transcribed region including a blastoderm-specific gene. J Mol Biol 186:149–166
Warner CK, Grell EH, Jacobson KB (1975) Mechanism of suppression in Drosophila. III. Phenoloxidase activity and the *speck* locus. Biochem Genet 13:353–357
Weintraub H (1985) Assembly and propagation of repressed and derepressed chromosomal states. Cell 42:705–711 (1985)
Woodruff RC, Ashburner M (1979) The genetics of a small autosomal region of *Drosophila melanogaster* containing the structural gene for alcohol dehydrogenase. II. Lethal mutations in the region. Genetics 92:133–149
Wright TRF (1977) The genetics of dopa decarboxylase and α-methyl dopa sensitivity in *Drosophila melanogaster*. Am Zool 17:707–721
Wright TRF, Wright EY (1978) Developmental effects of dopa decarboxylase deficient mutants, Ddc^{n}, in *Drosophila melanogaster*. XIV Int Congress of Genet, Moscow, Contributed Paper Sessions Abstracts, Part I, Continuation Sections 13–20, p 615
Wright TRF, Bewley GC, Sherald AF (1976a) The genetics of dopa decarboxylase in *Drosophila melanogaster*. II. Isolation and characterization of dopa decarboxylase deficient mutants and their relationship to the α-methyl dopa hypersensitive mutants: Genetics 84:287–310
Wright TRF, Hodgetts RB, Sherald AF (1976b) The genetics of dopa decarboxylase in *Drosophila melanogaster*. I. Isolation and characterization of deficiencies that delete the dopa-decarboxylase-dosage-sensitive region and the α-methyl-dopa-hypersensitive locus. Genetics 84:267–285
Wright TRF, Beermann W, Marsh JL, Bishop CP, Steward R, Black BC, Tomsett AD, Wright EY (1981a) The genetics of dopa decarboxylase in *Drosophila melanogaster*. IV. The genetics and cytology of the 37B10-37D1 region. Chromosoma 83:45–58
Wright TRF, Steward R, Bentley KW, Adler PN (1981b) The genetics of dopa decarboxylase in *Drosophila melanogaster*. III. Effects of a temperature sensitive dopa decarboxylase deficient mutation on female fertility. Dev Genet 2:223–235
Wright TRF, Black BC, Bishop CP, Marsh JL, Pentz ES, Steward R, Wright E (1982) The genetics of dopa decarboxylase in *Drosophila melanogaster*. V. *Ddc* and *l(2)amd* alleles: Isolation, characterization and intragenic complementation. Mol Gen Genet 188:18–26
Young MW, Judd BH (1978) Nonessential sequences, genes and the polytene chromosome bands of *Drosophila melanogaster*. Genetics 88:723–762
Zhimulev IF, Belyaeva ES, Semeshin VF (1981a) Information content of polytene chromosome bands and puffs. CRC Crit Rev Biochem 11:303–340
Zhimulev IF, Pokholkova GV, Bgatov AV, Semeshin VF, Belyaeva ES (1981b) Fine cytological analysis of the band 10A1.2 and the adjoining regions in *Drosophila melanogaster* X-chromosome. II. Genetical analysis. Chromosoma 82:25–40
Zimmerman JL, Petri W, Meselson M (1983) Accumulation of a specific subset of *D. melanogaster* heat shock mRNAs in normal development without heat shock. Cell 32:1161–1170

Heat Shock: Puffs and Response to Environmental Stress

M. L. PARDUE, W. G. BENDENA, and J. C. GARBE [1]

1 The Heat Shock Response Is a Complex Reaction to Environmental Factors

The heat shock response was identified by Ritossa (1962), who found that when *Drosophila* larvae were placed at temperatures above 36 °C a small number of puffs were induced on the polytene chromosomes. The puffs regressed when the larvae were allowed to recover at 25 °C. Beermann's (1952) analyses of the tissue and temporal specificity of puffing had already led to the hypothesis that puffs were morphological manifestations of gene activity. This hypothesis was strongly supported by the demonstration that a specific gene product, a secretory granule, was correlated with the presence of a puff at the site of that gene (Beermann 1961) and also by evidence that puffs are the sites of active RNA synthesis (Pelling 1964). Thus Ritossa concluded that the temperature shock was affecting the activity of a small set of genes.

Ritossa found that the set of puffs induced by heat shock was also induced by a number of agents that were believed to deplete cellular ATP (Ritossa 1964). Other workers have added to this list of inducers (see Ashburner and Bonner 1979, for review) and it now seems that the agents can best be categorized as stresses. Ritossa reported that heat treatment of isolated salivary glands induced the same set of puffs induced in intact larvae and that all of the larval polytene tissues that could be analyzed developed the same set of puffs (Ritossa 1962, 1964). These cytological studies strongly suggested that the response was a property of single cells, rather than an organismal response, and that the same set of genes might be induced in all types of cells. Both of these predictions were confirmed later when it became possible to study the response biochemically. Even cultured cell lines, diploid cells in continuous culture, responded to heat shock and other stresses, as did cells in the intact organism (Spradling et al. 1975; McKenzie et al. 1975).

As techniques for biochemical analysis became more powerful, Tissieres et al. (1974) showed that the appearance of heat shock puffs was correlated with the appearance of a small set of proteins, the heat shock proteins (hsp's). That work has been followed by a large number of studies on the hsp's, their mRNA's, and the genes of the heat shock puffs (see Atkinson and Walden 1985, for review). These studies have shown that the heat shock response is complex. The complexity is only partly understood. As implied by the heat shock puffs, there is an in-

[1] Massachusetts Institute of Technology, Cambridge, Massachusetts 02139, USA.

Results and Problems in Cell Differentiation 14
Structure and Function of Eukaryotic Chromosomes
Edited by W. Hennig

crease in the transcription of heat shock genes but there is also a decrease in the transcription of non-heat shock genes (reviewed by Bonner 1985). There is also a translational control which inhibits translation of all but the heat shock mRNA's (reviewed by Ballinger and Pardue 1985). In addition, some RNA processing is affected (Ellgaard and Clever 1971; Rubin and Hogness 1975; Yost and Lindquist 1986). Severe heat shocks involve changes of normal transcription, processing, and translation, but different stresses do not necessarily affect all of these processes together; slight stresses may induce production of hsp's without disrupting synthesis of most normal cell products.

The polytene puffs made it possible to observe that a variety of agents induce the same set of genes in *Drosophila*. It is not possible to do similar analyses on diploid organisms; however, after the products of the *Drosophila* heat shock puffs began to be characterized, it became apparent that heat shocks and other stresses induced related sets of proteins in animals, plants, and bacteria (reviewed by Craig 1985). The sequences of the major hsp's show remarkable evolutionary conservation. Other aspects of the response may vary with the life style of the organism. For example, in *Drosophila* hsp's are usually produced at temperatures above 30 °C; mammalian cells, which are normally at 37 °C, must be moved to still higher temperatures for the induction of hsp's. A *Drosophila* heat shock gene transfected into a mouse cell was not transcribed at 37 °C but was induced by conditions that induced the mouse heat shock genes (Corces et al. 1981). In yeast the preferential translation of hsp's is effected by a mechanism different from that used in *Drosophila* (Lindquist 1981). The evolutionary conservation of the heat shock response argues that the response is important to cells of all kinds. What is known about the biology of the response suggests that it is a homeostatic mechanism for coping with environmental change; low levels of induction of the response have been correlated with acquisition of thermotolerance in a number of organisms (see Craig 1985).

The polytene puffs of *Drosophila* have provided the initial clues to what is proving to be a highly conserved response of living cells. Much remains to be learned about both the function of the response and the mechanism(s) involved in its induction. The first five of the *D. melanogaster* heat shock loci to be cloned encoded the major hsp's (see Craig 1985), raising questions about the functions of the other members of the set of heat shock puffs. We have recently cloned sequences from a sixth puff, 93D (Garbe and Pardue 1986; Garbe et al. 1986). It appears that the role of this locus is quite different from that of the protein-encoding puffs (see below).

2 One Major *Drosophila* Heat Shock Puff Differs from the Others in Several Ways

Cytological studies of several *Drosophila* species have shown that each species has one major heat shock puff with some characteristics not shared with the other members of the heat shock set. In *D. melanogaster* this puff is at 93D on the polytene chromosomes. In other species the location of this puff is different (2-48B in *D. hydei,* 20 CD in *D. virilis,* 2-58 in *D. pseudoobscura*), but cytological analyses strongly suggest that in each species this puff is the 93D-equivalent puff. The 93D-equivalent puffs are induced as part of the heat shock set, but they are also specifically induced by several agents, including benzamide, vitamin B6 (although *D. melanogaster* is an exception), and colchicine (Lakhotia and Singh 1982). These puffs contain distinctive large ribonucleoprotein particles and share antigenic determinants not seen on other heat shock puffs (Dangli et al. 1983). Although these observations do not tell us the function of the 93D-equivalent puffs, they show that the puffs are very similar in phenotype.

In situ hybridization of heat shock RNA to polytene chromosomes has been used to study the *D. melanogaster* puff, 93D (Lengyel et al. 1980), and the *D. hydei* puff, 2-48B (Lubsen et al. 1978). These studies also reveal differences between the 93D-type loci and the other major heat shock loci. Significant amounts of RNA transcribed from 93D and 2-48B remain in the nucleus. Much of this nuclear transcript hybridizes to sequences that cannot be filled by cytoplasmic transcripts of the locus, suggesting either multiple transcripts or a large, relatively stable nuclear precursor from this locus. In contrast, transcripts of the other heat shock loci accumulated rapidly in the cytoplasm but were barely detectable in nuclear RNA (Lengyel et al. 1980).

In view of the evidence that 93D and 2-48B share many phenotypic characteristics, one conclusion from in situ hybridization experiments was surprising. *D. melanogaster* heat shock RNA hybridized with the hsp-encoding puffs of *D. hydei* but there was no hybridization with 2-48B (Peters et al. 1980). Furthermore, heat shock RNA from *D. hydei* did not cross-hybridize with puffs thought to be related to 2-48B in other more closely related species. These results suggested that the DNA of the 93D-like puffs might be evolving faster than the sequences encoding the hsp's, a suggestion supported by later sequence analyses (see below).

3 The 93D Locus Yields Multiple Transcripts

Analyses of cloned DNA show that the heat shock locus at 93D encodes several overlapping transcripts (Fig. 1), the three most abundant being approximately 10 kb, 1.9 kb, and 1.2 kb (Garbe and Pardue 1986). The abundance of all three major transcripts is significantly increased during heat shock; however, all are also found in non-heat shocked cells where they are also under developmental control (W. Bendena, unpublished). The appearance of individual minor transcripts may be more dependent on the specific conditions of the heat shock, but we have not yet completed analyzing this.

The transcribed region of 93D is approximately 10 kb (Fig. 2) and contains a region of unique sequence and a stretch of 280 base pair (bp) repeats (Walldorf et al. 1984; Garbe and Pardue 1986). All of the major transcripts are initiated in the unique region, approximately 2.5 kb from the 280 bp repeats. The 1.9 kb and the 1.2 kb transcripts terminate before the beginning of the repeats while the

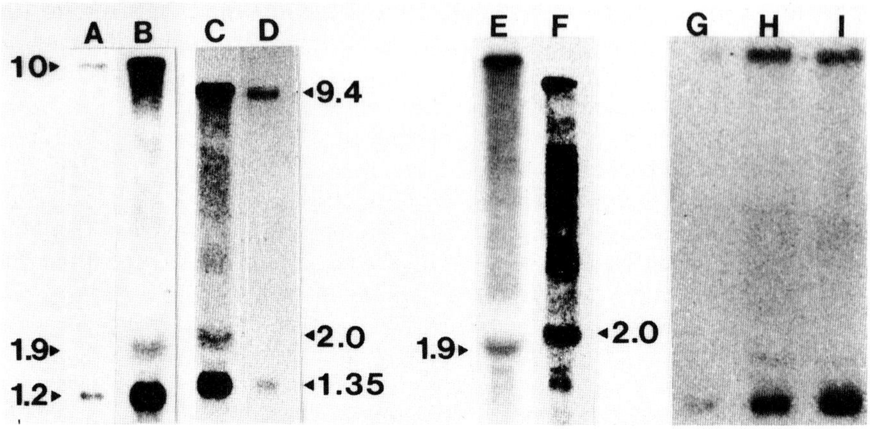

Fig. 1. 93D-type loci yield multiple transcripts. Three major transcripts are induced by heat shock (lanes *B, C, E, F*) and are also seen at a lower level in control cells (lanes *A* and *D*). The constitutive level of 93D transcripts increases in the later stages of embryogenesis (lanes *G–I*). RNA was separated on 1% agarose gels, transferred to filters, and hybridized with ^{32}P-labeled probe. An autoradiogram is shown. RNA's were: lane *A*, RNA from cultured *D. melanogaster* cells at 25 °C; lanes *B, E*, RNA from cultured *D. melanogaster* cells heat shocked for 1 h; lane *D, D. hydei* cells at 25 °C; lanes *C, F, D. hydei* cells heat shocked for 1 h; lanes *G, H, I*, non-heat-shocked *D. melanogaster* embryos, 0–6 h, 6–12 h, 12–18 h, respectively. Lanes *A–D* and *G–H* were probed with sequences of the 5′ exons of the transcript of the appropriate species. Lanes *E* and *F* were probed with intron sequences from the same species. Within each set of lanes equal numbers of cells are compared. Numbers indicate the size of the transcripts in kDa

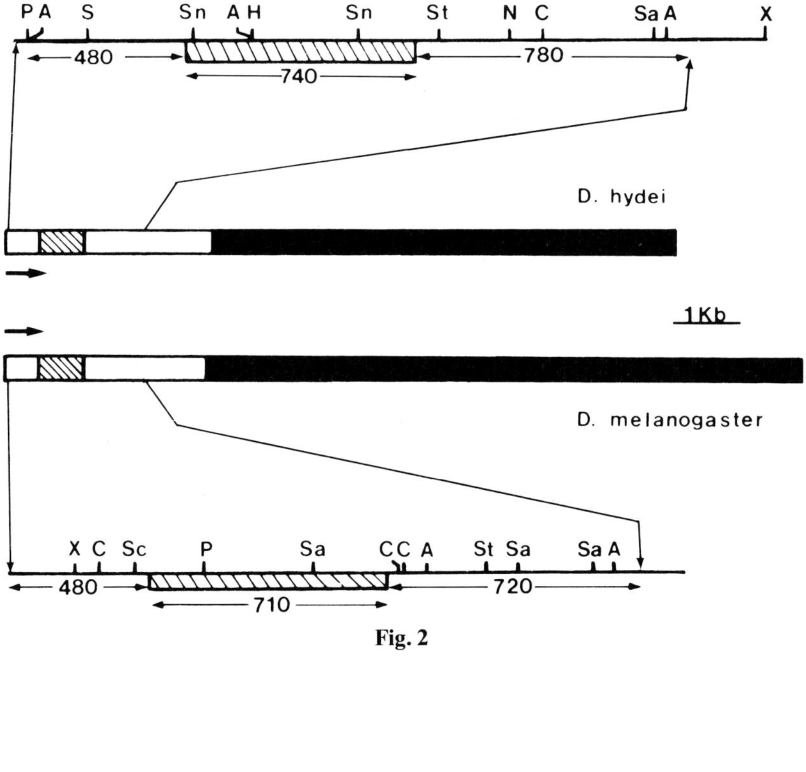

Fig. 2

10 kb transcript contains some 7.5 kb of these tandem repeats. Both the 10 kb and the 1.9 kb RNA's are limited to the nucleus. They are relatively abundant, suggesting that they may play a role in the nucleus rather than simply serving as precursors to the cytoplasmic species (Garbe et al. 1986).

The 1.2 kb RNA is found in the cytoplasm. It apparently differs from the 1.9 kb RNA by having a 710 bp intron spliced out (Garbe and Pardue 1986). Although much *Drosophila* mRNA is spliced, the only other *Drosophila* heat shock mRNA known to undergo this form of processing is the hsp82 mRNA (Holmgren et al. 1981). In addition to being spliced, the 1.2 kb transcript is polyadenylated

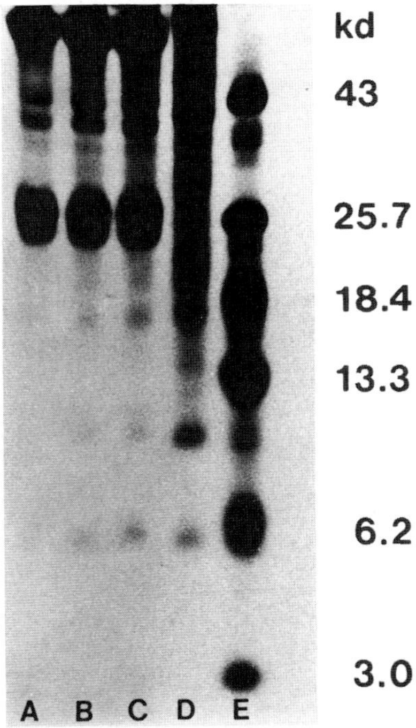

Fig. 3. Autoradiograph of low molecular weight polypeptides synthesized in *D. melanogaster* cultured cells at 25 °C or between 30 and 60 min of heat shock. Polypeptides predicted from the open reading frames of the cytoplasmic transcript are not seen although large amounts of the major hsps (e.g., those around 25 kDa) have accumulated. Control cells (lane *D*) were incubated in the presence of ^{3}H-labeled (uniformally labeled) amino acids for 30 min. Experimental cells (lanes *A–C*) were heat shocked for 30 min prior to labeling for an additional 30 min. Low molecular mass protein standards are shown in lane *E*. (Garbe and Pardue 1986)

Fig. 2. Molecular map of the *D. melanogaster* 93D locus and the *D. hydei* 2-48B locus. *Bars* represent the genomic organization of the transcription unit of each species. *Open bars* show the exon portions of the unique regions. *Cross-hatched bars* indicate the introns and *solid bars* the tandem repeats. *Short arrows* in the center indicate the direction of transcription. *Upper* and *lower figures* expand the regions that give rise to the two smaller transcripts for each species. Restriction enzyme sites are indicated as follows: A, Asu II; C, Cla I; H, Hind III; N, Nar I; P, Pst I; S, Sal I; Sa, Sac I; Sc, Sca I; Sn, Sna BI; St, Stu I; X, Xba I. *Numbers below the lines* indicate the approximate numbers of nucleotides in each block and thus has two of the features usually associated with mRNA. Surprisingly, the sequence of this RNA does not appear to encode a heat shock polypeptide. The longest open reading frame beginning with a methionine codon would encode only 34 amino acids. Open reading frames of that length would be expected to occur by chance in the *Drosophila* genome. In fact, the nontranscribed strand of the 1.2 kb sequence would yield longer open reading frames than found in the cytoplasmic RNA (Garbe et al. 1986).

Although we have not made an exhaustive search for the polypeptides predicted from the open reading frames in the 1.2 kb cytoplasmic RNA, we have looked carefully for such small polypeptides during the first hour of a 36 °C heat shock. During this time the 1.2 kb transcript shows a pattern of accumulation similar to that of the other hsp mRNA's. We detect abundant synthesis of the major hsp's but no small heat-induced polypeptides (Fig. 3). It seems most likely that the 1.2 kb mRNA does not function as an mRNA; however, it is possible that the translation product turns over too rapidly to be detected or that the RNA is translated only under special conditions.

4 93D-Equivalent Loci Have a Rapidly Diverging Sequence but a Conserved Structure

Studies of cloned DNA from 2-48B had shown that the locus has several kb of a 115 bp repeat (Peters et al. 1984) and a 9.4 kb transcript that included these repeats (Peters 1983). These similarities to the 93D locus prompted us to reconsider the lack of cross-hybridization reported for heat shock RNA (Peters et al. 1980). When the small subclones used to sequence the *D. melanogaster* locus were hybridized to DNA cloned from 2-48B, one subclone showed relatively strong hybridization, even at high stringency (Fig. 4). Surprisingly, this subclone was from the intron region, a region which has been reported to diverge more rapidly than exons, even in the hsp 82 gene (Blackman and Meselson 1986). Two exon subclones gave much lower hybridization. No detectable hybridization was seen with the rest of the subclones from the exon regions, even when hybridization was done at much reduced stringency.

The sequences homologous to the 93D intron were located about 2 kb from the 115 bp repeats of 2-48B, in a site analogous to the intron site of 93D. We have used subcloned fragments of this region of homology and of the surrounding sequences to show that 2-48B, like 93D, yields multiple transcripts (Fig. 1). We have concentrated on the three predominant transcripts. In both *D. melanogaster* and *D. hydei* these are two nuclear RNA's (9–10 kb and 1.9–2.0 kb) and one cytoplasmic RNA (1.2–1.35 kb). Sequence analysis confirms the earlier hybridization results (Fig. 5). The most striking region of homology is a 59 bp segment at the 3′ end of the intron, extending slightly into the 3′ exon. There are other much smaller stretches of homology around the 5′ splice site, the polyadenylation site, and near the 5′ end of the transcript. The regions at the 5′ end also show homology to conserved regions of the heat shock mRNA's (Garbe et al. 1986; Holmgren et al.

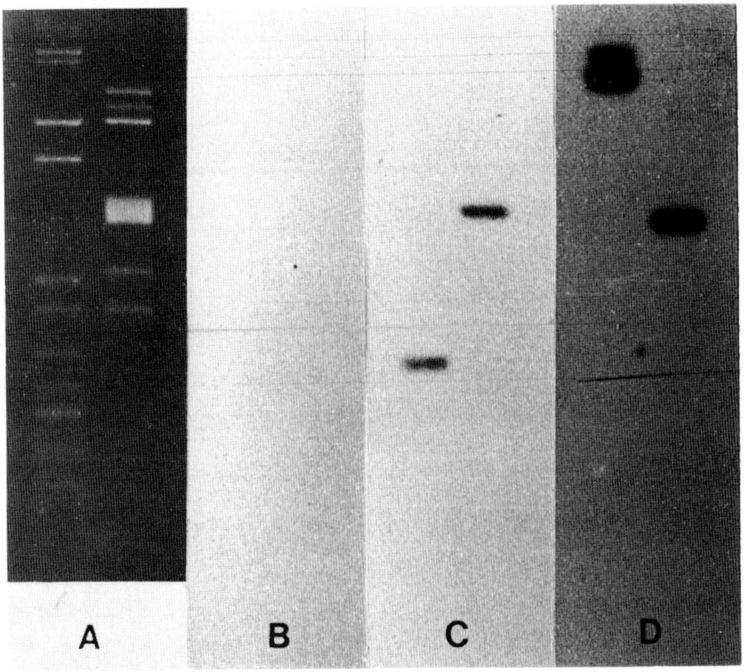

Fig. 4. Autoradiograph showing strong cross-hybridization of 93D intron sequence with DNA from *D. hydei* 2-48B. Weaker cross-homology is seen with fragments 3′ and 5′ to the *D. melanogaster* intron. Other regions of the 1.2 kb transcript show no hybridization to *D. hydei* DNA. *D. hydei* cosmid cDh171 was digested with Eco RI and Hind III (*right lane* of each pair) or with Eco RI and Xba I (*left lane*) and then separated on a 0.7% agarose gel (*A* shows ethidium stained sample of gel lanes). Seven replicas of the two gel lanes were transferred to membranes and hybridized with small fragments of the sequence of the *D. melanogaster* 1.9 kb transcript. *B, C,* and *D* are autoradiograms of 3 membranes after hybridization. Lane *D* was hybridized with the intron fragment. Lane *B* was hybridized with a cDNA fragment 200 bp 5′ to the intron. Lane *C* was hybridized with a fragment containing sequences flanking the intron. Hybridization stringency was set by washes in 0.3 M NaCl/0.03 M sodium citrate, pH 7.0 at 60°. (Garbe and Pardue 1986)

1981; Hultmark et al. 1986). The cytoplasmic transcript of 2-48B, like the analogous transcript of 93D, has only a few very short open reading frames and probably does not encode a protein.

The short tandem repeats are strikingly homogeneous within each species (Peters et al. 1984; Garbe et al. 1986) yet the repeats of 93D differ from those of 2-48B in both length and sequence. The longest region of homology that can be detected is nine nucleotides, ATAGGTAGG. Interestingly, this sequence is present once in the 115 bp repeat and twice in the 280 bp repeat, thus making the concentration and distribution of the sequence approximately the same on the two RNA's. If the sequence acts as some sort of a recognition site, the 93D transcript and the 2-48B transcript might be equivalent in function.

```
1
DRM  TATAAATAGAGCCGCCTCAGTCCGGTCACGTCACTCTCAAATGAAAAAGTGTTCAAGTGCATTCAAAGTGAAGCTGAAAAAATAACCAGT  90
DRH  TATAAATAAAGGTGACTGCCTGGACTGCAGTCATTCGAAAAA-AATAAGTGTTGAAGTGAAGTTCTCTAG----TGAAACTTAAAACTGT
91
DRM  TAAAAATAGTACAAAGAAATTTTCTTTCTTGCAATTTGCAAGCAGTAGCTACAACCAAAAATGGAAAAGTGTAAAAATCGTGTCCCAGCA  180
DRH  TTGAAGAATAAAATATATTTCTGGATAAGTGCTGTTTTGCATTGTTTCTACATAAATAATATACAACTGCAATGGAGAAGTGTACAGTAT
181
DRM  GACGAGCAGCAGCAGTACGAGTATTGCAAAATGCAGGGGCAAGGGCCCACGTAGTATTTTTCCACGTCGGGCATTTAATGCTCTCGAGTT  270
DRH  TTGTTGCGAAGGCGAGATTGCGGTAGCTATAACTCATGCTCACGTCGACTGAAATAACGTTTAGCGCACCG---TTCAAGCTGTTTTGCT
271
DRM  GGAAACAAAGAAACCATACGCAAACCCCCTGGAAAAGATGTGATTAGTCATCGATTTTGCTGGTCAGCGTCGGGGAATTTCGC-AATGCA  360
DRH  GTGAACAAAGAAAATAAAATCTGCGTCATCGTC---GTGGCCAGTTTTCGGCCGTTTTGCATTTGAGGAAATTGGGGCAACACGAAAGCA
361
DRM  GCAGGCAGTTTTCTCAA-ATTTGGCTAGAAAG-TGACCCACTAGGCAGTCTGAG---------GCAGTTATCCAGGATGTAAGGATGTGT  450
DRH  GCGGCCAAGTTTTTCAACATTTGGCTAGAAAAATTGCACTGTAGGCAGGCTGAGCAGATGCCAGGAGTTTGCGATAAATTGATGAGGCAG
451
DRM  GCAGTACTCTCTGTGTC---GAGTGGC-GATGGCCTGACCTGCGGATTTTGAAAGATTGAAATAGGAAGCCAGGTATACATACACAGCAA  540
DRH  GAAAGACTAGCCAGATCCCGGAGTGTATGTTGTCCTGACCTG-GACTTAAGGAGTTTGGAGTTAGGAAGCCAGGTATAAACCGAAA-CGG
541
DRM  CGAATGCAGCGTGTTCTATATTCTTAATTCGTAAGCCAAAC-CTATGTTTAAGACGTAGTTTATATAACCGTACATATGTATGTAGCGAA  630
DRH  CGAAACCAGCTTCGT--ACCTTATTA-TACGTAAGCGAAAATCTAAGCATAAGGCCTAGTTT-TATTACTCAAC---TATATGCAATTTA
631
DRM  ACACGTTTTCAAACGTCGCATCGAAGCCATTTCTTCTGCAGCCATTGCAATTTCATCGTGTTGCGTAATCGCAAATTGTAAATGTGTCCC  720
DRH  AAACGTTTTCAAGCCTGACGTCAGCGTAATTT-TTGTTTTTCGAGTAAACGTTTATACTCTGAAATTAGCGAATATTGTGTAATTCATTC
721
DRM  TACTACTTCACATTTGTGTGTATCCAAGTCGAAAGTTCTCGAACTATT-----CTCGTG---AGCAA-----CCCCCGGATCGAAAGGCC  810
DRH  GTA-ATTTTGTATTGCGTAAAATAATTTTCGAAGCTTCTAGCACTATTTCCACCCCATGCAAAGCATGTCTTCCCATCTATCTTTCGGCC
811
DRM  ACCAACTGG-----CTCGTTCTCTCTCTCCAGAAACAGCTTACTCTTACGTTTACAATTGACTCTCTACCTATCGGGTATACACATT---  900
DRH  ATTTCGAGTAAATTCTCGACTGTTCTTTGTTTAATCCCCTAAAAGGAGAGCAGCTGGCCTGCTCTAACAATAAATAGCATTGTATTTGCA
901
DRM  TTTATATATGTATACA-CTCAGAGACACCCCAATCCCCCACCATACATATGGCGAGATGTGTGAATGTATCGCAGTTGGACGAAAACGGC  990
DRH  TGTACATATATATAAATCTCGTATACACGTATTCTCTCTCGCTTGCACTTGCGCTTACATGCCGTTGCGTGGCA-TGGCAGCTGCGCGTA
991
DRM  AGCTGCGCAGAAAACGCTGCGGCGTACATACAAACGAGCTCTTTGTTTGCCTTTCCCCAGTT---TTGCTTCCCCATACGTATAACTGGG  1080
DRH  AACTCAGCCTCGTTGGTCGAGTTGTTCGACCCACCTTTGTCTTACAATGTATTTGTGTGTTTCCATCTCTTCTCTCTATACATATACGTA
1081
DRM  TGTTTATCAGCGGTAATCT-CGCCAGCCTGACTT-TGTCCCTGTCGCTTCTA-TTGACA------TTCTAAACTAGATTGTAGACGTTTT  1770
```

```
1171                                                                                          1260
DRM  AACATTT---ACCTAGCTGTTAATCTTATTTGTAAACATCTCTATATGCAACCTGCCATACCACTTAACCAAGCAATATGTATTTCTTTC
DRH  AGCCGTTTAAACCCTAGTGTTAATCTTATTTATAAAC-TGTCACT-TGATAATTGTTACACT--TTAACCAAGCAATATGTATTTCTTTC
1261                                                                                          1350
DRM  TCTAAACTTTATAGTTGGGCGTTGAAAGTTGATATCGATATCGATCCGTGAAAAGTCGATACCC-TGCGCAAGCATGGGGCGGCATATGG
DRH  TCTAAACTTTATAGTTGGGCGTTGAAAGTTGATA-CGCGATCGAACGATGTTAACTTGAACGCAATACGTGAGAAAAGGCCTTACAATTA
1351                                                                                          4440
DRM  GTGCTGAAAACGCACT---CGGCCCGATCCCGATTGCAGCGTTATTCGAAAGCTGTGTCTGCGACCGTGACTGAGATCATATGCGTACAT
DRH  GCTAAAAATGCTGATTGTTCGGTCCCATGTAGCCGTGAGAGTATATCTAAA-CTGCGACTAAATGTGTGTGTGTCGCGGCCTACACAGGG
1441                                                                                          1530
DRM  ATATCTAATGTCCGGGGTCGTGGGCCAGCCAGGGTGCTCGATTCTGTCAGATTGATTGTGCGGATTGTGTTATAGGAACACTGGTGTATC
DRH  GTA-CTAATCGTCGTTTTTATCAGATTGTGATTGCG-TGGATTGAATATATATAGGAACCTGTATCGACTTCTCTGCTCTACTAAGACTT
1531                                                                                          1620
DRM  GACTTCTCTGCTCCACTATGGGTGAAGGATACCCTACCGAAAAGGCCTTCTGTCGCTTAC-TATCATCGAACAAGTTCCGTAA-AGGGCA
DRH  GT-TCAAAAGAAAGGCGCCCAGTTACATATGTACTATATAAACGGTCGTCTCTTATTTACCTACCATCGAACCGTTATCTCAATACTGAT
1621                                                                                          1710
DRM  GACATACGTACA-CGTGGCAGCATATTACGTTCAATGACACATC--GTCTCTGGATTAGTAGTTGAACCAACGAGCTCG------AATAT
DRH  CGAATAGATACAACACTACTGACAACGATCGATATACACATATCTAGTATCTGGAGTAGCAACCGATTGCTCGTTCGTTTCACAAAGCAC
1711                                                                                          1800
DRM  ACCCTGCCAGTTGGTTGTAAATATTAATTTTGTTCTTATATTCAGTTGTAAAGTCAATACGGAAGACATTTCCACACCAAC-CATGTGTC
DRH  AAATTATAAGATACATCTTTCTTGTTATTTACTTAATATATTTTTTCTTTGTTGTTAACTTAAACATTTACAAGTACACGACAAGCGTG
1801                                                                                          1890
DRM  ACTTATGTTCTT-AAATACCAGAAACTGTTTAAAT--------------------TTTTGTTCTTATACACTTCACCAATTAAGAATTA
DRH  GCTTAAGTTCTTGAAATACCTGAAACTGTAAAAAAAAAAAAATACAAAAATTATTTCTTCTGTTTTTGTACATATCCAAGATTTAAGAAAA
1891                                                                                          1980
DRM  CCTGTACTGATGTAAGTGTACCTGTTGAATCACAACAAAAATAGAAAAAATGAATAACAAAAGAGCTCTATAACAGAAAAGCCACAGAAA
DRH  CTTGTGTCCATTTTTGTAATTACAATTGTCCAATTCAGTAATAGAAAATGAGAAGAAAAAAATAATACGAAAATGAAAAAGAAAAAGAAA
1981                                                                                  2064
DRM  AAGTGA-----------ATAAAAAAATTCAAAAATTCAAAAAA------------------------TAAATTTCGAAACATTTG
DRH  AAGAGAGCTCACCAAACAGAAAAGAAGCTACAATCAAAAAAAAGAATAAAAGAAATCTGAGACCAATAAATTTCGAAACAATTG
```

Fig. 5. Aligned sequences of the unique regions of the 93D and the 2-48B locus. DRM and DRH specific the *D. melanogaster* and *D. hydei* sequences, respectively. The two sequences were aligned for maximal homology using the programm ALIGN. The sequence alignment begins at the TATA box and ends at the polyadenylation site determined for *D. melanogaster*. Some of the significant regions of homology revealed by this alignment are enclosed in boxes. The 5′ splice junction (position 523) and the 3′ splice junction (position 1274) are indicated by *open arrows*. The polyadenylation site (position 2064) is indicated by a *solid arrow*. The start site and direction of transcription are indicated by the *arrow at the top* of the sequence. Sequences matching the heat shock mRNA leader consensus are marked with *solid triangles*. *Open triangle* indicates the nucleotide at +20 which does not match the consensus. The sequences enclosed within *brackets* represent a second region which is conserved within the hsp mRNA leader (Garbe et al. 1986)

5 Conclusions

Polytene puffs have been important for the study of the heat shock response because they allow the set of coordinately controlled genes to be visualized when they are induced. As expected, some of the puffs contain genes coding for the major polypeptides produced during the response. These loci have all been identified and cloned. One of the unanswered questions about the heat shock response concerns the functions of the remaining puffs. Our studies of one of these, 93D and its counterparts in other species, are revealing a locus for which evolutionary constraints appear to be somewhat different from those on protein-encoding loci. Only small segments of the nucleotide sequence are conserved but the number, size, and structure of the transcripts are conserved. The most striking sequence homologies between 93D and 2-48B occur in portions of the transcripts that remain in the nucleus. In addition, the nuclear transcripts are relatively more abundant than the typical precursor RNA. These observations suggest that transcripts of the 93D-type loci play roles both in the nucleus and in the cytoplasm. One intriguing possibility is that the locus might act to coordinate one or more events in the two cell compartments.

Acknowledgments. We are grateful to Drs. M. Wolfner and D. Hogness for clone padm129F5 and to Dr. N. Lubsen for clone cDh171 and for *D. hydei* cell lines. Will A. Gilbert and the Whitaker College Computing Facility have provided assistance with sequence analysis. Our work has been supported by a grant from the National Institutes of Health.

References

Ashburner M, Bonner JJ (1979) The induction of gene activity in Drosophila by heat shock. Cell 17:241–254

Atkinson BG, Walden DB (eds) (1985) Changes in eukaryotic gene expression in response to environmental stimuli. Academic Press, New York, pp 379

Ballinger D, Pardue ML (1985) Mechanism of translational control in heat-shocked Drosophila cells. In: Atkinson BG, Walden DB (eds) Changes in eukaryotic gene expression in response to environmental stress. Academic Press, New York, pp 53–70

Beermann W (1952) Chromomerenkonstanz und spezifische Modifikationen der Chromosomenstruktur in der Entwicklung und Organdifferenzierung von Chironomus tentans. Chromosoma 5:139–198

Beermann W (1961) Ein Balbiani-Ring als Locus einer Speicheldrüsen-Mutation. Chromosoma 12:1–25

Blackman RK, Meselson M (1986) Interspecific nucleotide sequence comparisons used to identify regulatory and structural features of the Drosophila hsp82 gene. J Mol Biol 188:499–515

Bonner JJ (1985) Mechanism of transcriptional control during heat shock. In: Atkinson BG, Walden DB (eds) Changes in eukaryotic gene expression in response to environmental stimuli. Academic Press, New York, pp 31–51

Corces V, Pellicer A, Axel R, Meselson M (1981) Integration, transcription, and control of a Drosophila heat shock gene in mouse cells. Proc Natl Acad Sci USA 78:7038–7042

Craig EA (1985) The heat shock response. CRC Crit Rev Biochem 18:239–280

Dangli A, Grond CJ, Kloetzel P, Bautz EKF (1983) Heat shock puff 93D from Drosophila melanogaster: accumulation of RNP specific antigen associated with giant particles of possible storage function. EMBO J 2:1747–1751

Ellgaard EG, Clever U (1971) RNA metabolism during puff induction in Drosophila melanogaster. Chromosoma 36:60–78

Garbe JC, Pardue ML (1986) Heat shock locus 93D of Drosophila melanogaster: a spliced RNA most strongly conserved in the intron. Proc Natl Acad Sci USA 83:1812–1816

Garbe JC, Bendena WG, Alfano M, Pardue ML (1986) A Drosophila heat shock locus with a rapidly diverging sequence but a conserved structure. J Biol Chem 261:16889–16894

Holmgren R, Corces V, Morimoto R, Blackman R, Meselson M (1981) Sequence homologies in the 5′ regions of four Drosophila heat shock genes. Proc Natl Acad Sci USA 78:3775–3778

Hultmark D, Klemenz R, Gehring W (1986) Translational and transcriptional control elements in the untranslated leader of the heat shock gene hsp22. Cell 44:429–438

Lakhotia SC, Singh AK (1982) Conservation of the 93D puff of Drosophila melanogaster in different species of Drosophila. Chromosoma 86:265–278

Lengyel JA, Ransom LJ, Graham ML, Pardue ML (1980) Transcription and metabolism of RNA from the Drosophila melanogaster heat shock puff site 93D. Chromosoma 80:237–252

Lindquist S (1981) Regulation of protein synthesis during heat shock. Nature 293:311–314

Lubsen NH, Sondermeijer PJA, Pages M, Alonso C (1978) In situ hybridization of nuclear and cytoplasmic RNA to locus 2-48B in Drosophila hydei. Chromosoma 65:199–212

McKenzie SL, Henikoff S, Meselson M (1975) Localization of RNA from heat-induced polysomes at puff sites in Drosophila melanogaster. Proc Natl Acad Sci USA 72:1117–1121

Pelling C (1964) Ribonukleinsäure-Synthese der Riesenchromosomen. Autoradiographische Untersuchungen an Chironomus tentans. Chromosoma 28:136–187

Peters FPAMN (1983) Genomic organization of heat shock locus 2-48B of Drosophila hydei. Ph D Thesis, Catholic Univ Nijmegen

Peters FPAMN, Lubsen NH, Sondermeijer PJA (1980) Rapid sequence divergence in a heat shock locus of Drosophila. Chromosoma 81:271–280

Peters FPAMN, Lubsen N, Walldorf U, Moormann RJM, Hovemann B (1984) The unusual structure of heat shock locus 2-48B in Drosophila hydei. Mol Gen Genet 197:392–398

Ritossa FM (1962) A new puffing pattern induced by heat shock and DNP in Drosophila. Experientia 18:571–573

Ritossa FM (1964) Experimental activation of specific puff loci in polytene chromosomes of Drosophila. Exp Cell Res 35:601–607

Rubin GM, Hogness DS (1975) Effect of heat shock on the synthesis of low molecular weight RNAs in Drosophila: accumulation of a novel form of 5S RNA. Cell 6:207–213

Spradling A, Penman S, Pardue ML (1975) Analysis of Drosophila mRNA by in situ hybridization: sequences transcribed in normal and heat shocked cultured cells. Cell 4:395–404

Tissieres A, Mitchell HK, Tracey UM (1974) Protein synthesis in salivary glands of Drosophila melanogaster: relation to chromosome puffs. J Mol Biol 84:389–398

Walldorf U, Richter S, Ryseck R-P, Steller H, Edstrom JE, Bautz EKF, Hovemann B (1984) Cloning of heat-shock locus 93D from Drosophila melanogaster. EMBO J 3:2499–2504

Yost HJ, Lindquist S (1986) RNA splicing is interrupted by heat shock and is rescued by heat shock protein synthesis. Cell 45:185–193

The Y Chromosomal Lampbrush Loops of *Drosophila*

W. HENNIG [1]

1 Introduction

In 1961, Meyer, Hess and Beermann published a description of "phase-specific functional structures in spermatocyte nuclei of *Drosophila melanogaster* and their dependence on the Y chromosome". This paper initiated extensive research work on the *Drosophila* Y chromosome, which is based on the discovery of lampbrush loop-like structures in primary spermatocyte nuclei. These nuclear structures in spermatocytes of *Drosophila* proved to be a general phenomenon, at least in the genus *Drosophila* (Hess and Meyer 1963 b; Hess 1967). They were recognized, on the basis of studies on spermatocytes of *Drosophila hydei,* as lampbrush loop-like active genetic loci related to the male fertility factors, which reside in the Y chromosome (Meyer 1963; Hess and Meyer 1963 b). By cytogenetic means Hess (1965) established a chromosome map of the Y chromosome of *Drosophila hydei* (Fig. 1). According to this map, one of the lampbrush loop-forming genes is located in the short arm of the Y chromosome, while four others reside in the long arm. These lampbrush loops in the long arm are clustered in two regions, one close to the kinetochore and one close to the end of the long arm (Hess 1965 a). The Y chromosome, in addition, carries two nucleolus organizer regions, one at each end (Meyer and Hennig 1974 a; Hennig et al. 1975).

The detailed genetic mapping of male fertility genes in the Y chromosome of *D. hydei* revealed that each lampbrush loop is correlated to only one complementation group (Hackstein et al. 1982). This is surprising, since each of the loops must accommodate large amounts of DNA (Hennig et al. 1974 a) and giant transcripts can be assigned to the loops (Hennig et al. 1974 b; Meyer and Hennig 1974 b; Glätzer and Meyer 1981; Grond et al. 1983; de Loos et al. 1984). Additional genetic loci were discovered in the Y chromosome, which are essential for male fertility, but which form no obvious lampbrush loops in the spermatocyte (Hackstein et al. 1982). Loci other than those required for male fertility have not been discovered in the Y chromosome.

[1] Department of Genetics, Faculty of Science, Katholieke Universiteit Nijmegen, Toernooiveld, 6525 ED Nijmegen, The Netherlands.

Results and Problems in Cell Differentiation 14
Structure and Function of Eukaryotic Chromosomes
Edited by W. Hennig

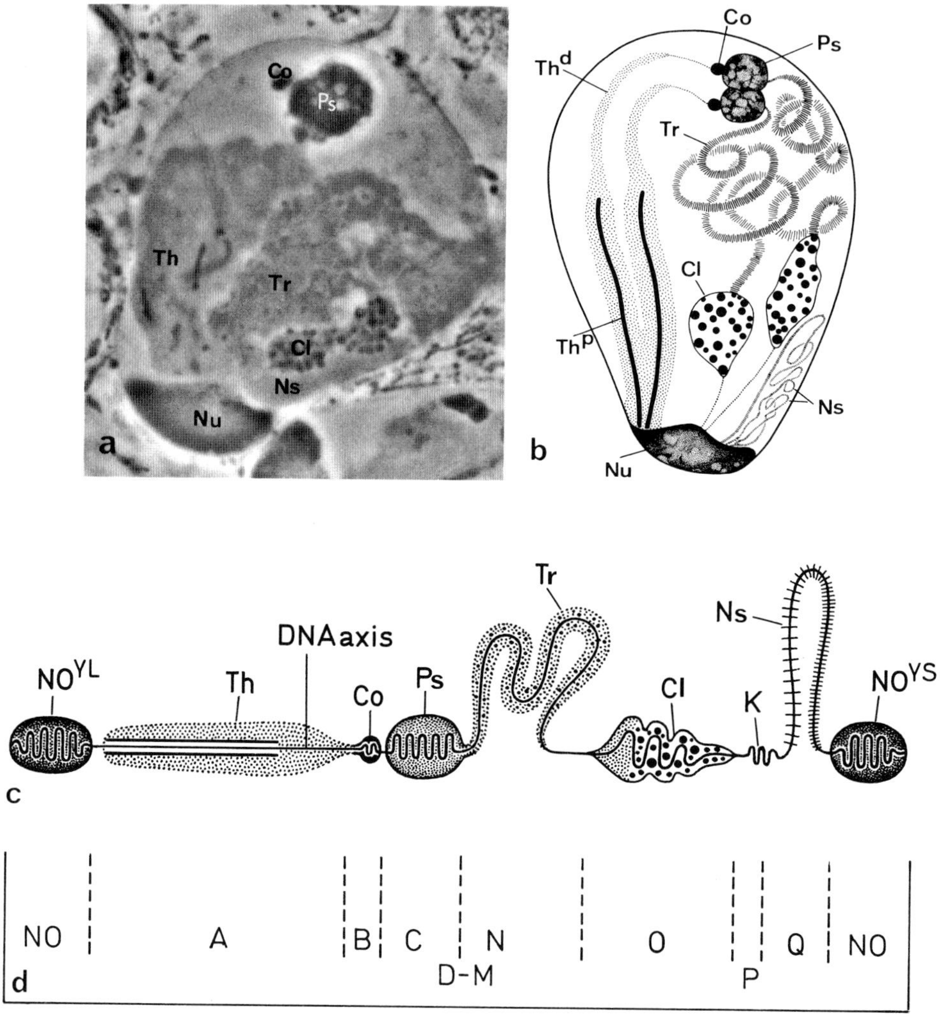

Fig. 1. Cytology and schematic explanation of a primary spermatocyte nucleus of *Drosophila hydei*. **a** Phase contrast picture of a wild-type nucleus. All visible material inside the nucleus except the nucleolus (*Nu*) belongs to Y chromosomal lampbrush loops. Autosomes and the X chromosome are not visible (see Fig. 2). **b** Diagramatic explanation of **a.** The structural relationship of the loops is indicated. Since the loops are formed during the first meiotic prophase, they are present in duplicates because the Y chromosome contains two chromatids. **c** Diagram of one chromatid (related to the scheme shown in **b**). The kinetochore (*K*) divides the Y chromosome into a long (*to the left*) and a short (*right*) arm. The long arm carries five loop-forming sites, the short arm only one. [Note: the cones (*Co*) have recently been identified as an independent transcription unit, which is most likely related to complementation group B. Huijser and Hennig, unpublished data]. **d** Location of the complementation groups in their relationship to the lampbrush loops (Hackstein et al. 1982). Only loci *A–C, N, O,* and *Q* are related to loop-forming sites. The residual complementation groups are indicated below in the approximate location relative to the loops. The locations of the loops are not shown in their correct positions on the chromosome. *Cl* clubs; *Co* cones; *K* kinetochore; *NO* nucleolus organizer region; *Ns* nooses; *Nu* nucleolus; *Ps* pseudonucleolus; *Th* threads (*d* distal part; *p* proximal part); *Tr* tubular ribbons. *A–Q* complementation groups. *YL* long arm, *YS* short arm of the Y chromosome. (Hackstein et al. 1982). Bar represents 10 µm

2 Structure of the Y Chromosomal Lampbrush Loops

2.1 Morphology of the Y Chromosome

The initial studies of the Y chromosomal lampbrush loops were concentrated on their ultrastructure and their cytology (see Hess and Meyer 1968). These studies revealed the close structural relationship to particular lampbrush loop pairs in amphibian oocytes, called the giant granular loops (Meyer 1963). A confirmation of this homology between the nuclear structures in *Drosophila* spermatocytes and oocyte lampbrush chromosomes was obtained from autoradiographic studies (Hennig 1967) which demonstrated that the loops incorporate ^{3}H-uridine, as must be expected for transcriptionally active loci. Consistent with the idea of an uncoiled DNA constitution in active chromosomal regions, the DNA concentration was found to be extremely low since studies of ^{3}H-thymidine incorporation yielded low levels of labeling in Y chromosomal regions (Hennig 1967). Several alternative models of the Y chromosome constitution in primary spermatocytes had been discussed by Meyer (1963). The observation of direct connections between most of the lampbrush loops in spermatocytes by Hennig (1967) and some other arguments are, however, in favor of the "open chromatid model" displayed in Fig. 1. This model has found substantial support by the observation that condensed Y chromosomal DNA cannot be detected in spermatocyte nuclei even with highly sensitive techniques such as staining with fluorescent dyes (Kremer

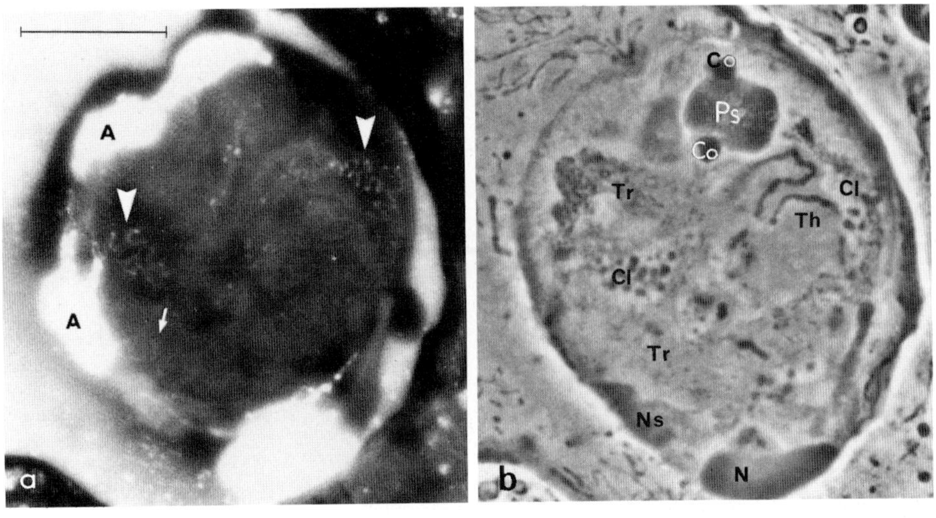

Fig. 2. The DNA in the spermatocyte nucleus can be visualized by staining with the fluorescent dye DAPI. The autosomes (*A*) are highly fluorescent and so is the nucleolus region (*N*). The autosomes are located close to the nuclear membrane, while the Y chromosomal loops engage most of the nuclear volume. Under the conditions of short exposure of the print, as shown in **a**, the DNA axis can be recognized within some of the lampbrush loops. *Large arrowheads* point toward the slightly coiled DNA axis of the clubs (*Cl*), the *small arrow* points toward the DNA axis of parts of the tubular ribbons (*Tr*). Designation of the loops as in Fig. 1. (Kremer et al. 1986). Bar represents 10 μm

et al. 1986). Recently, the exact location of DNA in some of the Y chromosomal lampbrush loops could be established by fluorescence techniques (Kremer et al. 1986) (Fig. 2). The "open chromatid model" proposed for the Y chromosome of *Drosophila hydei* (Hennig 1967) is, therefore, valid.

The Y chromosome of *D. hydei* contains 9.5% of the genomic DNA (approximately 40,000 kb DNA) (Zacharias et al. 1982). The genetic mapping of fertility genes in the Y chromosome (Hess 1965a) led to the conclusion that relatively large portions of the Y chromosome are devoid of genes. In this respect, the apparently complete decondensation of the Y chromosome during the first meiotic prophase is unexpected. Two explanations are possible. Either the decondensation is not of functional relevance or the decondensation displays functional activity of the region of the Y not involved in lampbrush loop formation. It is, at present, impossible to exclude this latter possibility.

2.2 Molecular Composition of the Loops

From the ultrastructural studies of Meyer (1963) and Grond et al. (1984), the following general picture of the lampbrush loop morphology emerges. Conventional histochemical methods are not sufficient to demonstrate the presence of the DNA. Therefore, only a negligible portion of the loop material consists of DNA. Consequently, the DNA must be highly decondensed. This is consistent with the structure established for oocyte lampbrush chromosome loops. The overwhelming amount of material in the loops is protein. Immunological investigations re-

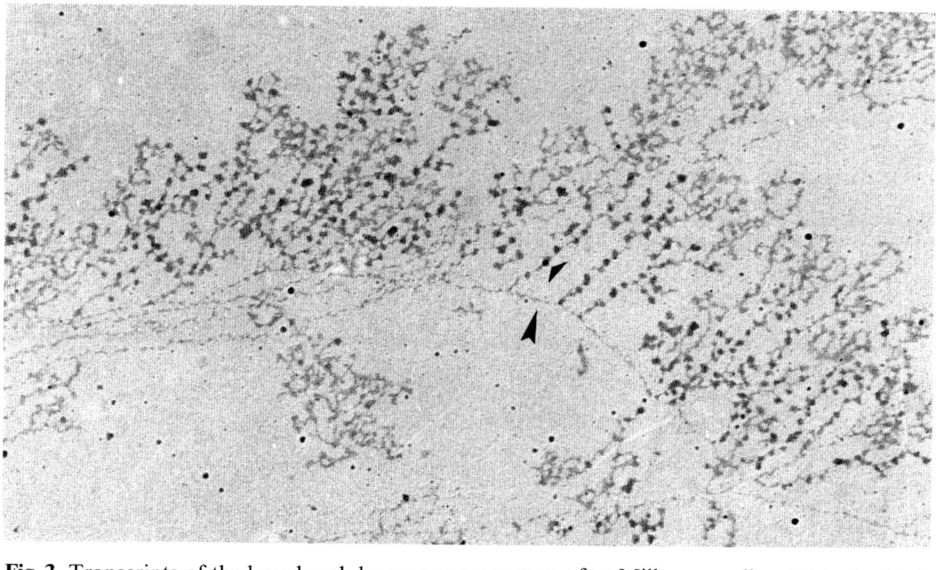

Fig. 3. Transcripts of the lampbrush loops nooses as seen after Miller spreading. Note the high degree of secondary structure and the association with protein granules (*short arrowhead*). The DNA axis is identified by a *large arrowhead*. (Electron micrographs of Siegmund and Hennig, unpublished). Bar represents 5 μm

vealed that different types of proteins must be distinguished. First, the loops contain proteins as they are associated with RNP particles. This has been demonstrated by Glätzer (1984), who found immunological cross reaction with antibodies directed against RNP-associated protein from *D. melanogaster* (Risau et al. 1983). These observations agree with the observations from ultrastructural studies which showed that RNP particle-like components are found in all the different lampbrush loop pairs (Meyer 1963; Grond et al. 1984).

The large amount of protein within the loops must, however, include other protein components, since most of the protein is not found in an RNP particle-like material (Meyer 1963; Grond et al. 1984). At least some of these proteins are basic proteins (Meyer 1963), which are loop-specific and may represent proteins important in the process of the substitution of chromosomal proteins, which is typical for male germ cell development (Hulsebos et al. 1984; Kremer et al. 1986; for discussion see Hennig 1985, 1987).

As active transcriptional sites, the Y chromosomal lampbrush loops should also contain RNA. The presence of newly synthesized RNA has been established

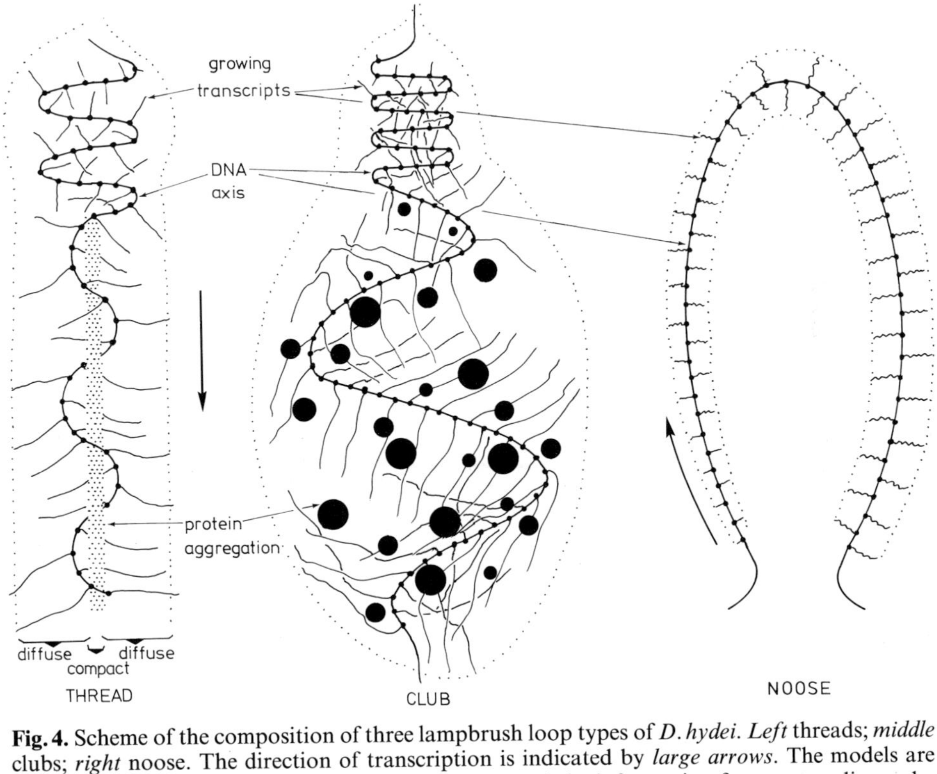

Fig. 4. Scheme of the composition of three lampbrush loop types of *D. hydei*. *Left* threads; *middle* clubs; *right* noose. The direction of transcription is indicated by *large arrows*. The models are based on the cytology of the loops as seen in Fig. 5 and the information from autoradiography after ^{3}H-uridine incorporation (Hennig 1967), transcript in situ hybridization experiments, ultrastructural analysis (Meyer 1963; Grond et al. 1984) and Miller spreading experiments (Grond et al. 1983; de Loos et al. 1984). The different secondary transcript structure within the loops is not shown. The accumulating proteins are indicated (cf. Fig. 6). For details of the general loop morphology compare Fig. 5

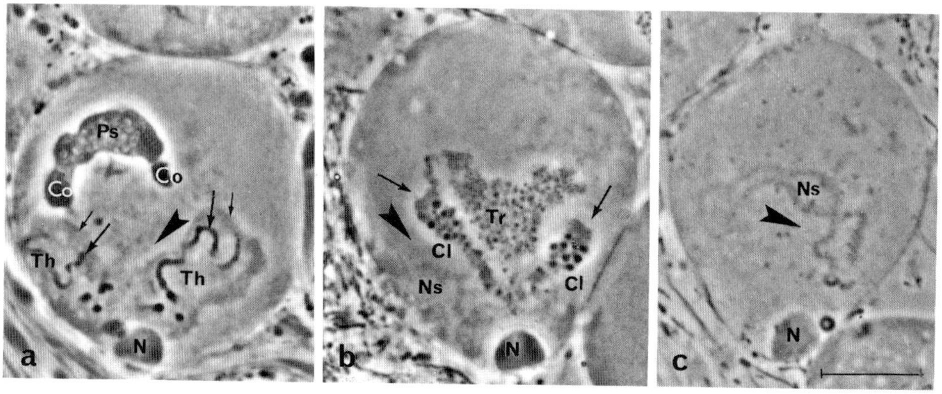

Fig. 5. Cytology of the lampbrush loops which are schematically shown in Fig. 4. The spermatocyte nuclei are from males with partial deletions of the Y chromosome to allow easier identification of details of the loops. **a** Only threads (*Th*), cones (*Co*) and pseudonucleolus (*Ps*) are present. The different parts of the threads (diffuse: *short arrow;* compact: *long arrow*) can be clearly seen (cf. Fig. 1 a). The direction of transcription of one of the loops is identified in all three nuclei by an *arrowhead.* **b** Clubs (*Cl*), tubular ribbons (*Tr*) and nooses (*Ns*) are present. Note the more diffuse distal sections (*arrows*) of the clubs which are connected to the tubular ribbons. **c** Only nooses (*Ns*) are present. Note the asymmetry. The nooses became only visible in aged cytological preparations since in vivo they are more diffuse (cf. Fig. 1). The refractivity is the result of some aggregation of the loop material. *N* nucleolus. (Photographs: Hennig, unpublished). Bar represents 10 μm

by autoradiography (Hennig 1967; Grond et al. 1984). Compared with the large amounts of protein, the proportion of RNA is, however, small. At the ultrastructural level, RNA has been identified in the RNP particle-like components of the loops (Grond et al. 1984). The most informative details on transcripts in the loops have, however, been obtained by Miller spreading experiments. The first studies revealed an extremely large size of the transcribing molecules (Meyer and Hennig 1974 b; Hennig et al. 1974 a). These data were confirmed by Glätzer and Meyer (1981). In further studies Grond et al. (1983) and de Loos et al. (1984) discovered a highly loop-specific transcript pattern. Not only the distances between the transcripts and the size pattern of transcripts turned out to be highly loop-specific, but also the secondary structure of the transcripts is characteristic for each loop. This is evident for the loops nooses (in the short arm of the Y chromosome), the pseudonucleolus and the threads (distal in the long arm, see Fig. 1). Also for the clubs (proximal in the long arm) a loop-specific transcript pattern has recently beeen established (Suijkerbuijk and Hennig, unpublished data). Thus, each loop is characterized by transcripts of giant size and highly loop-characteristic secondary structure (Fig. 3). It can be calculated from the cytological length of the loops (Hennig et al. 1974 a) that their sizes agree with the amounts of DNA expected to be active within the loops as estimated from the Miller spreading experiments. These experiments demonstrate, therefore, that the loops are transcribed in their entire length into giant primary transcripts (Figs. 4, 5). The length of these transcripts ranges between 260 kb in the smallest loop and 1500 kb in others. By current biochemical methods it is clearly not possible to identify or recover such

RNA molecules. It cannot, therefore, be excluded that comparably large transcripts for this reason have never been found in other genes, although they may be formed in some genomic sites.

3 Function of the Y Chromosomal Lampbrush Loops

3.1 Male Fertility Genes and Loops

Bridges (1916) and Stern (1927) had recognized that the Y chromosome of *D. melanogaster* is essential for sperm differentiation. The same holds true for *D. hydei* (Hess 1967). Males without Y chromosome (X/O males) are sterile due to a defective spermiogenesis. Relatively few developing germ cells pass through meiosis in X/O males of *D. hydei* (Hennig et al. 1974 b), dependent on the breeding temperature (Kremer unpublished). The Y chromosome is, hence, essential for the postmeiotic differentiation of the male germ cells. In cytogenetic and ultrastructural studies Hess and Meyer (see Hess and Meyer 1968) tried to obtain information on the biological function of the Y chromosomal fertility genes. All searching for the actual function of the Y chromosomal "fertility genes" has, however, up to now been unsuccessful.

3.2 No Major Structural Sperm Proteins Are Encoded in Loops

The difficulties in identifying the role of the Y chromosomal genes for spermatogenesis suggest that these problems are not based on purely methodological reasons. More likely, conceptual problems are involved. This becomes more evident in considering the observations on the Y chromosomal genes more closely. The morphology and molecular composition of the Y chromosomal lampbrush loops is not easily compatible with our conventional ideas on other eukaryotic genes. From the high level of transcriptional activity of the Y chromosomal lampbrush loops one would, for example, expect that the Y chromosomal fertility genes code for structural components of the spermatozoon. Such components should be required in large amounts since the size of the spermatozoa – more than 10 mm (Grond 1984) – is exceptional. However, so far no evidence for protein coding in the Y chromosome of *D. hydei* [and only questionable evidence (Goldstein et al. 1982) for the *D. melanogaster* Y chromosome: see Hennig 1987] has been obtained. Despite extended investigations, no proteins were found in testes which are encoded by the Y (Ingman-Baker and Candido 1980; Hulsebos et al. 1983). Also investigation of RNA species synthesized from the Y chromosome in testes in in vitro translation systems have so far provided no evidence for protein coding (B. Hennig, R. C. Brand, P. Huijser, W. Hennig, unpublished data). These data are, however, not sufficient to exclude a protein coding entirely.

Nevertheless, also other details are not compatible with the idea that major sperm protein fractions are encoded in the Y chromosome. Ultrastructural investigation of spermatids of males partially deficient in their Y chromosome (see Meyer 1968) provided no indications that any major structural component of the spermatid is absent, although drastic structural abnormalities in the spermatid differentiation are the rule in such males. Moreover, in *D. melanogaster* sperm differentiation proceeds considerably even in the absence of the Y chromosome. It is unlikely that this would be possible if major structural elements were encoded by Y chromosomal genes.

3.3 A Regulatory Role of the Fertility Genes?

What, then, is the function of Y chromosomal fertility genes? It has in the past repeatedly been postulated that the Y chromosomal genes play a regulative role for spermiogenesis (Hess and Meyer 1968; Beermann et al. 1967), although no concrete ideas on the character of the regulatory processes could be promoted. Some first experimental evidence for a regulatory role of the Y chromosome for processes during spermatogenesis was provided by Hulsebos et al. (1983). These authors recognized that in X/O testes of *D. hydei* three major protein fractions are expressed in reduced amounts. These proteins, including tubulins, are, however, clearly not encoded by Y chromosomal DNA sequences. This supports the conclusion that the Y chromosome is involved in the control of the expression of these proteins. Recent molecular studies demonstrated that such a control must be exerted at the translational level, at least for tubulins, since in X/O testes normal amounts of tubulin mRNA are synthesized (Brand et al., in preparation).

Comparable conclusions can be drawn from the observations on the *Stellate* locus of *D. melanogaster* (Hardy et al. 1984; Livak 1984). The *Stellate* locus in the X chromosome encodes a protein of M_r 17,000, but homologous DNA sequences were found in the Y chromosome. These Y chromosomal gene copies cannot be assigned to any of the known fertility genes but are located between fertility loci kl2 and kl3 (Hardy et al. 1984). Whenever the Y chromosome is absent, excessive amounts of the M_r 17,000 protein are synthesized which lead to the formation of the crystals typical for X/O testes of *D. melanogaster* (Meyer et al. 1961). Also here, the Y chromosome has a controlling effect on testes proteins encoded elsewhere in the genome, although in an opposite sense than in *D. hydei,* namely by suppressing X chromosomal gene expression.

3.4 DNA Structure in the Lampbrush Loops

Before further considering the potential function of Y chromosomal genes, I must briefly discuss the molecular structure of the Y chromosomal lampbrush loop-forming genes. The recovery of Y chromosomal DNA turned out to be more difficult than one should have expected from the consideration that 9.5% of the genomic DNA of males must be Y chromosomal. The explanation is that the Y chromosomal DNA shares homologies extensively with DNA sequences outside the Y chromosome (Vogt and Hennig 1983; Hennig et al. 1983). For this class of Y chromosomal DNA ("Y-associated" DNA according to our nomenclature),

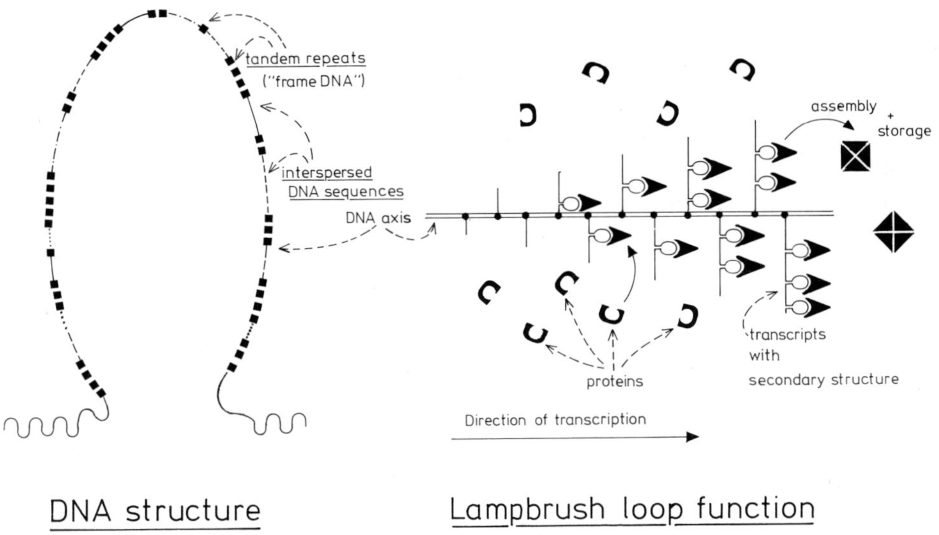

Fig. 6. Model of the molecular structure of Y chromosomal lampbrush loops (**a**) and of their proposed function (**b**). The loop DNA is composed of tandem repeat clusters ("Y-specific" type) (*black boxes*) with other repetitive DNA sequences interspersed between these clusters. These interspersed DNA sequences are only in part homologous to one another, but are also homologous to DNA outside the Y chromosome ("Y-associated" type). **b** The transcripts select, on the basis of their secondary structure, nuclear proteins which are subsequently deposited within the loop (cf. Fig. 5)

the application of microcloning has been proven to be almost obligatory (Hennig et al. 1983; see Huijser and Hennig 1987). Another, "Y-specific", class of DNA is more easy to detect since these sequences are exclusively located in the Y chromosome (Lifschytz 1979; Vogt et al. 1982; Vogt and Hennig 1983; Awgulewitsch and Bünemann 1986).

The insight into the arrangement and properties of the Y chromosomal DNA sequences permits us to design a model of the DNA sequence arrangement of the lampbrush loop nooses (Fig. 6), which most likely applies to all Y chromosomal lampbrush loop pairs (Hennig et al. 1986; Vogt and Hennig 1986a, b; Vogt et al. 1986). Each loop is constructed of a family of repeated DNA sequences which occur in several tandem repeat clusters distributed all over the loop. These clusters must result from several subsequent and independent amplification steps of an original DNA sequence of a few hundred nucleotide pairs during evolution. Inserted between these tandem repeat clusters are repeated DNA sequences of a lower repetition frequency but of higher complexity (up to several kb in length) ("Y-associated" DNA). These interspersed sequences occur also in various locations outside the Y chromosome. All evidence available so far points toward a transposable nature of these DNA sequences.

3.5 Transcripts of the Lampbrush Loops and Their Potential Functions

From Miller spreading experiments it is clear that the lampbrush loops are transcribed into primary transcripts representing most if not all of the loop DNA. Consequently, both types of DNA sequences – Y-specific and Y-associated ones – are transcribed into one primary transcript. The fate of these giant primary transcripts is, however, still entirely unclear. Testis RNA contains a mixture of Y chromosomal transcripts heterogeneous in size (Vogt et al. 1982; Vogt and Hennig 1983; Lifschytz et al. 1983). RNA fractions of discrete molecular weight were found to be homologous to Y-associated DNA sequences (Brand et al., in preparation). Unfortunately, the actual genomic origin of such RNA species can only partially be determined by transcript-in situ hybridization. Such experiments display Y chromosomal transcription, as must be expected from the interspersed location of these sequences within the loops. This, however, does not prove that the cytoplasmic RNA species are indeed derived from the primary Y chromosomal transcripts by splicing. In some instances, transcripts homologous to Y-associated DNA sequences have been found in X/O testes, which suggests an autosomal or X chromosomal transcription of the homologous DNA sequences; but in other cases such transcripts are absent in X/O testes. This does not prove an exclusive Y chromosomal origin of the discrete-sized RNA molecules, since the transcription outside the Y chromosome could depend on the presence of the Y chromosome in the genome. The genomic origin of RNA species homologous to the Y-associated DNA sequence type is, therefore, unresolved. It might, however, be of fundamental importance for our understanding of the function of Y chromosomal genes to explore the actual sites of transcription of RNA species of the discrete-size class.

Some data suggested that Y chromosomal transcripts remain restricted to the nucleus (Lifschytz et al. 1984). While this might hold true for parts of the primary transcripts, other sections of the transcripts occur, without doubt, in the cytoplasm, as in situ hybridization experiments with tissue sections demonstrated (Huijser and Hennig 1987; Hennig et al., unpublished data). Similarly, earlier DNA/RNA hybridization experiments gave evidence that Y chromosomal RNA species are still present in spermatids (Hennig 1968; Hennig et al. 1974a). RNA of Y chromosomal origin is, therefore, not restricted to the spermatocyte nucleus, but is released into the cytoplasm of primary spermatocytes and is retained up to postmeiotic spermatid stages.

What is the function of this RNA? The DNA sequencing data argue against a protein-coding character of most of the loop DNA (Vogt and Hennig 1986a, b; Huijser and Hennig 1987). The occurrence of transcripts in the cytoplasm of spermatids suggests a functional role during the postmeiotic development. I mentioned before evidence which implies a regulatory function of the Y chromosomal fertility genes on protein synthesis. In combination, both observations suggest that the cytoplasmic RNA fractions derived from the Y chromosome interact with genes from other genomic locations by activating (or inactivating) their mRNA for translation. The homologies between the Y chromosomal interspersed DNA sequences and other genomic loci may be important in this respect. It is, for example, possible, that in sites outside the Y chromosome the complementary DNA strands are transcribed. This would facilitate interaction of RNA of a Y chromosomal origin with RNA of protein coding genes. Examples for such regulatory interactions of genes are known from prokaryotes.

This interpretation of the biological role of the Y chromosomal lampbrush loops does not explain why such spectacular lampbrush loops are formed. An answer to this question can be derived if their specific features such as the high amounts of proteins, which are in part loop-specific, their loop-specific deposition (see, for example, the cytology of the threads and clubs in Figs. 1 and 4), the extreme size of the transcripts and their high degree of secondary structure are taken into account. In our working hypothesis, the specific secondary structure of the transcripts is required to bind particular protein species in the loops (Fig. 6) (see Hennig et al. 1986). The possibility of loop-specific interactions between transcript structure and associated proteins had been considered much earlier in order to explain the loop-specific morphology of the ribonucleoprotein in the loops (Hennig 1967).

What is the possible biological reason for binding proteins in Y chromosomal loops? We proposed that the lampbrush loops are required to assemble nuclear proteins inside the nucleus. Such proteins may be involved in rearrangements in the chromatin as they occur during and after meiosis (see Hennig et al. 1986). The cytological investigation of the chromatin constitution during the germ cell development revealed a complicated sequence of chromatin condensation and decondensation steps after meiosis (Kremer et al. 1986). This study suggested that the lampbrush loops accumulate proteins required in this process. Consistent with this idea is the observation that antigenic determinants of a DNA-binding protein, recognized with the aid of an antibody from *D. melanogaster*, Bv96 (Frasch 1985), are in the primary spermatocyte nucleus enriched in the lampbrush loop nooses, but can later be detected in the heads of maturing spermatids. Alternatively, nuclear proteins could be involved in postmeiotic regulation processes.

4 Conclusions

The data reviewed in this article show that the cytological and molecular structure of the Y chromosomal lampbrush loops in *D. hydei* is understood in its basic features. Future work will have to be focused on the molecular basis of structural modifications of Y chromosomal lampbrush loops that can result from mutations (Hess 1965 b; Leoncini 1977; Hackstein et al. 1982; Hackstein et al. 1987). To understand the molecular basis of such mutations it will be necessary to recover the mutated sections of the DNA sequences. Prior to this, the reconstruction of major parts of the respective lampbrush loop at the DNA level is required, since the reasons for structural modifications of a loop can probably only be understood if the mutated DNA sequence can be placed into its sequence context within the loop.

The main question on the biological function of the Y chromosomal fertility genes can still not be answered. Evidently, they display a highly abnormal molecular structure and unusual features during their transcription. Based on a number of indirect arguments, we were, however, able to propose a working hypothesis for the functional role of these genes.

Acknowledgment. For critical reading the manuscript I am grateful to Dr. Ingrid Hennig.

References

Awgulewitsch A, Bünemann (1986) Isolation of Y-chromosomal repetitive DNA sequences of *Drosophila hydei* via enrichment of chromosome-specific sequences by heterogeneous hybridization between female and male DNA. J Biochem Biophys Method 12:37–50

Beermann W, Hess O, Meyer GF (1967) Lampbrush Y chromosomes in spermiogenesis of *Drosophila*. In: Wolff E (Ed) The relationship between experimental embryology and molecular biology. Gordon and Breach, New York, pp 61–81

Bridges CB (1916) Non-disjunction as proof of the chromosome theory of heredity. Genetics 1:1–52 and 107–162

Frasch M (1985) Charakterisierung chromatinassoziierter Kernproteine von *Drosophila melanogaster* mit Hilfe monoklonaler Antikörper. Diss, Univ Tübingen

Glätzer KH (1984) Preservation of nuclear RNP antigens in male germ cell development of *Drosophila hydei*. Mol Gen Genet 196:236–243

Glätzer KH, Meyer GF (1981) Morphological aspects of the genetic activity in primary spermatocyte nuclei of *Drosophila hydei*. Biol Cell 41:165–172

Goldstein LSB, Hardy RW, Lindsley DL (1982) Structural genes on the Y chromosome of *Drosophila melanogaster*. Proc Natl Acad Sci USA 79:7405–7409

Grond CJ (1984) Spermatogenesis in *Drosophila hydei*. Ph D Thesis, Univ Nijmegen

Grond CJ, Siegmund I, Hennig W (1983) Visualization of a lampbrush loop-forming fertility gene in *Drosophila hydei*. Chromosoma 88:50–56

Grond CJ, Rutten RGJ, Hennig W (1984) Ultrastructure of the Y chromosomal lampbrush loops in primary spermatocytes of *Drosophila hydei*. Chromosoma 89:85–95

Hackstein JHP, Leoncini O, Beck H, Peelen G, Hennig W (1982) Genetic fine structure of the Y chromosome of *Drosophila hydei*. Genetics 101:257–277

Hackstein JHP, Hennig W, Siegmund I (1987) Y chromosome-specific mutations induced by a giant transposon in *Drosophila hydei*. Mol Gen Genet 207:455–465

Hardy RW, Lindsley DL, Livak KJ, Lewis B, Silversten AL, Joslyn GL, Edwards J, Bonaccorsi S (1984) Cytogenetic analysis of a segment of the Y chromosome of *Drosophila melanogaster*. Genetics 107:591–610

Hennig W (1967) Untersuchungen zur Struktur und Funktion des Lampenbürsten-Y-Chromosoms in der Spermatogenese von *Drosophila*. Chromosoma 22:294–357

Hennig W (1968) Ribonucleic acid synthesis of the Y chromosome of *Drosophila hydei*. J Mol Biol 38:227–239

Hennig W (1985) Y chromosome function and spermatogenesis in *Drosophila hydei*. Adv Genet 23:179–234

Hennig W (1987) Spermatogenesis. In: Malacinsky G (ed) Developmental genetics. Macmillan, New York, in press

Hennig W, Meyer GF, Hennig I, Leoncini O (1974a) Structure and function of the Y chromosome of *Drosophila hydei*. Cold Spring Harbor Symp Quant Biol 38:673–683

Hennig W, Hennig I, Leoncini O (1974b) Some observations on spermatogenesis of *Drosophila hydei*. Drosophila Information Service 51:127

Hennig W, Link B, Leoncini O (1975) The location of the nucleolus organizer region in *Drosophila hydei*. Chromosoma 51:57–63

Hennig W, Huijser P, Vogt P, Jäckle H, Edström J-E (1983) Molecular cloning of microdissected lampbrush loop DNA sequences of *Drosophila hydei*. EMBO J 2:1741–1746

Hennig W, Brand RC, Hackstein J, Huijser P, Kirchhoff C, Kremer H, Lankenau DH, Vogt P (1986) Structure and function of Y chromosomal genes in *Drosophila*. Chromosomes Today 9 (in press)

Hess O (1965 a) Strukturdifferenzierungen im Y-Chromosom von *Drosophila hydei* und ihre Beziehungen zu Gen-Aktivitäten. III. Sequenz und Lokalisation der Schleifenbildungsorte. Chromosoma 16:222–248

Hess O (1965 b) Strukturdifferenzierungen im Y-Chromosom von *Drosophila hydei* und ihre Beziehungen zu Gen-Aktivitäten. I. Mutanten der Funktionsstrukturen. Verh Dtsch Zool Ges 156–163

Hess O (1967) Morphologische Variabilität der chromosomalen Funktionsstrukturen in den Spermatocytenkernen von *Drosophila*-Arten. Chromosoma 21:429–445

Hess O, Meyer GF (1963 a) Artspezifische funktionelle Differenzierungen des Y-Heterochromatins bei *Drosophila*-Arten der *D. hydei*-Subgruppe. Port Act Biol Ser A VII 1–2:29–46

Hess O, Meyer GF (1963 b) Chromosomal differentiations of the lampbrush type formed by the Y chromosome in *Drosophila hydei* and *D. neohydei*. J Cell Biol 16:527–539

Hess O, Meyer GF (1968) Genetic activities of the Y chromosome in *Drosophila* during spermatogenesis. Adv Genet 14:171–223

Huijser P, Hennig W (1987) Ribosomal DNA-related sequences in a Y chromosomal lampbrush loop of *Drosophila hydei*. Mol Gen Genet 206:441–451

Hulsebos TJM, Hackstein JHP, Hennig W (1983) Involvement of Y chromosomal loci in the synthesis of *Drosophila hydei* sperm proteins. Dev Biol 100:238–243

Hulsebos TJM, Hackstein JHP, Hennig W (1984) Lampbrush loopspecific protein of *Drosophila hydei*. Proc Natl Acad Sci USA 81:3404–3408

Ingman-Baker J, Candido EPM (1980) Proteins of the *Drosophila melanogaster* male reproductive system: Two-dimensional gel patterns of protein synthesized in the X/O, X/Y and X/Y/Y testis and paragonial gland and evidence that the Y chromosome does not code for structural sperm protein. Biochem Genet 18:809-828

Kremer H, Hennig W, Dijkhof R (1986) Chromatin organization in the male germ line of *Drosophila hydei*. Chromosoma 94:147–161

Leoncini O (1977) Temperatursensitive Mutanten im Y Chromosom von *Drosophila hydei*. Chromosoma 63:329–357

Lifschytz E (1979) A procedure for the cloning and identification of Y-specific middle repetitive sequences in *Drosophila hydei*. J Mol Biol 133:267–277

Lifschytz E, Hareven D, Azriel A, Brodsly H (1983) DNA clones and RNA transcripts of four lampbrush loops from the Y chromosome of *Drosophila hydei*. Cell 32:191–199

Livak KJ (1984) Organization and mapping of a sequence on the *Drosophila melanogaster* X and Y chromosomes that is transcribed during spermatogenesis. Genetics 107:611–634

Loos F de, Dijkhof R, Grond CJ, Hennig W (1984) Lampbrush loop-specificity of transcript morphology in spermatocyte nuclei of *Drosophila hydei*. EMBO J 3:2845–2849

Meyer GF (1963) Die Funktionsstrukturen des Y-Chromosoms in den Spermatocytenkernen von *Drosophila hydei, D. neohydei, D. repleta* und einigen anderen *Drosophila*-Arten. Chromosoma 14:207–255

Meyer GF (1968) Spermiogenese in normalen und Y-defizienten Männchen von *Drosophila melanogaster* und *D. hydei*. Z Zellforsch Mikrosk Anat 84:141–175

Meyer GF, Hennig W (1974 a) The nucleolus in primary spermatocytes of *Drosophila hydei*. Chromosoma 46:121–144

Meyer GF, Hennig W (1974 b) Molecular aspects of the fertility factors in *Drosophila*. In: Afzelius BA (ed) The functional anatomy of the spermatozoon. Pergamon Press, Oxford New York, pp 69–75

Meyer GF, Hess O, Beermann W (1961) Phasenspezifische Funktionsstrukturen in Spermatocytenkernen von *Drosophila melanogaster* und ihre Abhängigkeit vom Y-Chromosom. Chromosoma 12:676–716

Risau W, Symmons P, Saumweber H, Frasch M (1983) Nonpackaging and packaging proteins of hnRNA in *Drosophila melanogaster*. Cell 33:529–541

Stern C (1927) Ein genetischer und zytologischer Beweis für Vererbung im Y-Chromosom von *Drosophila melanogaster*. Z Indukt Abstamm Vererbungsl 44:187–231

Vogt P, Hennig W (1983) Y chromosomal DNA of *Drosophila hydei*. J Mol Biol 167:37–56

Vogt P, Hennig W (1986 a) Molecular structure of the lampbrush loop nooses of the Y chromosome of *Drosophila hydei*. I. The Y chromosome-specific repetitive DNA sequence family ay1 is dispersed in the loop DNA. Chromosoma 94:449–458

Vogt P, Hennig W (1986 b) Molecular structure of the lampbrush loop nooses of the Y chromosome of *Drosophila hydei*. II. DNA sequences with homology to other genomic locations are a major constituent of the loop. Chromosoma 94:459–467

Vogt P, Siegmund I, Hennig W (1982) Identification of cloned Y chromosomal DNA sequences from a lampbrush loop of *Drosophila hydei*. Proc Natl Acad Sci USA 79:5132–5136

Vogt P, Hennig W, ten Hacken D, Verbost P (1986) Evolution of Y chromosomal lampbrush loop DNA sequences of *Drosophila*. Chromosoma 94:367–376

Zacharias H, Hennig W, Leoncini O (1982) Microspectrophotometric comparison of the genome sizes of *Drosophila hydei* and some related species. Genetica 58:153–157

Contributions of Electron Microscopic Spreading Preparations ("Miller Spreads") to the Analysis of Chromosome Structure

U. SCHEER [1]

1 Introduction

Before introduction of the chromatin spreading technique by Miller and Beatty in 1969 (Miller and Beatty 1969 a–d), relatively limited information was available concerning the submicroscopic organization of chromosomes in mitosis and interphase. Thus, at the end of the 1960s the general picture that emerged from inspection of ultrathin sections of cell nuclei, isolated chromatin, and of whole mount preparations was that mitotic chromosomes and interphase chromatin consisted of a complex meshwork of irregularly sized and knobby fibers with diameters ranging from 20 to 30 nm (for reviews see Ris 1969; DuPraw 1970; Solari 1974; Ris and Korenberg 1979; Hozier 1979). Although the internal organization of chromatin fibers was particularly suitable for study by surface spreading on an air-water interface, originally introduced by Gall (1963), neither a discrete subunit organization was recognized nor could the "thick" chromatin fibers be reproducibly unfolded into thinner fibers which were believed to be transcriptionally active (Gall 1966). Hence, electron microscopic methods were unavailable to study features of chromosomes during interphase, i.e., their functional subdivision into transcriptionally active and inactive domains.

A technical breakthrough came with the chromatin spreading technique developed by O. Miller and co-workers (Miller and Beatty 1969 a–d; Miller and Hamkalo 1972; Miller and Bakken 1972; Hamkalo and Miller 1973; Hamkalo et al. 1973). This spreading procedure was designed to unfold chromatin gently, but extensively by exposure to a medium of very low ionic strength at slightly alkaline pH and to avoid mechanical stress due to surface tensions. The "Miller spreading", as the preparation is often referred to, unraveled chromosomes and interphase chromatin to the first level of the nucleoprotein organization and permitted visualization of transcript fibrils which remained associated with the extended chromatin filaments. In fact, the first papers published by Miller and Beatty (1969 a–d) described the morphological features of transcriptionally active genes while only later, in 1973, Olins and Olins and independently Woodcock drew attention to repeating globular subunit structures present in transcriptionally inactive chromatin fibers prepared from a variety of cells according to the Miller procedure (Olins and Olins 1973; Woodcock 1973).

[1] Institute of Zoology I, University of Würzburg, Röntgenring 10, 8700 Würzburg, FRG.

Results and Problems in Cell Differentiation 14
Structure and Function of Eukaryotic Chromosomes
Edited by W. Hennig

It is obvious that one essential step of the Miller spreading technique is the extensive unfolding of chromatin by hypotonic media in order to disrupt all higher levels of chromosomal arrangements as they exist in the living cell. Although modifications of the Miller procedure (e.g., inclusion of salt in the dispersal medium, very brief exposure to low salt medium) have also been successfully applied to the study of higher order, supranucleosomal packaging of chromatin and to identify intermediate, transitory states during experimentally induced relaxation (see, e.g., Hozier 1979; Zentgraf et al. 1980a, b, 1981; Hamkalo et al. 1981; Labhart et al. 1982), such aspects will not be considered in the present chapter. Rather, this overview focuses on molecular aspects of chromatin organization with special emphasis on structural modifications that are concerned with transcription. Furthermore, more recent developments of the Miller chromatin spreading method will be discussed which combine immunolocalization and in situ hybridization techniques. Such applications of the chromatin spreading method are of great potential value for identification of DNA- and RNA-associated proteins and the genetic content of specific transcription units.

2 The Nucleosome Chain: The Elementary Structural Component of Chromosomes

When mitotic chromosomes or interphase chromatin preparations are unfolded by exposure to a buffer of very low ionic strength (e.g., 0.1 mM Na-borate buffer, pH 8–9) and centrifuged onto an electron microscopic grid essentially as described in detail by Miller and Hamkalo (1972), Miller and Bakken (1972), and Hamkalo and Miller (1973), the bulk of the transcriptionally inactive chromatin displays the characteristic linear arrays of closely packed particles, the nucleosomes, connected by thin filaments representing linker DNA (Fig. 1a, b). Such extended beaded chains were first described by Olins and Olins (1973, 1974) and Woodcock (1973; Woodcock et al. 1976); the term nucleosome was introduced by Oudet et al. (1975). The beaded chromatin fiber or primary nucleofilament organization has been found in all eukaryotic species so far studied with the notable exception of some dinoflagellates which do not contain histones (Herzog and Soyer 1981). Thus, chromatin of plant and animal species is morphologically indistinguishable in Miller spreads (e.g., chromatin from human HeLa cells and amphibian oocytes; Fig. 1a, b; amphibian and avian erythrocytes, *Drosophila* polytene chromosomes, *Chlamydomonas:* Woodcock et al. 1976; rat and chicken: Olins et al. 1975; Zentgraf et al. 1980a; sea urchin sperm: Zentgraf et al. 1980b; *Physarum polycephalum:* Scheer et al. 1981; *Dytiscus marginalis:* Scheer and Zentgraf 1978; *Drosophila melanogaster:* McKnight and Miller 1976; *Oncopeltus fasciatus:* Foe 1978; *Cyclops furciper:* Beermann 1984; *Zea mays:* Zentgraf et al. 1981; Greimers and Deltour 1981; *Allium cepa* and *Sambucus nigra:* Lutz and Nagl 1980; cultured mammalian cells of various origin: Howze et al. 1976; Franke et al. 1978; Hamkalo et al. 1981; Labhart et al. 1982).

The packing of DNA into a repeating subunit structure has also been established by biochemical and physicochemical methods and it is fascinating to follow

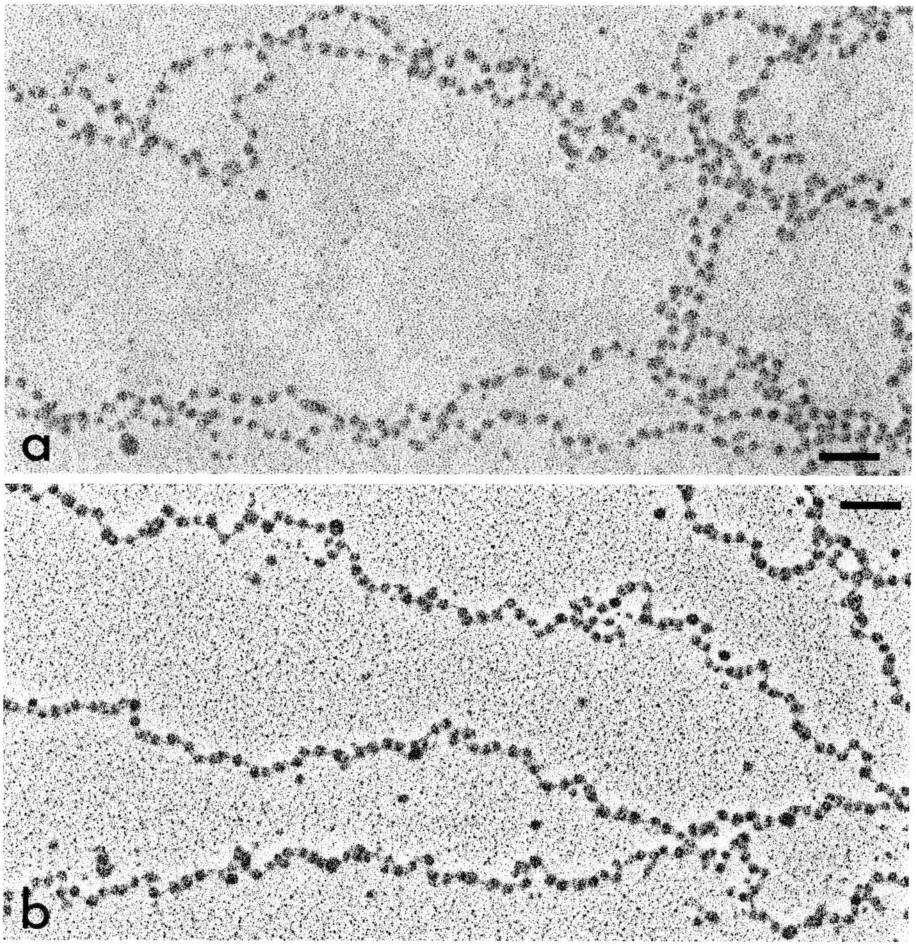

Fig. 1 a, b. Electron microscopic appearance of transcriptionally inactive chromatin in Miller spreads. Chromatin from human HeLa cells (**a**) and oocytes of the salamander *Pleurodeles waltlii* (**b**) was unfolded by exposure to very low salt conditions prior to centrifugation onto electron microscopic grids. The chromatin fibers are composed of linear arrays of globular subunits, the nucleosomes. × 80 000; *bars* = 0.1 μm

how the data, obtained by various methods, were eventually integrated into the nucleosome model (for reviews see McGhee and Felsenfeld 1980; Igo-Kemenes et al. 1982; Kornberg 1977). Since each nucleosome contains about 200 base pairs (bp) of DNA (with an extended length of the B-form of 200 × 3.4 Å = 680 Å), the coiling of the DNA around the nucleosome core with a diameter of approximately 110 Å results in an apparent foreshortening of the DNA by a factor of about 6. In Miller spreads nucleosomes are usually somewhat distantly spaced (Fig. 1 a, b) so that the apparent contraction ratio of DNA is only about two to three.

3 Visualization of Transcribing rRNA Genes

The amplified ribosomal RNA (rRNA) genes of amphibian oocytes were the first genes visualized by electron microscopy (Miller and Beatty 1969a–d; Figs. 2 and 3). Subsequently numerous investigators employed the Miller technique to

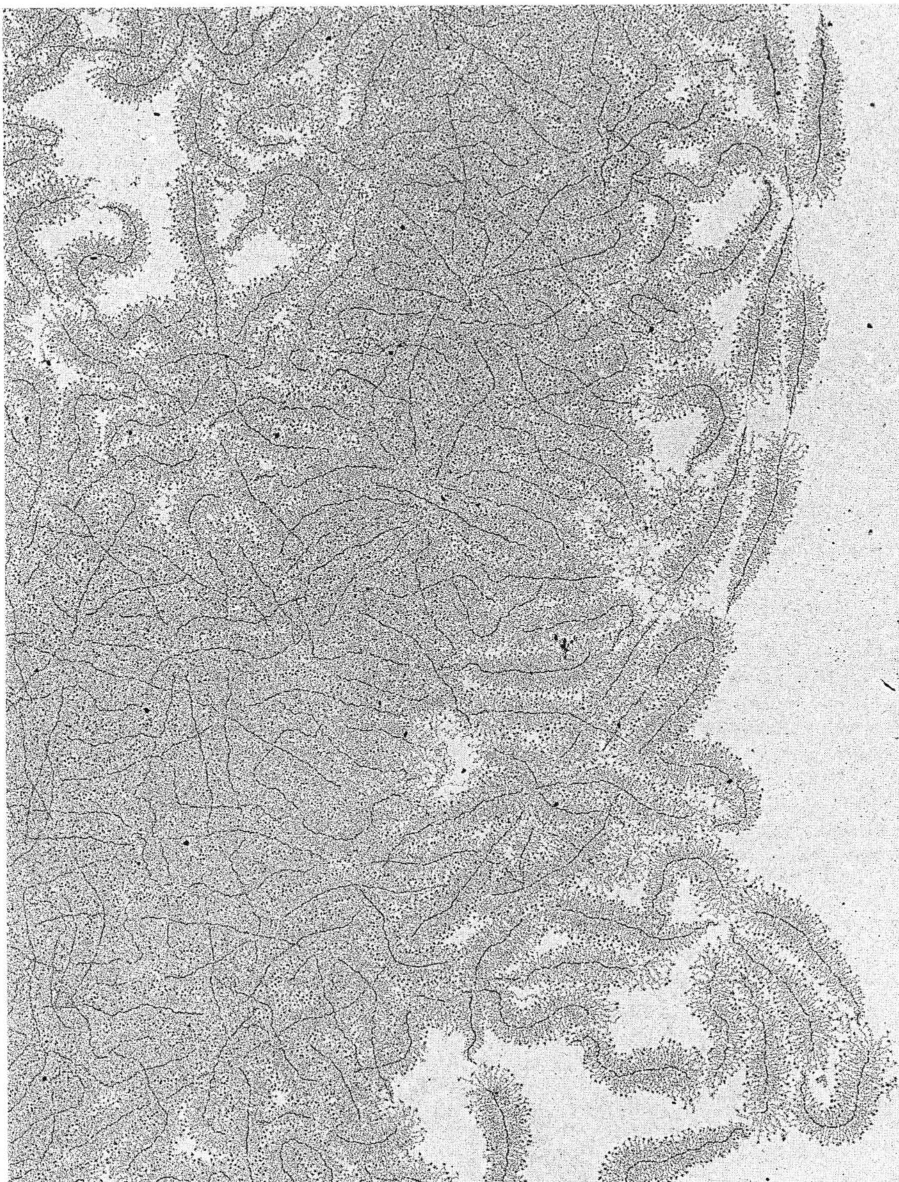

Fig. 2. Survey electron micrograph of spread nucleolar chromatin from a mid-sized *Pleurodeles* oocyte. This oogenic stage is characterized by intense synthesis of rRNA. ×9000; *bar* = 1 μm

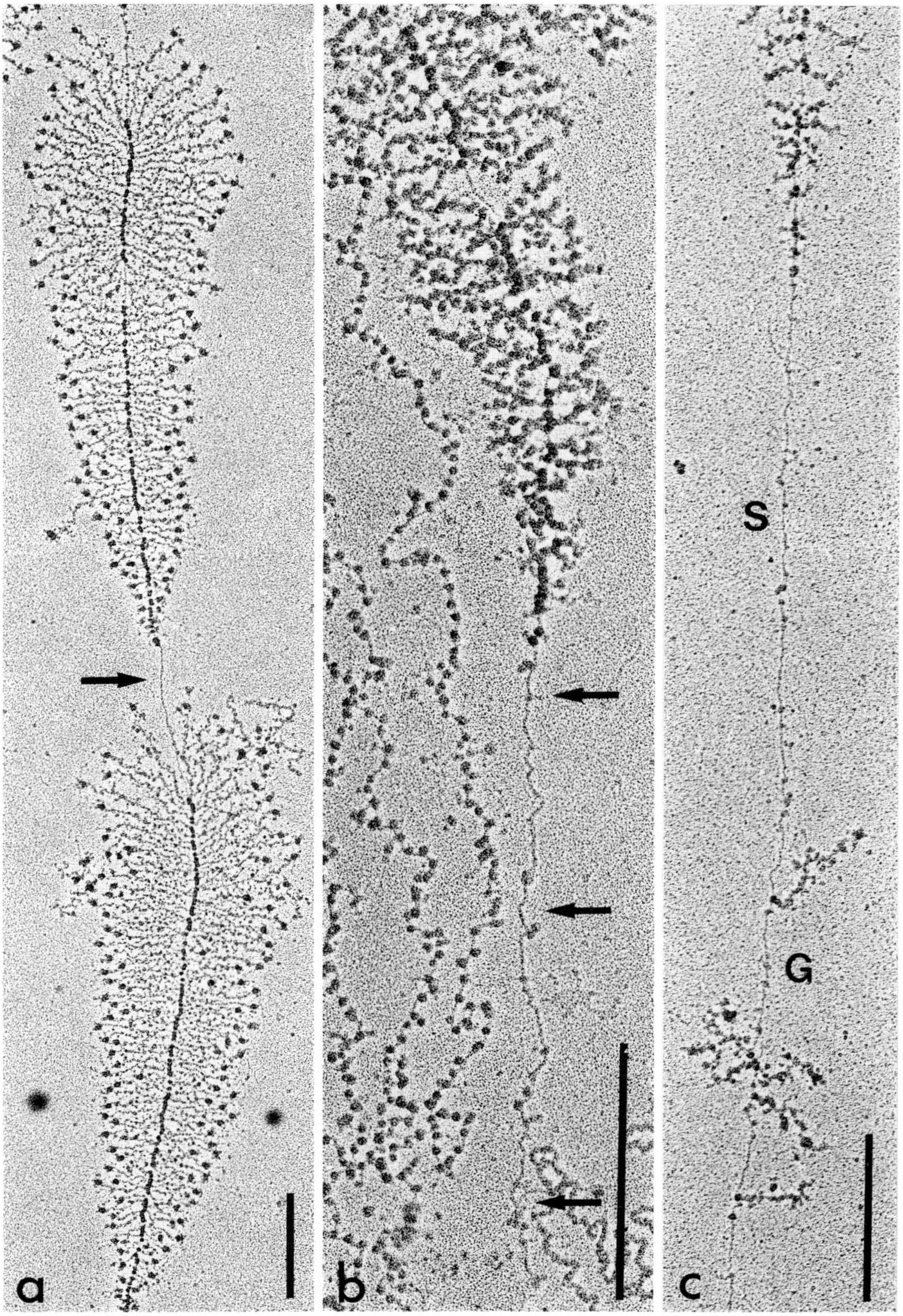

Fig. 3a–c. Structural details of transcribed rRNA genes as seen in Miller spreads. **a** Two tandemly arranged rRNA genes from a *Pleurodeles* oocyte. Note the high packing density of the RNA polymerases and the smooth configuration of the spacer region (*arrow*). **b** Higher magnification of the spacer region (*arrows*) and the proximal portion of an rRNA gene from an oocyte of *Triturus cristatus*. The spacer chromatin is clearly different from adjacent inactive chromatin fibers which reveal the characteristic nucleosome arrangement. **c** The smooth, nonnucleosomal chromatin conformation is maintained in the spacer (*S*) and the gene (*G*) regions in stages of reduced transcriptional activity (spread preparation from a mature oocyte of *Triturus cristatus*). **a** ×28000; **b** ×70000; **c** ×45000; *bars* = 0.5 μm

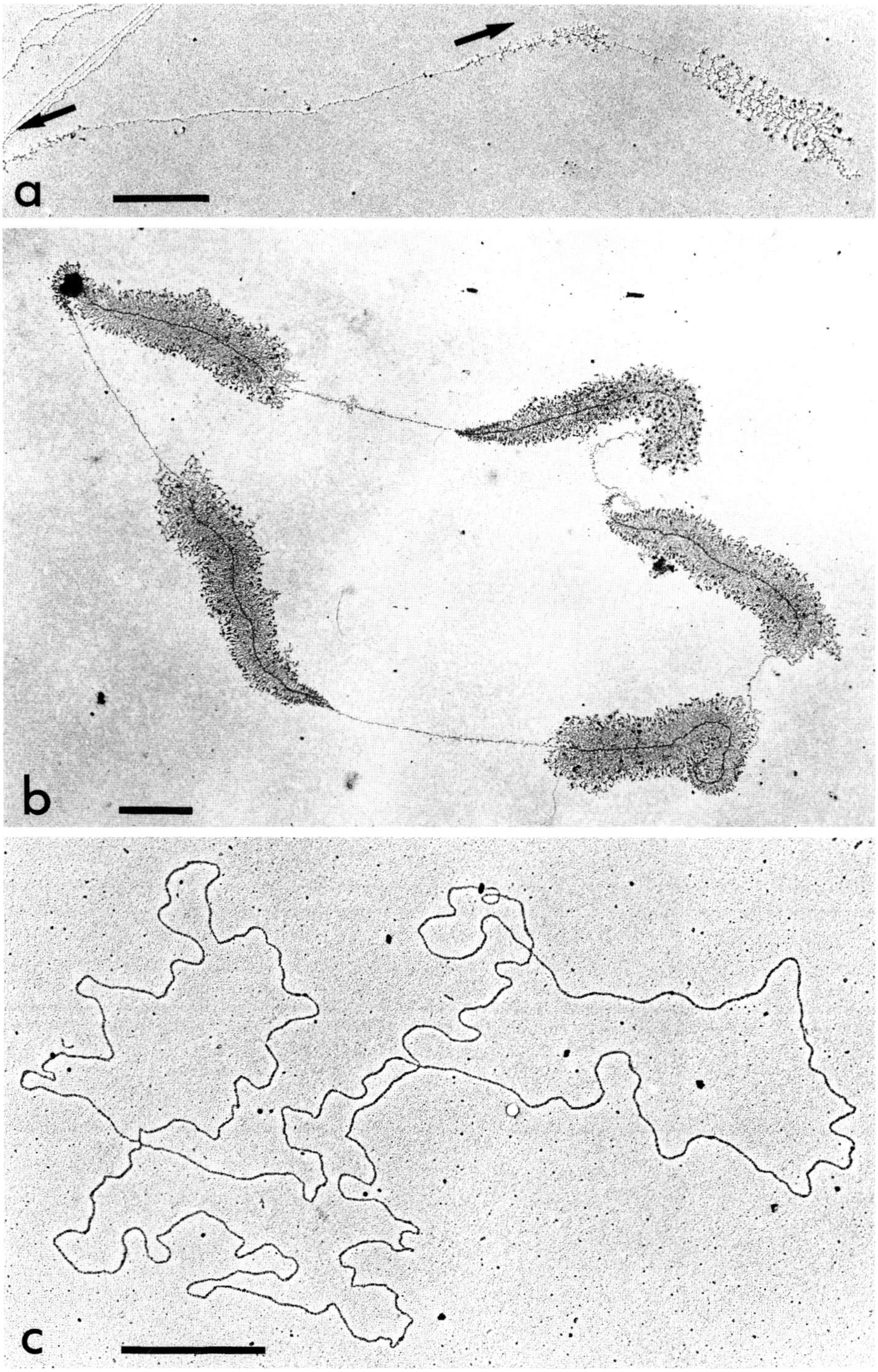

Fig. 4 a–c

analyze transcribing rRNA genes from a variety of plant and animal cells (for reviews see Miller and Hamkalo 1972; Miller and Bakken 1972; Hamkalo and Miller 1973; Hamkalo et al. 1973; Chooi 1976; Franke and Scheer 1978; Franke et al. 1979; McKnight et al. 1979; Miller 1981, 1984; Scheer and Zentgraf 1982; Berger and Schweiger 1982; Puvion-Dutilleul 1983; Trendelenburg 1983). Figures 2 to 4 illustrate the appearance of rRNA genes from amphibian and insect oocytes and from the slime mold, *Physarum polycephalum.* Since the rDNA of *Xenopus laevis* has been completely sequenced, the ribosomal transcription units of this species can be correlated precisely with specific DNA sequences (e.g., Sollner-Webb et al. 1982; Bakken et al. 1982).

In spite of extensive interspecies differences in the lengths of transcription units, the interspersed nontranscribed spacers and the resulting repeat units (for a compilation of quantitative data see Franke et al. 1979), several common characteristic features of rRNA genes emerged from these studies.

1. The chromatin axis of active rRNA genes is usually densely covered by RNA polymerase I-containing particles from which the transcript fibrils extend laterally. The lateral fibrils, containing nascent pre-rRNA chains, gradually increase in length in the direction of transcription so that the characteristic "Christmas trees" are formed (Figs. 2 and 3a). The free ends of the transcripts are often accentuated by a terminal knob. In most transcription units the lateral fibrils increase in length as a function of the distance from the transcription initiation site, although the length increment seems to be lower or almost zero for the more distal transcripts (Franke et al. 1976a; Laird et al. 1976; Puvion-Dutilleul et al. 1977; Angelier et al. 1979). In *Dictyostelium,* the fibril length gradient is interrupted, suggesting that processing of the nascent pre-rRNA takes place at a specific period during transcription (Grainger and Maizels 1980).
2. In most species, multiple rRNA genes are arranged in tandem with identical polarities along a given chromatin strand and are separated from each other by nontranscribed spacers (Fig. 3a). It should be noted, however, that at least in amphibians, the so-called nontranscribed spacer might, in fact, be transcribed. Spacer transcripts arranged in defined transcription units are not infrequently found in Miller spreads of *Xenopus laevis* oocytes and seem to originate from the reduplicated promotor-like sequences in the spacer DNA (cf. Trendelenburg 1981; Morgan et al. 1983). There are, however, notable exceptions to the familiar gene-spacer-gene pattern. The rRNA genes of certain unicellular green algae are tightly clustered without detectable spacers or are arranged with alternating polarities along the chromatin axis (Berger and Schweiger 1982). Opposite polarities

Fig. 4. a Linear nucleolar chromatin strand from the slime mold *Physarum polycephalum* with two rRNA genes at either end in opposite polarities (the direction of transcription is indicated by the *arrows;* only the proximal part of the left transcription unit is shown). The intergenic nontranscribed region is compacted into nucleosomes. In such Miller spreads the intergene region has an average length of 4.4 μm as compared to the extended length of the corresponding DNA (B-form) of 8.5 μm (cf. Seebeck and Braun 1982). **b** Nucleolar chromatin ring with five repeating units from an oocyte of the water beetle, *Dytiscus marginalis.* The contour length of the ring shown is 35 μm. **c** rDNA circle isolated from *Dytiscus* nucleolar chromatin with a contour length of 34.3 μm (cytochrome c-surface spreading). Molecules of this size class represent 5-gene rings. **a** ×13000; **b** ×9800; **c** ×19800; *bars* = 1 μm of adjacent rRNA genes are also present in the amplified nucleolar chromatin of *Physarum, Dictyostelium,* and *Tetrahymena* which form linear strands with two rRNA genes at either end in a palindromic arrangement (Fig. 4a; cf. Grainger and Ogle 1978; Grainger and Maizels 1980; Seebeck and Braun 1982; Vavra et al. 1982; Engberg 1985).

3. Transcribed nucleolar chromatin, as well as the intervening spacers are devoid of nucleosomes. The absence of nucleosomes is compatible with the full extension of the rDNA to the length of its B-form in Miller spreads. This has been shown by the congruence of the lengths of rDNA repeating units at the DNA and chromatin level (Trendelenburg et al. 1976; Scheer et al. 1977; Reeder et al. 1978) and by estimation of the DNA length contained in ribosomal transcription units by gel electrophoretic determinations of the molecular weight of the pre-rRNAs (e.g., Meyer and Hennig 1974; Scheer et al. 1977; Spring et al. 1976).

In Miller spreads the chromatin structure of active rDNA is readily distinguished from inactive chromatin. Providing that the transcripts are more distantly spaced, thus allowing visualization of the underlying chromatin axis, active rDNA has a thin, nonbeaded conformation (Fig. 3b, c; see also Franke et al. 1976b, 1978; Scheer 1978; Foe 1978; Busby and Bakken 1980; Labhart and Koller 1982; Scheer and Zentgraf 1982; Puvion-Dutilleul 1983). A similarly extended, nonnucleosomal organization has also been found for the nontranscribed spacer of several species (Fig. 3a–c; Trendelenburg et al. 1976; Scheer et al. 1977; Scheer 1980; Scheer and Zentgraf 1982; Labhart and Koller 1982). That the nonnucleosomal conformation of nucleolar chromatin is maintained even when transcriptional activity is reduced suggests that the altered chromatin conformation is due to an intrinsic property of rDNA chromatin and not the transcription process per se (Fig. 3c; Foe 1978; Scheer 1978). Only in stages of complete inactivation does rDNA-containing chromatin assume the characteristic nucleosomal conformation (see next section). The most compelling evidence for a stable extension of the DNA of both the gene and the spacer of transcribing nucleolar chromatin is found in studies of the circular rRNA genes of *Dytiscus* oocytes (Fig. 4b; see also Trendelenburg et al. 1976). These circles contain one or more repeating units and, therefore, fall into discrete size classes. The rDNA can be isolated from the chromatin rings and surface spread by the cytochrome c-method (Fig. 4c). The identical contour lengths of both the chromatin rings and the corresponding rDNA circles indicate that the rDNA is fully extended in the chromatin (see also Trendelenburg et al. 1976; Scheer and Zentgraf 1978).

The central nontranscribed chromatin regions separating the palindromically arranged rRNA genes of *Physarum, Dictyostelium,* and *Tetrahymena* are exceptional in that they reveal a bead-like pattern, indicative of a nucleosomal organization (Fig. 4a). In fact, the DNA of this intergene region is compacted by a factor of about 2 relative to the length of the corresponding DNA (Fig. 4a; see also Grainger and Ogle 1978) and reveals the characteristic canonical nucleosome repeat after nuclease digestion (Borchsenius et al. 1981; Prior et al. 1983; Palen and Cech 1984).

3.1 rRNA Genes in Different States of Activity

Ultrastructural studies of several biological systems have provided considerable insight into the regulation of rRNA synthesis. During insect embryogenesis and amphibian oogenesis, the rate of rRNA accumulation is extensively modulated (Foe 1978; McKnight and Miller 1976; Scheer et al. 1976a). Using these systems, it is possible to examine rRNA genes by the Miller spreading technique in states of increasing and decreasing rRNA synthesis. Such studies have shown that rRNA synthesis is regulated at the level of transcription since reduced rRNA synthesis is correlated with a decreased number of transcripts per gene (Fig. 3c) and/or a decreased number of active genes. Furthermore, it has been demonstrated that rRNA gene activation proceeds via a two-step mechanism. First, there is a structural transition of rDNA chromatin from a beaded to a smooth form, which is then followed by transcription (Foe 1978; Scheer 1978). Each rRNA gene appears to be regulated individually since adjacent genes reveal variable degrees of transcript densities. The same pattern of chromatin reorganization is seen after inactivation of rRNA genes, both in natural stages of inactivation (e.g., Trendelenburg and McKinnell 1979) or after inhibition of transcription by drugs, such as actinomycin D (Scheer 1978; Franke et al. 1979). Long-term inactivated nucleolar chromatin is structurally indistinguishable from the bulk chromatin. Not only is it condensed into nucleosomes, but it is also organized into thick, i.e., higher order, fibrils (Scheer and Zentgraf 1978).

4 Visualization of Transcriptionally Active Nonnucleolar Genes

The electron microscopic chromatin spreading method has also provided new approaches to analyze the structural organization of transcriptionally active nonnucleolar genes and, in a few cases, to identify the transcription units of specific protein coding genes. The first nonnucleolar genes to be visualized in the electron microscope were the transcriptional units of the lateral loops of lampbrush chromosomes from amphibian oocytes (Miller and Beatty 1969d). Amphibian lampbrush chromosomes are especially well suited for such studies since their numerous lateral loops represent sites of intense transcriptional activities. As shown originally by Miller and co-workers and subsequently by other authors, loop transcription units are characterized by a high packing density of RNA polymerases (of type II, see Bona et al. 1981) and by nascent RNP transcripts reaching considerable lengths (Miller and Beatty 1969d; Miller and Hamkalo 1972; Miller et al. 1972; Miller and Bakken 1972; Miller 1981; Angelier and Lacroix 1975; Franke et al. 1976b; Scheer et al. 1976b, 1979a; Hill 1979; Hill and McGregor 1980).

The light microscopic appearance of isolated lampbrush chromosomes from a salamander oocyte is presented in Fig. 5a. Usually one loop represents a single transcription unit. Occasionally, however, a single loop contains more than one

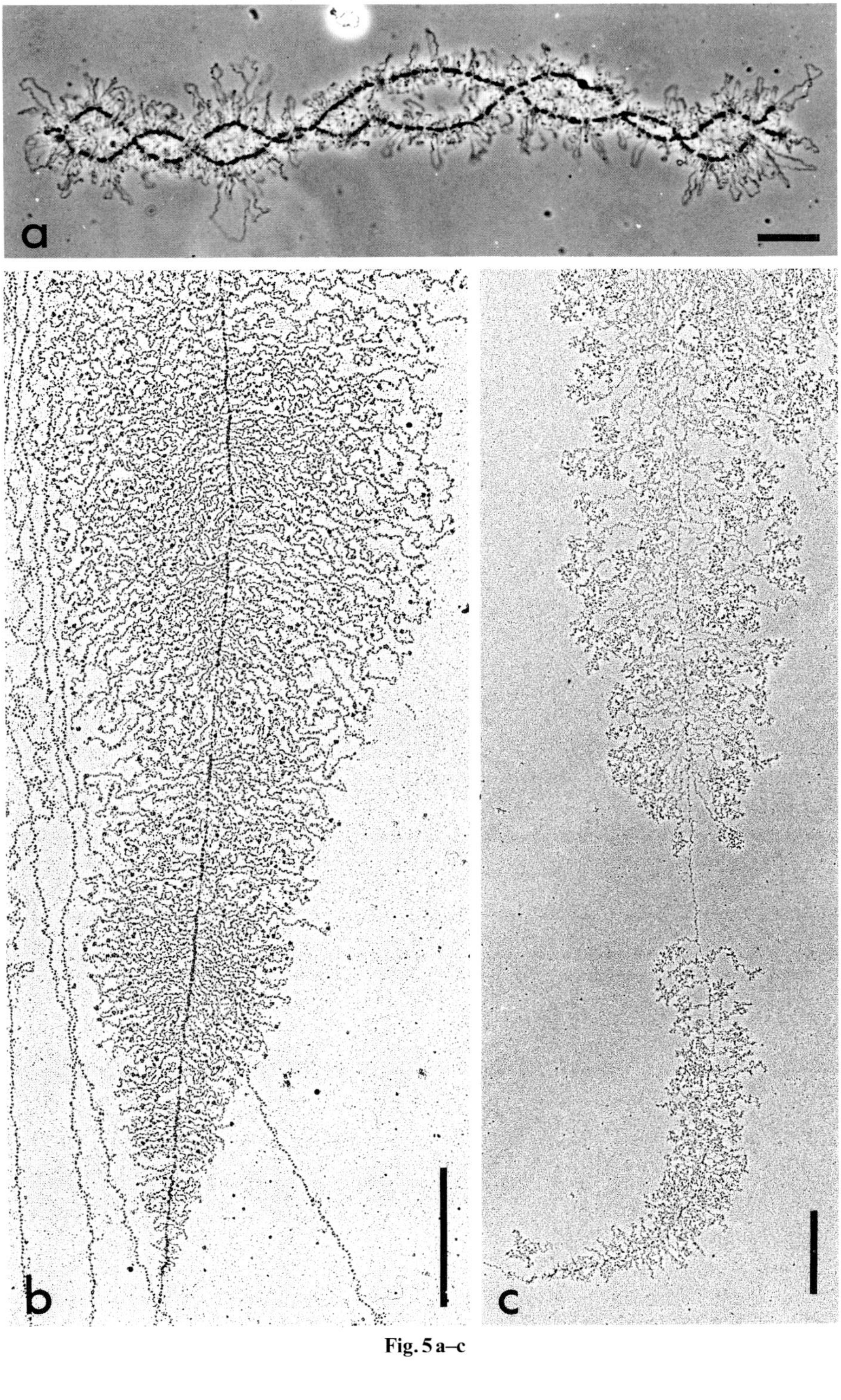

Fig. 5 a–c

transcription unit arranged with identical, opposite, or alternating polarities (Scheer et al. 1976b). Often the transcriptional polarity is recognized by the increase in thickness of the loop matrix between both ends of a loop (for a detailed account of lampbrush chromosome features see Callan 1982). In spread preparations transcriptional units of the lampbrush chromosomes (or fragments therefrom) are identified by their large size, the close spacing of transcripts, and the

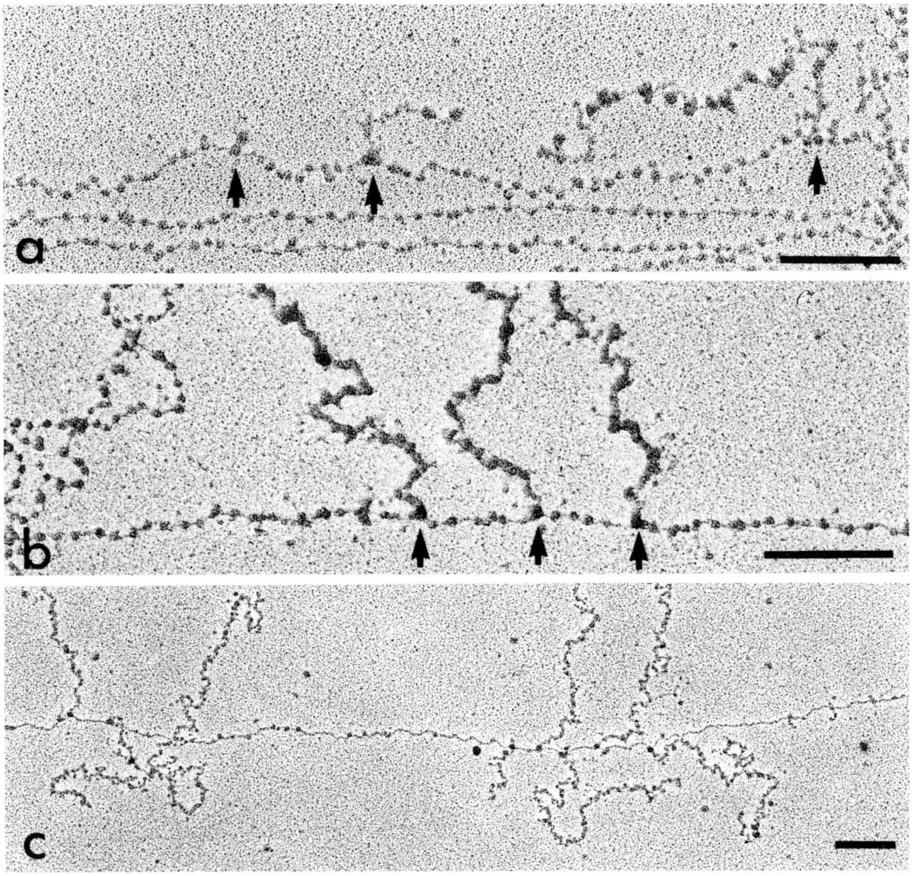

Fig. 6 a–c. Nucleosomal organization of moderately transcribed nonnucleolar chromatin from cultured *Xenopus laevis* kidney cells (**a**) and oocytes of *Pleurodeles waltlii* (**b**). The attachment sites of the transcripts to the chromatin axis are denoted by *arrows*. Inclusion of the anionic detergent Sarkosyl in the spreading medium releases the histones from the DNA and destroys the beaded organization (**c**). The transcripts remain stably bound to the DNA under such conditions. **a** ×80000; **b** ×85000; **c** ×40000; *bars* = 0.2 μm

◄

Fig. 5. a Light microscopic appearance of a lampbrush chromosome bivalent from a *Pleurodeles waltlii* oocyte. The lateral loops represent loci of intense transcriptional activity. **b–c** Miller spreads of lampbrush chromosomes from *Pleurodeles waltlii* oocytes. Note the different morphological aspects of the nascent transcripts in **b** and **c. a** ×430; **b** ×20000; **c** ×12000; *bars* = 20 μm (**a**) and 1 μm (**b, c**)

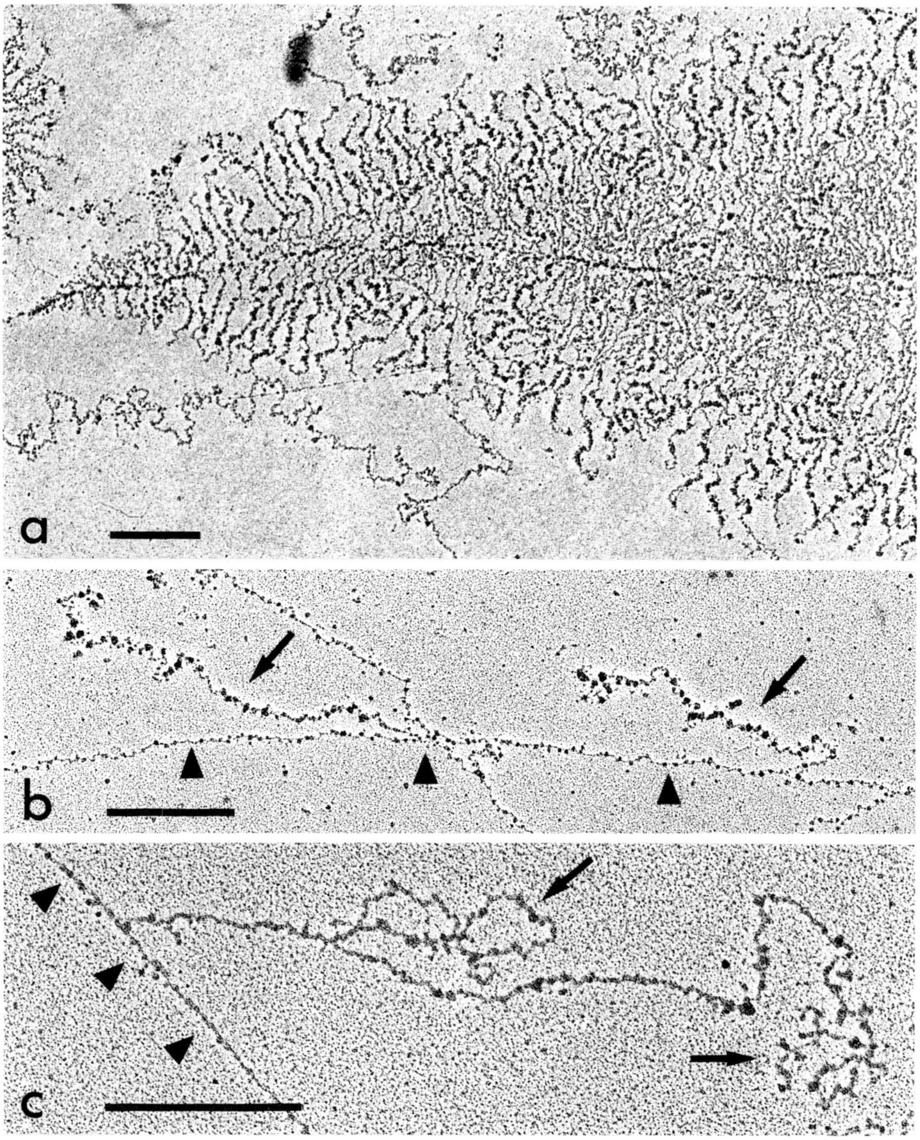

Fig. 7 a–c. Various aspects of transcript organization in Miller spreads of lampbrush chromosomes of *Triturus cristatus* (**a**) and cultured kidney cells of *Xenopus laevis* (**b, c**). Each transcript of the transcription unit shown in (**a**) reveals a subterminal thickening at corresponding positions. The two distantly spaced transcripts in **b** (*arrows*) are condensed into arrays of globular particles. Secondary structures characteristic of nonnucleolar transcripts (ring- and bushlike formations) are shown in **c** (*arrows*). The chromatin axis is indicated by *arrowheads* in **b** and **c**. **a** × 24 000; **b** × 34 000; **c** 54 000; *bars* = 0.5 μm

Fig. 8 a, b. Part of a lampbrush loop transcription unit from a *Triturus helveticus* oocyte after spreading for electron microscopy according to the Miller procedure. Each lateral fibril carries a conspicuous "loop-stem" structure at corresponding positions (*arrows* in **b**). A transcript fibril from the terminal region of the same transcription unit is shown in (**a**). In addition to the loop-stem structure indicated by the *arrow* a second fold-back structure of the nascent RNA is visible (*arrowhead*). Note that the double-stranded RNA "stem" structures are almost devoid of a protein coat. × 60 000; *bars* = 0.2 μm

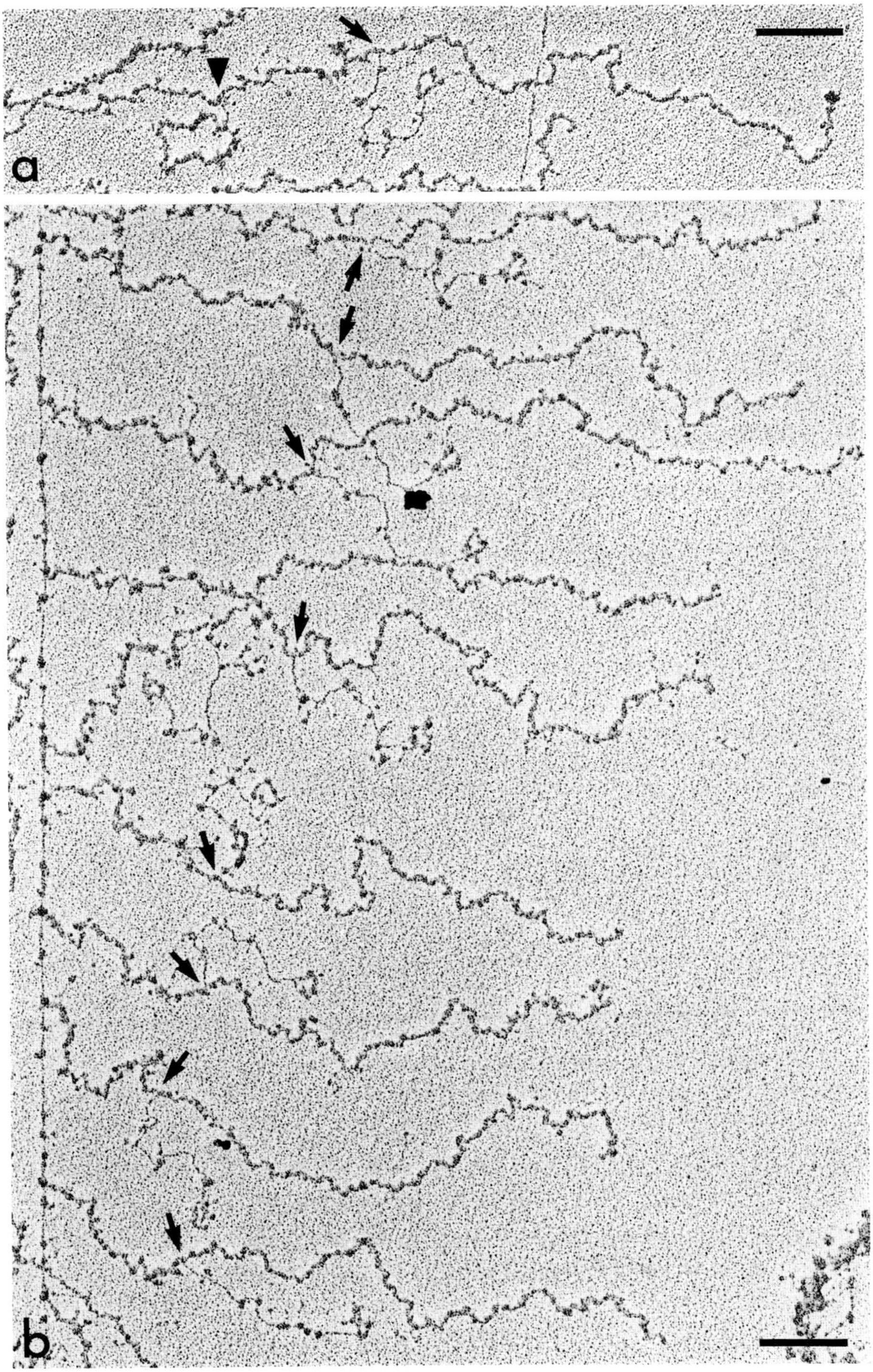

Fig. 8a, b

extensive secondary structures of the transcript fibrils which often give rise to complex bush- and ring-like configurations (Figs. 5 b, c, 7 a, 8 a, b; for reviews see Scheer et al. 1979 a; Sommerville 1981).

Similar high packing densities of nascent transcripts are also found in lampbrush chromosomes of the unicellular green algae, *Acetabularia* (Spring et al. 1975; Franke et al. 1976 b; Scheer et al. 1979 a), in certain lampbrush loops of the Y-chromosome of *Drosophila* spermatocytes (Grond et al. 1983; Glätzer and Meyer 1981), in autosomal chromosomes of *Drosophila* primary spermatocytes (Glätzer 1975), in silk fibroin genes of *Bombyx mori* (McKnight et al. 1976), in chorion protein genes of *Drosophila* (Osheim and Miller 1983; Osheim et al. 1985), and in Balbiani ring genes of polytene chromosomes from *Chironomus tentans* salivary glands (Lamb and Daneholt 1979; Widmer et al. 1984). Usually, however, nonnucleolar transcription units of a variety of cell types and species are characterized by relatively few transcript fibrils, with extended unoccupied chromatin regions, indicative of a low level of ongoing transcriptional activity (Figs. 6 a, 7 b, c; see also Miller and Bakken 1972; Hamkalo and Miller 1973; Hamkalo et al. 1973; Kierszenbaum and Tres 1975; Laird and Choii 1976; Laird et al. 1976; Foe et al. 1976; Amabis and Nair 1976; Oda et al. 1977; Villard and Fakan 1978; McKnight et al. 1979; Harper and Puvion-Dutilleul 1979; Busby and Bakken 1979, 1980; Hughes et al. 1979; Cotton et al. 1980; Petrov et al. 1980; Beyer et al. 1980, 1981; Scheer et al. 1981; Greimers and Deltour 1981; Bachvarova et al. 1982; Rattner et al. 1982; Puvion-Dutilleul 1983).

4.1 Chromatin Structure in Different States of Transcriptional Activity

Intensely transcribed nonnucleolar chromatin regions with maximum packing densities of RNA polymerases are devoid of nucleosome-sized particles (Franke et al. 1976 b). Furthermore, the DNA appears to be extended to almost the length of the B-conformation. This has been shown by a comparison of the length of the transcription units of the highly active silk fibroin genes of *Bombyx mori* and the chorion protein genes of *Drosophila melanogaster* follicle cells with the corresponding DNA (McKnight et al. 1976; Osheim et al. 1985). However, in stages of reduced activity and more distantly spaces transcripts, the chromatin axis not engaged in transcription generally assumes a beaded, nucleosomal configuration. The rapid reformation of nucleosomes after the passage of a transcriptional event is seen in lampbrush chromosomes in stages of experimentally induced or naturally occurring reduced transcriptional activity (Fig. 6 b, c; see also Angelier and Lacroix 1975; Scheer 1978; Bona et al. 1981), as well as in transcription units of somatic cells (Figs. 6 a, 7 b, c; see also Kierszenbaum and Tres 1975; Foe et al. 1976; Laird and Choii 1976; Laird et al. 1976; Oda et al. 1977; McKnight et al. 1979; Villard and Fakan 1978; Busby and Bakken 1979, 1980; Hughes et al. 1979; Cotton et al. 1980; Petrov et al. 1980; Greimers and Deltour 1981). Thus, in contrast to transcribing rDNA chromatin, nonnucleolar chromatin appears to assume the nucleosomal organization relatively fast, provided that the RNA polymerases are sufficiently apart to allow reformation of nucleosomes (for identification of nucleosomes in Miller spreads see below).

4.2 Ultrastructure of Nascent RNP Transcripts

The specific low salt chromatin spreading conditions induce the nascent RNP fibrils to unravel from a condensed into a largely extended state. This is best illustrated by comparing the in situ conformation of nascent RNP transcripts of amphibian lampbrush chromosomes with their appearance in spread preparations. When lampbrush chromosomes are isolated in the presence of physiological salt concentrations followed by fixation and flat embedding for electron microscopy, ultrathin sections reveal that the majority of the transcripts are arranged in linear arrays consisting of 25 to 40 nm globular subunits (Mott and Callan 1975; Spring and Franke 1981; N'Da et al. 1986). In contrast, in Miller spreads much thinner fibrils of the transcripts predominate (Figs. 5 b, c, 6 b, c) which apparently are derived by relaxation of the higher order globular packaging structures (see also Sommerville 1981; Scheer and Dabauvalle 1985). An ordered packaging of the basic 10 nm transcript fibril into a RNP particle of defined morphology and size has also been described for the transcripts of the Balbiani ring genes in *Chironomus tentans* (Skoglund et al. 1983). It should be noted that the extended transcripts as seen in Miller spreads are much thicker than naked RNA, indicating that the nascent RNA chains become coated with proteins immediately following their synthesis.

Not all RNP transcripts unravel into extended fibrils when spread for electron microscopy. Some retain extensive secondary structures and/or display a distinctly beaded morphology due to the presence of 20 nm particles. Examples are presented in Figs. 5 c, 6 a, 7 a–c, and 8 a, b, both from amphibian lampbrush chromosomes and cultured *Xenopus laevis* kidney cells (for further examples see, e.g., Kierszenbaum and Tres 1975; Hughes et al. 1979; Beyer et al. 1980, 1981; Greimers and Deltour 1981; Osheim et al. 1985). A remarkably high degree of secondary structure occurs along the transcripts of the Y chromosome in spermatocytes of *Drosophila hydei*. The transcripts possess a folding pattern that is characteristic for a specific transcription unit (Glätzer 1975; Glätzer and Meyer 1981; Grond et al. 1983; de Loos et al. 1984; Hennig 1985).

To what extent intramolecular base pairing of the nascent RNA chain or protein-protein interactions are involved in establishing the secondary structures of transcript fibrils seen in Miller spreads is not known (for discussion see Sommerville 1981). Several authors have analyzed the distribution of secondary structures, such as loop formations, bush-like ramifications, or RNP particles, in the multiple transcripts of a given transcription unit. The conclusion stemming from such studies is that secondary structures are not randomly distributed. Rather, they occur at specific sites indicating a relationship to specific RNA sequences (e.g., Glätzer 1980; Beyer et al. 1980; Hennig 1985; Osheim and Beyer 1985).

Two transcription units of *Pleurodeles* lampbrush chromosomes with different types of transcript organization are shown in Fig. 5 b and c. The transcripts are either fully extended (Fig. 5 b) or display, specifically at their termini, a particulate morphology and are folded up into complex bush-like structures (Fig. 5 c). Another example is shown in Fig. 7 a where the subterminal region of each consecutive transcript carries an array of closely spaced particles at corresponding positions.

A clear example that the RNA of nascent transcripts can form intramolecular duplex structures, probably by a fold-back mechanism and base pairing of inverted repeat sequences, is provided in Fig. 8. Two different loop-stem structures are recognized, a large loop with a short stem and a small loop with a long stem (Fig. 8a). The loops presumably contain single-stranded RNA, whereas the stems are formed by duplex regions of the RNA (cf. Sommerville 1981). It is noteworthy that the loops display the same ultrastructural morphology as the principal RNP fibril. However, the stem connecting the loop to the transcript fibril is very thin and is apparently deficient of proteins. This finding is in agreement with biochemical studies showing that double-stranded RNA regions of hnRNA, in comparison with single-stranded regions, are relatively devoid of proteins in living cells (Calvet and Pederson 1979). The loop-stem structures form at corresponding positions of the multiple transcripts shown in Fig. 8b. The micrograph depicts only a relatively small area close to the transcription initiation site where the second loop has not yet been formed. It is obvious that the distance between the hairpin structures and the anchoring sites of the fibrils into the chromatin axis increase as a function of the growth of the transcripts. A quantitative analysis of the total transcription unit with 48 traceable fibrils is presented in Fig. 9.

Whether the foldback structures of the type depicted in Fig. 8 have a function in the process of RNA sequence excision is presently not known. However, RNA particles frequently seen on transcripts of a given gene, at sequence-specific sites, might be functionally involved in splicing reactions by bringing together two splice junctions with the intron transcript looping out (Osheim et al. 1985; Osheim

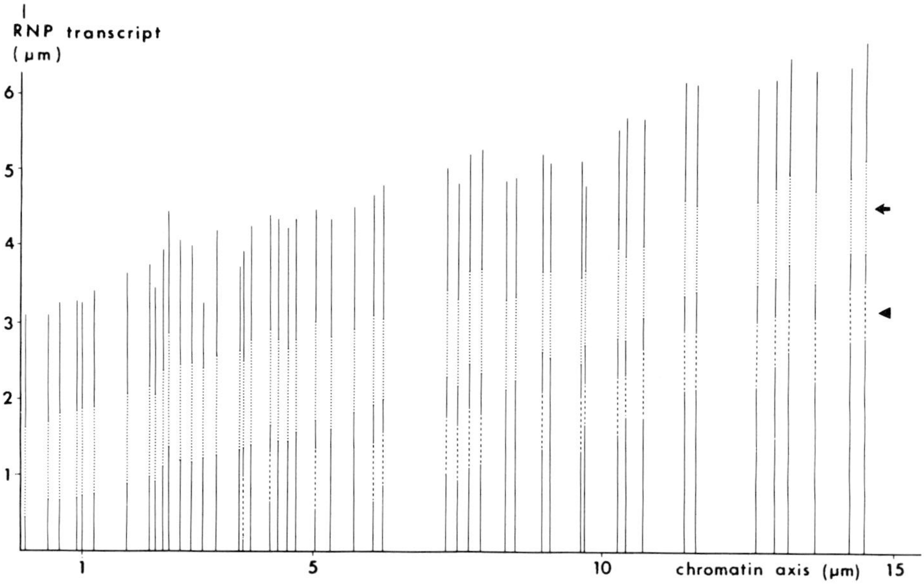

Fig. 9. Quantitative evaluation of 48 traceable transcripts from the transcription unit shown in Fig. 8. Each transcript is represented in its linearized form, i.e., loop-stem structures are linearized and indicated as *dotted lines* (at the *arrow* and *arrowhead;* the corresponding structures are shown in Fig. 8a). The graphic representation clearly demonstrates that the hairpin structures are related to specific sequences of the nascent RNA chains

and Beyer 1985). Such splicing reactions might well occur during transcription. In fact, numerous nonnucleolar transcription units from a variety of species analyzed by the chromatin spreading method reveal interruptions in the RNP fibril length gradients. This indicates that cleavage of transcripts occurs during transcription (e.g., Laird and Choii 1976; Scheer et al. 1979a; Beyer et al. 1981). In addition, a considerable proportion of the RNA contained in terminal transcript fibrils of amphibian lampbrush chromosomes apparently does not consist of contiguous chains, but of nicked molecules (Scheer and Sommerville 1982).

5 Application of Localization Techniques to Miller Spreads

An advantage of the chromatin spreading technique is that it is potentially capable of elucidating the ultrastructure and biochemical composition of transcription units of defined genes in various states of activity. So far only a few protein-coding genes have been identified in Miller spreads (silk fibroin genes of *Bombyx mori:* McKnight et al. 1976; Balbiani ring genes of salivary glands of *Chironomus tentans:* Lamb and Daneholt 1979; and chorion protein genes of *Drosophila melanogaster* follicle cells: Osheim and Miller 1983; Osheim et al. 1985). With the availability of specific nucleic acid probes it should be feasible, by using modified protocols for in situ hybridization, to correlate transcription units in Miller spreads with specific genes. In this way one may analyze the distribution of specific proteins within the transcription unit, in the flanking regions, and along the nascent RNP transcripts by immunochemical approaches.

5.1 In Situ Hybridization of Nucleic Acids

Transcription units with numerous attached lateral fibrils represent a favorable situation for in situ hybridization of DNA or RNA probes to the nascent RNA chains because of the multiplication of the DNA coding sequence at the RNA level. Hybridization of nucleic acid probes to nascent RNA has already been used, at the light microscopic level, to localize histone genes to specific loops of amphibian lampbrush chromosomes (Old et al. 1977; Callan 1982; Gall et al. 1981; Diaz et al. 1981; Diaz and Gall 1985).

Similar procedures may also be adapted to investigations at the electron microscopic level. Visualization of the hybridized DNA or RNA probe could be accomplished by autoradiographic or nonradioactive methods, such as biotinylation or chemical modification of the probe which are then detected by specific antibodies (e.g., Fostel et al. 1984; Landegent et al. 1985).

When recombinant plasmids containing rDNA sequences of *Xenopus laevis* are hybridized to Miller spreads of amphibian oocyte nuclei, the DNA probe binds selectively to the pre-rRNA of the lateral fibrils of the rRNA genes (for examples see Scheer and Zentgraf 1982). We are confident that further refinements of this approach will permit a better ultrastructural preservation of the transcription units and a higher hybridization efficiency so that protein coding genes can be identified as well.

An additional approach to identify transcribed genes in the electron microscope is based on restriction endonuclease cleavage prior to chromatin spreading in order to produce identifiable fragments of transcription units (e.g., Reeder et al. 1976; Pruitt and Reeder 1984).

5.2 Biochemical and Immunological Techniques

It is known that transcriptional complexes are extremely stable to high salt concentrations and ionic detergents, i.e., conditions which lead to the dissociation of the majority of chromatin proteins including histones (for details see Scheer 1978). This selective stability of transcriptional complexes has been exploited to demonstrate that the beaded morphology of a chromatin fiber is dependent on the presence of histones and does not reflect, e.g., arrays of template-bound RNA polymerases without nascent transcripts. When chromatin is spread in the presence of the anionic detergent Sarkosyl or the commercial dishware detergent "Joy" (originally introduced by Miller and Bakken 1972) at concentrations which release most chromatin proteins from the DNA, the beaded organization of chromatin is no longer visible (Scheer 1978). Figure 6b and c illustrate the effect of Sarkosyl on the ultrastructure of lampbrush chromosome loops transcribed with moderate efficiency. Without detergent the chromatin axis displays a beaded morphology (Fig. 6b); addition of Sarkosyl results in the disappearance of the beads while the transcripts remain attached to the smoothly contoured chromatin fiber (Fig. 6c), indicating that the beads in fact represent nucleosomes.

A direct approach to localize specific proteins in Miller spreads is based on the use of immunolocalization techniques. As shown by several authors, antibodies to core histones react specifically with nucleosomal beads of inactive chromatin (e.g., Bustin et al. 1976; Ghose et al. 1983). Antibodies to the core histones H2B and H3 also bind to the axis of nonnucleolar genes which are sparsely covered with transcripts (McKnight et al. 1978). The identification of histones by this immuno-electron microscopic approach in conjunction with the beaded morphology of the chromatin regions between distantly spaced transcripts clearly indicates that a transient nucleosomal packaging of the chromatin template takes place between successive transcriptional events.

Whether histones remain associated also with intensely transcribed chromatin regions, such as the lampbrush chromosome loops or nucleolar chromatin, is as yet an open question (for discussion see Mathis et al. 1980; Karpov et al. 1984; Sargan and Butterworth 1985). Indirect evidence based on the effect of histone antibodies microinjected into nuclei of living amphibian oocytes suggests that some histones are present on the heavily transcribed lampbrush loops. However, these experiments did not allow quantitative estimates and also did not permit the precise localization of histones (Scheer et al. 1979b; Scheer 1986).

Immuno-electron microscopy has also been used to study, at a very high level of resolution, the distribution of nonhistone proteins in chromatin. T-antigen has been mapped to a specific region of SV40 minichromosomes, the nucleosome-

and histone-free gap containing the origin of replication and promotor elements (Harper et al. 1984; Weiss et al. 1985).

Do specific ribosomal proteins bind to nascent pre-rRNA of rRNA genes? This question has been studied by Choii and Leiby (1981) by combining Miller spreads with immunolocalization techniques. They used two different protocols for binding of the antibodies. Either the antibodies were added directly to the lysed *Drosophila* embryonic cells followed by centrifugation of the chromatin onto the electron microscopic grid or rRNA genes were first immobilized on the grid followed by incubation with the antibodies. Their results show that individual ribosomal proteins bind to specific sequences of the nascent pre-rRNA. In this connection it should be emphasized that not all ribosomal proteins are involved in the early steps of ribosomal assembly, but become integrated into preribosomal particles only at more advanced stages of maturation (Hügle et al. 1985).

Another approach to localize nucleolus-specific proteins in Miller spreads was taken by Angelier et al. (1982). These authors adapted the silver staining method to spread preparations of rRNA genes of amphibian oocytes. They showed that silver staining proteins were preferentially associated with the transcribed rDNA and/or with the basal portions of the nascent RNP transcripts.

In conclusion, the tools and the probes necessary for identification and analysis of transcription units at the molecular level are presently available. Their application, in conjunction with the Miller chromatin spreading technique, should lead to a more detailed understanding of the conformational and compositional differences between active and inactive chromatin and the complex processes involved in generation and maturation of RNP transcripts.

Acknowledgments. I thank Dr. Frank Longo and Dr. Werner W. Franke for critical reading and correcting the manuscript and Dr. Marie-Christine Dabauvalle for quantitative evaluation of some micrographs. The work presented here received financial support from the Deutsche Forschungsgemeinschaft.

References

Amabis JM, Nair KK (1976) Ultrastructure of gene transcription in spermatocytes of *Trichosia pubescens* Morgante, 1969 (diptera: sciaridae). Z Naturforsch 31c:186–189

Angelier N, Lacroix JC (1975) Complexes de transcription d'origines nucléolaire et chromosomique d'ovocytes de *Pleurodeles waltlii* et *P. poireti* (amphibiens, urodèles). Chromosoma 51:323–335

Angelier N, Hemon D, Bouteille M (1979) Mechanisms of transcription in nucleoli of amphibian oocytes as visualized by high-resolution autoradiography. J Cell Biol 80:277–290

Angelier N, Hernandez-Verdun D, Bouteille M (1982) Visualization of Ag-NOR proteins on nucleolar transcriptional units in molecular spreads. Chromosoma 86:661–672

Bachvarova R, Burns JP, Spiegelman I, Choy J, Chaganti RSK (1982) Morphology and transcriptional activity of mouse oocyte chromosomes. Chromosoma 86:181–196

Bakken A, Morgan G, Sollner-Webb B, Roan J, Busby S, Reeder RH (1982) Mapping of transcription initiation and termination signal on *Xenopus laevis* ribosomal DNA. Proc Natl Acad Sci USA 79:56–60

Beermann S (1984) Circular and linear structures in chromatin diminution of *Cyclops*. Chromosoma 89:321–328

Berger S, Schweiger H-G (1982) Characterization and species differences of rRNA in algae. In: Busch H, Rothblum L (eds) The cell nucleus, vol 10. Academic Press, New York, pp 31–65
Beyer AL, Miller OL, McKnight SL (1980) Ribonucleoprotein structure in nascent hnRNA is nonrandom and sequence-dependent. Cell 20:75–84
Beyer AL, Bouton AH, Miller OL (1981) Correlation of hnRNP structure and nascent transcript cleavage. Cell 26:155–165
Bona M, Scheer U, Bautz EKF (1981) Antibodies to RNA polymerase II (B) inhibit transcription in lampbrush chromosomes after microinjection into living amphibian oocytes. J Mol Biol 151:81–99
Borchsenius S, Bonven B, Leer JC, Westergaard O (1981) Nuclease-sensitive regions on the extrachromosomal r-chromatin from *Tetrahymena pyriformis*. Eur J Biochem 117:245–250
Busby S, Bakken A (1979) A quantitative electron microscopic analysis of transcription in sea urchin embryos. Chromosoma 71:249–262
Busby S, Bakken A (1980) Transcription in developing sea urchins: electron microscopic analysis of cleavage, gastrula and prism stages. Chromosoma 79:85–104
Bustin M, Goldblatt D, Sperling R (1976) Chromatin structure visualization by immunoelectron microscopy. Cell 7:297–304
Callan HG (1982) Lampbrush chromosomes. Proc R Soc Lond B 214:417–448
Calvet JP, Pederson T (1979) Heterogeneous nuclear RNA double-stranded regions probed in living HeLa cells by crosslinking with the psoralen derivative aminomethyltrioxsalen. Proc Natl Acad Sci USA 76:755–759
Choii WY (1976) RNA transcription and ribosomal protein assembly in *Drosophila melanogaster*. In: King RC (ed) Handbook of genetics, vol 5. Plenum, New York, pp 219–265
Chooi WY, Leiby KR (1981) An electron microscopic method for localization of ribosomal proteins during transcription of ribosomal DNA: a method for studying protein assembly. Proc Natl Acad Sci USA 78:4823–4827
Cotton RW, Manes C, Hamkalo BA (1980) Electron microscopic analysis of RNA transcription in preimplantation rabbit embryos. Chromosoma 79:169–178
de Loos F, Dijkhof R, Grond CJ, Hennig W (1984) Lampbrush chromosome loop-specificity of transcript morphology in spermatocyte nuclei of *Drosophila hydei*. EMBO J 3:2845–2849
Diaz MO, Gall JG (1985) Giant readthrough transcription units at the histone loci on lampbrush chromosomes of the newt *Notophthalmus*. Chromosoma 92:243–253
Diaz MO, Barsacchi-Pilone G, Mahon KA, Gall JG (1981) Transcripts from both strands of a satellite DNA occur on lampbrush chromosome loops of the newt *Notophthalmus*. Cell 24:649–659
DuPraw EJ (1970) DNA and chromosomes. Holt, Rinehart and Winston, New York
Engberg J (1985) The ribosomal RNA genes of *Tetrahymena:* structure and function. Eur J Cell Biol 36:133–151
Foe VE (1978) Modulation of ribosomal RNA synthesis in *Oncopeltus fasciatus:* an electron microscopic study of the relationship between changes in chromatin structure and transcriptional activity. Cold Spring Harbor Symp Quant Biol 42:723–740
Foe VE, Wilkinson LE, Laird CD (1976) Comparative organization of active transcription units in *Oncopeltus fasciatus*. Cell 9:131–146
Fostel J, Narayanswami S, Hamkalo B, Clarkson SG, Pardue ML (1984) Chromosomal location of a major tRNA gene cluster of *Xenopus laevis*. Chromosoma 90:254–260
Franke WW, Scheer U (1978) Morphology of transcriptional units at different states of activity. Philos Trans R Soc Lond B 283:333–342
Franke WW, Scheer U, Spring H, Trendelenburg MF, Krohne G (1976a) Morphology of transcriptional units of rDNA. Exp Cell Res 100:233–244
Franke WW, Scheer U, Trendelenburg MF, Spring H, Zentgraf H (1976b) Absence of nucleosomes in transcriptionally active chromatin. Cytobiologie 13:401–434
Franke WW, Scheer U, Trendelenburg MF, Zentgraf H, Spring H (1978) Morphology of transcriptionally active chromatin. Cold Spring Harbor Symp Quant Biol 42:755–772
Franke WW, Scheer U, Spring H, Trendelenburg MF, Zentgraf H (1979) Organization of nucleolar chromatin. In: Busch H (ed) The cell nucleus, vol 7. Academic Press, New York, pp 49–95

Gall JG (1963) Chromosome fibers from an interphase nucleus. Science 139:120–121
Gall JG (1966) Chromosome fibers studied by a spreading technique. Chromosoma 20:221–233
Gall JG, Stephenson EC, Erba HP, Diaz MO, Barsacchi-Pilone G (1981) Histone genes are located at the sphere loci of newt lampbrush chromosomes. Chromosoma 84:159–171
Ghose D, Weiss E, Homo JC, Van Regenmortel MHV, Oudet P (1983) Visualization of antihistone antibodies on SV40 minichromosomes by scanning transmission electron microscopy (STEM). Exp Cell Res 147:135–142
Glätzer KH (1975) Visualization of gene transcription in spermatocytes of *Drosophila hydei.* Chromosoma 53:371–379
Glätzer KH (1980) Regular substructures within homologous transcripts of spread *Drosophila* germ cells. Exp Cell Res 125:519–523
Glätzer KH, Meyer GF (1981) Morphological aspects of the genetic activity in primary spermatocyte nuclei of *Drosophila hydei.* Biol Cell 41:165–172
Grainger RM, Maizels N (1980) *Dictyostelium* ribosomal RNA is processed during transcription. Cell 20:619–623
Grainger RM, Ogle RC (1978) Chromatin structure of the ribosomal RNA genes in *Physarum polycephalum.* Chromosoma 65:115–126
Greimers R, Deltour R (1981) Organization of transcribed and nontranscribed chromatin in isolated nuclei of *Zea mays* root cells. Eur J Cell Biol 23:303–311
Grond CJ, Siegmund I, Hennig W (1983) Visualization of a lampbrush loop-forming fertility gene in *Drosophila hydei.* Chromosoma 88:50–56
Hamkalo BA, Miller OL (1973) Electronmicroscopy of genetic activity. Annu Rev Biochem 42:379–396
Hamkalo BA, Miller OL, Bakken AH (1973) Ultrastructure of active eukaryotic genomes. Cold Spring Harbor Symp Quant Biol 38:915–919
Hamkalo BA, Goldsmith MR, Rattner JB (1981) Higher order structure in chromosomes. In: Schweiger HG (ed) International cell biology 1980–1981. Springer, Berlin Heidelberg New York, pp 152–161
Harper F, Puvion-Dutilleul F (1979) Non-nucleolar transcription complexes of rat liver as revealed by spreading isolated nuclei. J Cell Sci 40:181–192
Harper F, Florentin Y, Puvion E (1984) Localization of T-antigen on simian virus 40 minichromosomes by immunoelectron microscopy. EMBO J 3:1235–1241
Hennig W (1985) Y chromosome function and spermatogenesis in *Drosophila hydei.* Adv Genet 23:179–234
Herzog M, Soyer M-O (1981) Distinctive features of dinoflagellate chromatin. Absence of nucleosomes in a primitive species *Prorocentrum micans* E. Eur J Cell Biol 23:295–302
Hill RS (1979) A quantitative electron-microscope analysis of chromatin from *Xenopus laevis* lampbrush chromosomes. J Cell Sci 40:145–169
Hill RS, MacGregor HC (1980) The development of lampbrush chromosome-type transcription in the early diplotene oocytes of *Xenopus laevis*: an electron-microscope analysis. J Cell Sci 44:87–101
Howze GB, Hsie AW, Olins AL (1976) v-bodies in mitotic chromatin. Exp Cell Res 100:424–428
Hozier JC (1979) Nucleosomes and higher levels of chromosomal organization. In: Taylor JH (ed) Molecular genetics, part 3. Academic Press, New York, pp 315–385
Hügle B, Hazan R, Scheer U, Franke WW (1985) Localization of ribosomal protein S1 in the granular component of the interphase nucleolus and its distribution during mitosis. J Cell Biol 100:873–886
Hughes ME, Bürki K, Fakan S (1979) Visualization of transcription in early mouse embryos. Chromosoma 73:179–190
Igo-Kemenes T, Hörz W, Zachau HG (1982) Chromatin. Annu Rev Biochem 51:89–121
Karpov VL, Preobrazhenskaya OV, Mirzabekov AD (1984) Chromatin structure of hsp 70 genes, activated by heat shock: selective removal of histones from the coding region and their absence from the 5′ region. Cell 36:423–431
Kierszenbaum AL, Tres LL (1975) Structural and transcriptional features of the mouse spermatid genome. J Cell Biol 65:258–270

Kornberg RD (1977) Structure of chromatin. Annu Rev Biochem 46:931–954
Labhart P, Koller T (1982) Structure of the active nucleolar chromatin of *Xenopus* laevis oocytes. Cell 28:279–292
Labhart P, Koller T, Wunderli H (1982) Involvement of higher order chromatin structures in metaphase chromosome organization. Cell 30:115–121
Laird CD, Chooi WY (1976) Morphology of transcription units in *Drosophila melanogaster*. Chromosoma 58:193–218
Laird CD, Wilkinson LE, Foe VE, Chooi WY (1976) Analysis of chromatin-associated fiber arrays. Chromosoma 58:169–192
Lamb MM, Daneholt B (1979) Characterization of active transcription units in Balbiani rings of *Chironomus tentans*. Cell 17:835–848
Landegent JE, Jansen in de Wal N, van Ommen GJB, Baas F, de Vijlder JJM, van Duijn P, van der Ploeg M (1985) Chromosomal localization of a unique gene by non-autoradiographic in situ hybridization. Nature 317:175–177
Lutz C, Nagl W (1980) A reliable method for preparation and electron microscopic visualization of nucleosomes in higher plant. Planta 149:408–410
Mathis D, Oudet P, Chambon P (1980) Structure of transcribing chromatin. Prog Nucleic Acid Res Mol Biol 24:1–55
McGhee JD, Felsenfeld G (1980) Nucleosome structure. Annu Rev Biochem 49:1115–1156
McKnight SL, Miller OL (1976) Ultrastructural patterns of RNA synthesis during early embryogenesis of *Drosophila melanogaster*. Cell 8:305–319
McKnight SL, Sullivan NL, Miller OL (1976) Visualization of the silk fibroin transcription unit and nascent silk fibroin molecules on polyribosomes of *Bombyx mori*. Prog Nucleic Acid Res Mol Biol 19:313–318
McKnight SL, Bustin M, Miller OL (1978) Electron microscopic analysis of chromosome metabolism in the *Drosophila melanogaster* embryo. Cold Spring Harbor Symp Quant Biol 42:741–754
McKnight SL, Martin KA, Beyer AL, Miller OL (1979) Visualization of functionally active chromatin. In: Busch H (ed) The cell nucleus, vol 7. Academic Press, New York, pp 97–122
Meyer GF, Hennig W (1974) The nucleolus in primary spermatocytes of *Drosophila hydei*. Chromosoma 46:121–144
Miller OL (1981) The nucleolus, chromosomes, and visualization of genetic activity. J Cell Biol 91:15s–27s
Miller OL (1984) Some ultrastructural aspects of genetic activity in eukaryotes. J Cell Sci Suppl 1:81–93
Miller OL, Bakken AH (1972) Morphological studies of transcription. Acta Endocrinol Suppl 168:155–177
Miller OL, Beatty BR (1969a) Visualization of nucleolar genes. Science 164:955–957
Miller OL, Beatty BR (1969b) Extrachromosomal nucleolar genes in amphibian oocytes. Genetics Suppl 61:133–143
Miller OL, Beatty BR (1969c) Nucleolar structure and function. In: Lima-de-Faria A (ed) Handbook of molecular cytology. North-Holland, Amsterdam, pp 605–619
Miller OL, Beatty BR (1969d) Portrait of a gene. J Cell Physiol 74: Suppl 1:225–232
Miller OL, Hamkalo BA (1972) Visualization of RNA synthesis on chromosomes. Int Rev Cytol 33:1–25
Miller OL, Beatty BR, Hamkalo BA (1972) Nuclear structure and function during amphibian oogenesis. In: Biggers JD, Schuetz AW (eds) Oogenesis. Univ Park Press, Baltimore; Butterworths, London, pp 119–129
Morgan GT, Reeder RH, Bakken AH (1983) Transcription in cloned spacers of *Xenopus laevis* ribosomal DNA. Proc Natl Acad Sci USA 80:6490–6494
Mott MR, Callan HG (1975) An electron-microscope study of the lampbrush chromosomes of the newt *Triturus cristatus*. J Cell Sci 17:241–261
N'Da E, Bonnanfant-Jais ML, Penrad-Mobayed M, Angelier N (1986) Size uniformity of ribonucleoprotein matrix particles in loops of *Pleurodeles waltlii* lampbrush chromosomes visualized by electron microscopy. J Cell Sci 81:17–27

Oda T, Nakamura T, Watanabe S (1977) Transcription complexes of chromatin showing a beads-like structure of heterogeneous ribonucleoprotein chains. J Electron Microsc 26:203–207
Old RW, Callan HG, Gross KW (1977) Localization of histone gene transcripts in newt lampbrush chromosomes by in situ hybridization. J Cell Sci 27:57–79
Olins AL, Olins DE (1973) Spheroid chromatin units (v bodies). J Cell Biol 59:252a
Olins AL, Olins DE (1974) Spheroid chromatin units (v bodies). Science 183:330–332
Olins AL, Carlson RD, Olins DE (1975) Visualization of chromatin substructure: v bodies. J Cell Biol 64:528–537
Osheim YN, Beyer AL (1985) Nascent ribonucleoprotein structure of polymerase I, II, and III gene transcripts. In: Smuckler EA, Clawson GA (eds) Nuclear envelope structure and RNA maturation. Liss, New York, pp 277–295
Osheim YN, Miller OL (1983) Novel amplification and transcriptional activity of chorion genes in *Drosophila melanogaster* follicle cells. Cell 33:543–553
Osheim YN, Miller OL, Beyer AL (1985) RNP particles at splice junction sequences on *Drosophila* chorion transcripts. Cell 43:143–151
Oudet P, Gross-Bellard M, Chambon P (1975) Electron microscopic and biochemical evidence that chromatin structure is a repeating unit. Cell 4:281–300
Palen TE, Cech TR (1984) Chromatin structure at the replication origins and transcription-initiation regions of the ribosomal RNA genes of *Tetrahymena*. Cell 36:933–942
Petrov P, Raitcheva E, Tsanev R (1980) Nucleosomes and nonribosomal RNA transcription in early mouse embryo. An electron microscopic study. Eur J Cell Biol 22:708–713
Prior CP, Cantor CR, Johnson EM, Littau VC, Allfrey VG (1983) Reversible changes in nucleosome structure and histone H3 accessibility in transcriptionally active and inactive states of rDNA chromatin. Cell 34:1033–1042
Pruitt SC, Reeder RH (1984) Effect of topological constraint on transcription of ribosomal DNA in *Xenopus* oocytes. J Mol Biol 174:121–139
Puvion-Dutilleul F (1983) Morphology of transcription at cellular and molecular levels. Int Rev Cytol 84:57–101
Puvion-Dutilleul F, Bernadac A, Puvion E, Bernhard W (1977) Visualization of two different types of nuclear transcriptional complexes in rat liver cells. J Ultrastruct Res 58:108–117
Rattner JB, Sauners C, Davie JR, Hamkalo BA (1982) Ultrastructural organization of yeast chromatin. J Cell Biol 92:217–222
Reeder RH, Higashinakagawa T, Miller OL (1976) The 5′-3′ polarity of the *Xenopus* ribosomal RNA precursor molecule. Cell 8:449–454
Reeder RH, McKnight SL, Miller OL (1978) Contraction ratio of the nontranscribed spacer of *Xenopus* rDNA chromatin. Cold Spring Harbor Symp Quant Biol 42:1174–1177
Ris H (1969) The molecular organization of chromosomes. In: Lima-de-Faria A (ed) Handbook of molecular cytology. North-Holland, Amsterdam, pp 221–250
Ris H, Korenberg J (1979) Chromosome structure and levels of chromosome organization. In: Prescott DM, Goldstein L (eds) Cell biology, vol 2. Academic Press, New York, pp 267–361
Sargan DR, Butterworth PHW (1985) Eukaryotic ternary transcription complexes: transcription complexes of RNA polymerase II are associated with histone-containing, nucleosome-like particles in vivo. Nucl Acids Res 13:3805–3822
Scheer U (1978) Changes of nucleosome frequency in nucleolar and non-nucleolar chromatin as a function of transcription: an electron microscopic study. Cell 13:535–549
Scheer U (1980) Structural organization of spacer chromatin between transcribed ribosomal RNA genes in amphibian oocytes. Eur J Cell Biol 23:189–196
Scheer U (1986) Injection of antibodies into the nucleus of amphibian oocytes: an experimental means of interfering with gene expression in the living cell. J Embryol Exp Morph 97 Suppl:223–242
Scheer U, Dabauvalle M-C (1985) Functional organization of the amphibian oocyte nucleus. In: Browder LW (ed) Developmental biology, vol 1. Plenum, New York, pp 385–430
Scheer U, Sommerville J (1982) Sizes of chromosome loops and hnRNA molecules in oocytes of amphibia of different genome sizes. Exp Cell Res 139:410–416

Scheer U, Zentgraf H (1978) Nucleosomal and supranucleosomal organization of transcriptionally inactive rDNA circles in *Dytiscus* oocytes. Chromosoma 69:243–254

Scheer U, Zentgraf H (1982) Morphology of nucleolar chromatin in electron microscopic spread preparations. In: Busch H, Rothblum L (eds) The cell nucleus, vol 11. Academic Press, New York, pp 143–176

Scheer U, Trendelenburg MF, Franke WW (1976a) Regulation of transcription of genes of ribosomal RNA during amphibian oogenesis. A biochemical and morphological study. J Cell Biol 69:465–489

Scheer U, Franke WW, Trendelenburg MF, Spring H (1976b) Classification of loops of lampbrush chromosomes according to the arrangement of transcriptional complexes. J Cell Sci 22:503–519

Scheer U, Trendelenburg MF, Krohne G, Franke WW (1977) Lengths and patterns of transcriptional units in the amplified nucleoli of oocytes of *Xenopus laevis*. Chromosoma 60:147–167

Scheer U, Spring H, Trendelenburg MF (1979a) Organization of transcriptionally active chromatin in lampbrush chromosome loops. In: Busch H (ed) The cell nucleus, vol 7. Academic Press, New York, pp 3–47

Scheer U, Sommerville J, Bustin M (1979b) Injected histone antibodies interfere with transcription of lampbrush chromosome loops in oocytes of *Pleurodeles*. J Cell Sci 40:1–20

Scheer U, Zentgraf H, Sauer H (1981) Different chromatin structures in *Physarum polycephalum*. Chromosoma 84:279–290

Seebeck T, Braun R (1982) Organization of rDNA in chromatin: *Physarum*. In: Busch H, Rothblum L (eds) The cell nucleus, vol 11. Academic Press, New York, pp 177–191

Skoglund U, Andersson K, Björkroth B, Lamb MM, Daneholt B (1983) Visualization of the formation and transport of a specific hnRNP particle. Cell 34:847–855

Solari AJ (1974) The molecular organization of the chromatin fiber. In: Busch H (ed) The cell nucleus, vol 1. Academic Press, New York, pp 493–535

Sollner-Webb B, Wilkinson JK, Miller KG (1982) Transcription of Xenopus ribosomal RNA genes. In: Busch H, Rothblum L (eds) The cell nucleus, vol 12. Academic Press, New York, pp 31–67

Sommerville J (1981) Immunolocalization and structural organization of nascent RNP. In: Busch H (ed) The cell nucleus, vol 8. Academic Press, New York, pp 1–57

Spring H, Franke WW (1981) Transcriptionally active chromatin in loops of lampbrush chromosomes at physiological salt concentration as revealed by electron microscopy sections. Eur J Cell Biol 24:298–308

Spring H, Scheer U, Franke WW, Trendelenburg MF (1975) Lampbrush-type chromosomes in the primary nucleus of the green alga *Acetabularia mediterranea*. Chromosoma 50:25–43

Spring H, Krohne G, Franke WW, Scheer U, Trendelenburg MF (1976) Homogeneity and heterogeneity of sizes of transcriptional units and spacer regions in nucleolar genes of *Acetabularia*. Biol Cell 25:107–116

Trendelenburg MF (1981) Initiations of transcription at distinct promoter sites in spacer regions between pre-rRNA genes in oocytes of *Xenopus laevis*: an electron microscopic analysis. Biol Cell 42:1–12

Trendelenburg MF (1983) Progress in visualization of eukaryotic gene transcription. Hum Genet 63:197–215

Trendelenburg MF, McKinnell RG (1979) Transcriptionally active and inactive regions of nucleolar chromatin in amplified nucleoli of fully grown oocytes of hibernating frogs, *Rana pipiens* (amphibia, anura). Differentiation 15:73–95

Trendelenburg MF, Scheer U, Zentgraf H, Franke WW (1976) Heterogeneity of spacer lengths in circles of amplified ribosomal DNA of two insect species, *Dytiscus marginalis* and *Acheta domesticus*. J Mol Biol 108:453–470

Vavra KJ, Colavito-Shepanski M, Gorovsky MA (1982) Organization of rDNA in chromatin: *Tetrahymena*. In: Busch H, Rothblum L (eds) The cell nucleus, vol 11. Academic Press, New York, pp 193–223

Villard D, Fakan S (1978) Visualisation des complexes de transcription dans la chromatine étalée de mammifères: étude en autoradiographie à haute résolution. C R Acad Sci Paris 286D:777–780

Weiss E, Ghose D, Schultz P, Oudet P (1985) T-antigen is the only detectable protein on the nucleosome-free origin region of isolated simian virus 40 minichromosomes. Chromosoma 92:391–400

Widmer RM, Lucchini R, Lezzi M, Meyer B, Sogo JM, Edström J-E, Koller T (1984) Chromatin structure of a hyperactive secretory protein gene (in Balbiani ring 2) of *Chironomus*. EMBO J 3:1635–1641

Woodcock CLF (1973) Ultrastructure of inactive chromatin. J Cell Biol 59:368a

Woodcock CLF, Safer JP, Stanchfield JE (1976) Structural repeating units in chromatin. Exp Cell Res 97:101–110

Zentgraf H, Müller U, Franke WW (1980 a) Reversible in vitro packing of nucleosomal filaments into globular supranucleosomal units in chromatin of whole chick erythrocyte nuclei. Eur J Cell Biol 23:171–188

Zentgraf H, Müller U, Franke WW (1980 b) Supranucleosomal organization of sea urchin sperm chromatin in regularly arranged 40 to 50 nm large granular subunits. Eur J Cell Biol 20:254–264

Zentgraf H, Müller U, Scheer U, Franke WW (1981) Evidence for the existence of globular units in the supranucleosomal organization of chromatin. In: Schweiger HG (ed) International cell biology 1980–1981. Springer, Berlin Heidelberg New York, pp 139–151

Replication of DNA in Eukaryotic Chromosomes

J. H. TAYLOR [1]

1 Introduction

Studies of DNA replication in chromosomes first clearly revealed the compartments of the cell cycle and provided a means for measuring the parameters of the S-phase in the cell cycle. Radioisotopic labeling, especially with [^{3}H]-thymidine, provided a specific label easily detected by autoradiography, as well as by liquid scintillation counting for biochemical studies. The research utilizing these tools soon centered on what seemed to be two separate areas. One that occupied our attention for many years deals with the structural properties of chromosomes and the exchanges and aberrations that could be revealed. Most of these problems are now understood, at least in broad outline. The discovery of fluorescence and staining procedures to distinguish bromodeoxyuridine (BrdU) labeled segments from those with thymidine in chromosomal DNA stimulated the structural studies, particularly in the last 10–12 years. In this review I will spend little time discussing this area; for an extensive review of some aspects see Tice and Hollaender (1984).

The second area of study deals with what I choose to call the time compartments in the S-phase. Although this aspect of replication has stimulated much research activity, the full significance of the regulatory systems associated with these compartments is only now being appreciated. I briefly reviewed the topic not long ago (Taylor 1984), but several developments now make the concepts clearer. However, as I will outline below the structural units of replication in chromosomes, the replicons, still elude us. Yet, it is difficult to imagine the degree of control so far observed to operate without replicons and origins. Origins have been identified in the replicons of prokaryotes, in viral DNAs, in mitochondrial DNAs, and in chromosomal DNA of at least one eukaryote, yeast. Origins may have been identified in other eukaryotic chromosomes, but as will be shown below we lack the appropriate assay to verify these reports.

There is enough evidence to indicate that replication of DNA in many eukaryotes is compartmentalized even though the nucleus may have no morphological compartments. The compartments to which I refer are separated by time rather than structural barriers. After briefly reviewing the structural deductions from DNA replication studies, I will consider the time compartments and the regulatory aspects at more length.

[1] Institute of Molecular Biophysics and Department of Biological Science, Florida State University, Tallahassee, Florida 32306, USA.

Results and Problems in Cell Differentiation 14
Structure and Function of Eukaryotic Chromosomes
Edited by W. Hennig

2 Structural Deductions from Chromosomal DNA Replication

Chromosomes contain two longitudinal subunits with opposite polarity as originally discovered by labeling with [^{3}H]-thymidine (Taylor et al. 1957; Taylor 1958). The two subunits are the two chains of a DNA helix which separate during replication and form the two chromatids, each of which consists of a DNA helix and the associated proteins. Since the two subunits have opposite polarity, exchanges between them are prohibited, but exchanges occur between chromatids (sister chromatid exchanges) at a low frequency. The frequency can be enhanced by a variety of agents that break chromosomes and some others that induce very few exchanges between nonsister chromatids. The discovery of ways of distinguishing by fluorescence or staining chromatids with different amounts of bromodeoxyuridine (BrdU) substituted for thymidine has made the study of sister chromatid exchanges and the time of replication of different segments or bands easier to detect than with radiolabeling and autoradiography (see Tice and Hollaender 1984). The BrdU staining, which has also been demonstrated with antibodies to BrdU (Morstyn et al. 1983; Vogel et al. 1986; Speit and Vogel 1986), increases resolution and makes feasible the analysis of sister chromatid exchanges involving smaller segments and smaller chromosomes than had been possible by autoradiography. Some of the early controversy over apparent exceptions to the semiconservative distribution of subunits could be shown to be due to closely spaced sister chromatid exchanges as Callan (1972) had proposed. Segregation is regular and the assortment of labeled chromatids to daughter nuclei is random at the second and subsequent divisions after one round of replication in labeled thymidine.

3 Regulation of DNA Replication in Chromosomes

Two major unsolved problems of replication relate to regulation, i.e., control of rounds of replication per S-phase and time of replication over the S phase. Two types of regulatory mechanisms are evident. One insures a single replication of each locus per S-phase with rare exceptions. The other regulates the time in the cycle when a particular chromosomal segment will be replicated. Both regulatory mechanisms would appear to require specific origins from which replication is initiated. The degree to which the timing is controlled is subject to some debate, but the original studies of animal cells in culture by pulse labeling with [^{3}H]-thymidine indicated that one part of the complement is replicated in the first half of S-phase, while the remainder is replicated in the last half (Taylor 1960). Among the parts replicating in the last half are the inactive segments of one X chromosome in cells of mammalian females, most of the Y chromosome and about one-half of the segments of the autosomes, the G-bands (Dutrillaux et al. 1976). In human lymphocytes the whole of one X chromosome is replicated in the first 3 h of S-phase, while most of the other X chromosome is replicated in the last half of S phase (Petersen 1964). In 1961 Lyon published evidence that the heterochromatic X, which proved to be the late replicating X, was genetically inactive in the mouse. Many subsequent studies showed this correlation to be a general one for mammals. As I will argue below, the inactive X is a special case evolved to inactivate one allele of many of the X-linked genes and I think it utilizes a general regulatory system of eukaryotes that was probably established long before it was utilized for dosage compensation of X-linked genes.

The second replication of a region under the control of an origin could in principle be prevented by one of two general mechanisms, either by the formation of a stable complex at the origin during replication, which prevents a second round, or by the modification of the DNA at the origin. A suggestion that a modification could result from a delay in methylation at the origin after replication (Taylor 1978) is a possible mechanism except that some DNAs may lack methylation. The hemimethylated state at CpG sites in mammals does direct mismatch repair enzymes to replace the bases on the new, unmethylated chains (Hare and Taylor 1985). This evidence demonstrates that the hemimethylated state could provide a signal sufficient to prevent a second replication during an S-phase.

3.1 Viral Origins for Replications

Unlike the chromosomal origins, the origins of SV40 and other similar DNA viruses have evolved to be utilized many times in one S phase. However, when these genomes are integrated in the chromosomes, the origin is either rendered functionless or comes under a cis-acting control mechanism of the type that keeps replication to one round per S phase in chromosomes. In the replicative state, the SV40 genome is not methylated at its CpG sites, most of which are in the origin of replication or nearby in the adjacent promoter for the late genes. If these sites become methylated in the integrated state, that modification might be sufficient to inactivate the origin. However, some recent experiments by Roberts and Weintraub (1986) suggest that the regulation could involve cis-acting sequences in the chromosomes. Two cis-acting and at least one trans-acting factor can modulate the rate of replication and convert the SV40 origin in a plasmid to one that replicates once per cell cycle.

The bovine papalloma virus (BPV) genome is a small circular DNA of 7–8 kilobases (kb) that can replicate as a plasmid in mouse cells in culture. All of the replicative functions are contained in a 5.4 kb Hind III-BamH I fragment (Lusky and Botchan 1984). This segment contains the BPV replication origin, several transcription promoters, two known enhancers, and eight open reading frames, three of which encode proteins that modulate replication (Lusky and Botchan 1985). This portion of the BPV genome is maintained at about 150 copies per cell by replicating once each cell cycle under the control of two plasmid maintenance sequences. One is contained within a 541 bp segment (PMS I) separated by about 2.6 kb from the second maintenance sequence (PMS II) with 140 bp.

When Roberts and Weintraub joined a 550 bp segment of SV40, containing the enhancer, the origin, and the early and late promoters to the 5.4 kb fragment of BPV, they obtained a plasmid with a modulated replication cycle in Cos-7 cells. Cos-7 is a monkey cell line with an integrated partial copy of the SV40 genome that constitutively produces T antigen, but is nonpermissive for BPV replication. It will support the replication of plasmids that contain the SV40 origin many times per cell cycle. They compared the replication of the fused genome segments by co-transfecting the constructs (SV-BPV) with SV40 plasmids (pSV) that replicate uncontrolled in the Cos-7 cells. Equimolar amounts of pSV and SV-BPV were co-transfected into Cos-7 cells by calcium phosphate precipitation and total cellular DNA was isolated at 16, 24, and 48 h after transfection. The supercoiled plasmids were measured in Southern blots. The BPV segment reduced the replication of the attached SV40 origin about 20-fold. Even though it still replicated 1000–5000 copies, they think that is because the unregulated T-antigen-driven replication predominates until the BPV-encoded, trans-acting proteins accumulate to levels sufficient for regulation.

They used the above described technique to locate the cis- and trans-acting factors by making a variety of constructs containing deleted portions of the BPV segment linked to the SV40 origin and testing these against the uncontrolled pSV plasmids. Two cis-acting negative control (NCOR) sequences were located in the BPV genome that were active on the SV40 origin. One is located within the PMS I and the other is adjacent to the PMS II. A 24 bp repeated sequence within each NCOR was found which has twofold symmetry across 20 nucleotides (two helical turns), suggesting that these two sequences could serve as a recognition site for a protein dimer.

They also obtained some evidence that the 5′ end of the open reading frame E1 of the BPV genome codes for a trans-acting negative regulator for the hybrid genome as it does for the BPV genome (Berg et al. 1986). It is already known that the 3′ end defines a positive trans-acting factor necessary for replication of BPV (Lusky and Botchan 1985). Using this information Roberts and Weintraub (1986) were able to construct a plasmid, with the SV40 origin, the BPV cis- and trans-acting loci and a neomycin resistance factor, that was maintained as a multiple copy plasmid in a drug resistant Cos-7 cell line with a regulated replication of the hybrid plasmid. Although they have not grown the cells free of the antibiotic to see if the cell line is stable, the plasmid appears to replicate once per cell cycle. This model system of origin regulation may give some ideas of how the chromosomal origins operate. However, the mechanism by which the trans-acting factors interact differentially with the unreplicated and the newly replicated origins in an S-phase is not revealed.

3.2 The Nature of Chromosomal Origins for Replication

Although the origins of viral genomes may give clues about the chromosomal origins, they have evolved to support run-away replication in the proper cellular environment and therefore must escape any trans-acting factors produced by chromosomal genes. We must isolate and study the chromosomal origins to learn their characteristics. The lack of an appropriate assay for origins has frustrated attempts to isolate and clone origins except in yeast. There, the existence of a small plasmid with an origin similar to those of the chromosomes has made the isolation of functional origins from a number of sources possible. These sequences, called autonomous replication sequences (ARS), contain a concensus sequence (A/TTTTATA/GTTTA/T) (Broach et al. 1983), but may lack regulatory sequences of the type described above. Those isolated from other organisms have not been shown to act as origins in the donor cells.

A suggestion that the large family of repeats (Alu-type) that are dispersed throughout the genome of mammals (Jelinek et al. 1980) may be origins has some experimental support. They have maintained a highly conserved region of 9–14 bp homologous to the SV40 origin. Johnson and Jelinek (1986) were able to demonstrate a limited replication of a plasmid containing one of the Alu repeats, BLUR 8, when transferred to Cos cells which have a constitutive production of T antigen known to be essential for the replication of SV40 DNA. The vector plasmid, pBR322, failed to replicate in the same cells. These experiments do not indicate necessarily that Alu repeats serve as chromosomal origins, but provide a possible assay for a cloned chromosomal origin that can be transferred by techniques used for DNA transformation.

The major problem in isolating chromosomal origins has been the lack of a suitable assay. Several researchers have injected plasmids into *Xenopus* eggs, but the egg appears to be too permissive (Mechali and Kearsey 1984); any DNA will replicate and the differences in efficiencies reported (Watanabe and Taylor 1980; Hirago et al. 1982; Chambers and Taylor 1982) may not be indicative of functional origins. Recent attempts to demonstrate replication of plasmids containing the repeats found in the presumptive origins cloned from the *Xenopus* genome in similar transient transformation experiments failed (Riggs and Taylor, unpublished). However, without a positive control, the adequacy of this test remains uncertain. The problem of studying the replication of plasmids in higher eukaryotic cells is their tendancy to integrate or be lost at division because of their failure to move on the spindle. An exception is the circular genome of the bovine papilloma virus, which replicates in cultured mouse cells as a plasmid. The genome has two maintenance sequences which may be equivalent to origins. If either of the maintenance sequences is removed, the BPV plasmid can still replicate but it becomes subject to integration. The intact plasmid is maintained as a separate small ring of DNA at a copy number of about 150 (Lusky and Botchan 1984). It replicates in synchrony with the chromosomes once each S phase. However, in certain skin cells in vivo it produces virus in what must be a run-away type of replication like other DNA viruses. As indicated above this plasmid may be useful as a model for chromosomal origins.

A potential origin of replication has been identified by Burhans et al. (1986) near the dihydrofolate reductase (DHFR) gene in Chinese hamster cells. In a cell line with about 1000 copies of the gene they have identified a region downstream of the gene that replicates first in the early S-phase. Among several cloned segments from this region, one hybridizes preferentially with pulse-labeled Okazaki fragments produced during the first few minutes of S-phase. Since the adjacent cloned segments hybridize to a greater extent with fragments produced later in S phase they think the origin is localized in this segment of about 1500 bp, which replicates first. However, no other functional test has been devised to verify the hypothesis.

4 The Timing of DNA Replication over the S-Phase

Early attempts to demonstrate that replicons exhibit a precise sequence in the S-phase were unsuccessful (Amaldi et al. 1972; Adegoke and Taylor 1977). These results led me to the conclusion that replicons may not have a precise time for replication except that all early replicating DNA must finish before the pause (Fig. 1) that usually occurs before the transition to late S. Origins probably have different affinities for the replication complex of proteins. Those with the highest affinities will replicate very early, but even some of these may not follow a precise sequence. I am aware of two studies that could be an exception and may indicate a more precise control. One is a study of the timing of replication in the slime mold, *Physarum,* which has a multinucleate plasmodium. The nuclei divide in synchrony and transcription appears to begin exclusively, or preferentially, in the replication eyes. Three actin loci replicate during the first 8–10 min of S-phase while the fourth replicates 80–90 min into S-phase, which in this species is late S. The chronology is invariant in consecutive cell cycles (Pierron et al. 1984). It is possible that in this species mRNAs for transacting initiation factors depends on the activation of a previously replicated set of replicons as S-phase proceeds. The other study by Schildkraut (personal communication) shows that in several mouse cell lines a cluster of immunoglobulin, IgC_H, genes replicate in the following order: α, ε, γ2a, γ2b, γ1, γ3, δ, and μ, followed by J and D segments. The genes with that order in the chromosome replicate during the first half of S-phase in erythroleukemia (MEL) cells, in fibroblasts, hepatoma cells, and a T-cell lymphoma. In contrast, in two plasmacytoma lines and two leukemia transformed pre-B cell lines, all genes replicate in early S with no obvious temporal order. The whole cluster comprises about 300 kb DNA that could be a huge replicon or possibly a cluster of replicons that fire in sequence in some cells.

All of these examples can be accomodated in a model with an early S-phase (S_E) and a late S-phase (S_L) with a gap in synthesis between as indicated by pulse labeling across S-phase (Stubblefield et al. 1967; Collins 1978; Schempp and Vo-

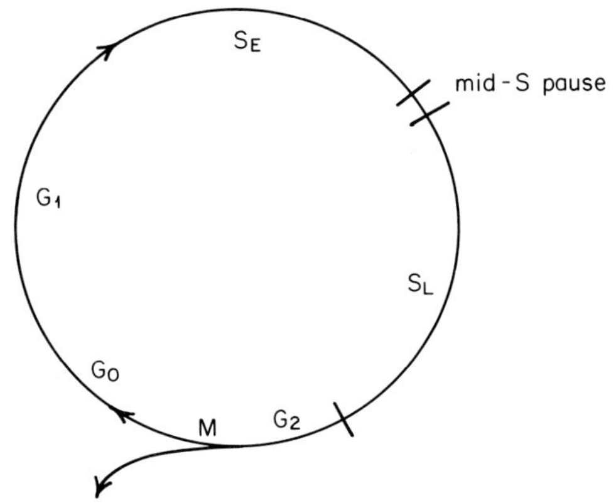

Fig. 1. Cell cycle showing the hypothetical pause at mid-S-phase as indicated by evidence presented in the text

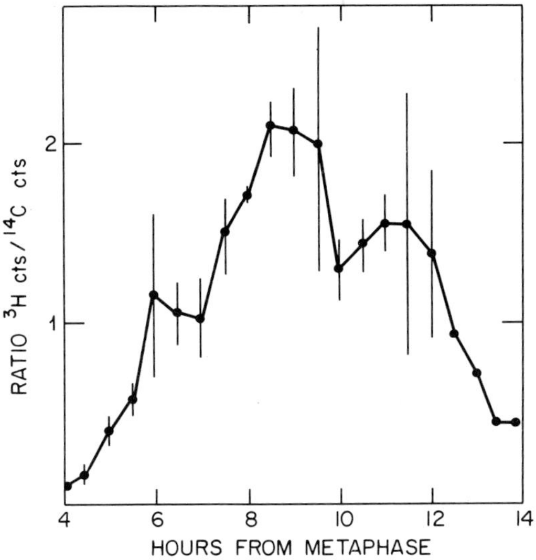

Fig. 2. Graph showing incorporation of [^{3}H]-thymidine into DNA of Chinese hamster cells over the S-phase. Data were obtained by prelabeling the DNA with [^{14}C]-thymidine and selecting cells by shaking off dividing cells. No other treatment was used to synchronize cells after their removal from the original cultures. Cells were distributed to small flasks and pulse labeled for 10 min with [^{3}H]-thymidine at 30 min intervals beginning 4 h from metaphase. Data from four independent experiments are combined and the *vertical lines* indicate standard errors of the means. The S-phase was measured to be about 7 h in other experiments; the longer interval indicated by these data represents asynchrony of the transit through S-phase. The dip in the rate of synthesis that reaches a maximum at 10 h after division represents the pause in mid-S. The small dip on the ascending curve at 7 h has not been regularly observed and considering the large standard error of the preceding point it probably does not represent a second pause. (From unpublished data collected in my lab by N. M. Straubing in 1969)

gel 1978; Schmidt 1980). The first two studies indicated a pause between the first half of S-phase (S_E) and the second half (S_L) when synchronized cells in culture were pulse labeled with [^{3}H]-thymidine, and collected at division over short intervals for counting of the radioactivity in the DNA (see Fig. 2 for the results of a similar experiment). In the last two studies synchronized cells were labeled for 1 h with BrdU and dividing cells collected at hourly intervals. The chromosomes were examined by microscopy after staining for replication bands by the 33258-Hoechst-Giemsa method (Perry and Wolf 1974; D'Andrea et al. 1973). Those bands that have replicated in the presence of BrdU lose their dark staining capacity after this treatment. Although there is some replication during the mid-S pause, the bands that are typically either early or late do not replicate in that interval.

The above studies indicate at least two time compartments, which would allow a gene or segment of a chromosome to form complexes with certain DNA-binding proteins if replicated in one compartment. If, in another cell, that same segment replicates in another time compartment, the complexes formed could be different if the proteins available at that time were not the same. One complex might be compatible with transcription of the segment, while the other would not be. As long as replication of the chromosomal segment remains in the same compartment, each cycle will regenerate either the potentially functional or the repressed state of the segment.

The X chromosomes of mammals would represent a specialized use of these two systems of general regulation to suppress one allele of most of the X-linked genes in cells of females. In this way dosage compensation can occur when the males have one set of the genes and females have two.

5 Possible Factors that Suppress Late Replicating X Chromosome Genes

As mentioned above, I would like to propose that S-phase in most cells consists of two compartments, early replicating (S_E) and late replicating (S_L) periods usually interrupted by a short interval. This idea has considerable evidence to support it, but there appear to be exceptions and it should be considered a working hypothesis. There may be satellite DNAs that can be programmed outside these two principal compartments, since they may not have promoters for transcription. In addition, the rapidly dividing cells in some early embryos probably do not have these compartments. The only regulation necessary at that time may be the one that prevents two rounds of replication per S-phase. Probably origins are preprogrammed to replicate either early or late when the specific factors that differentially bind origins are first produced and cells begin to differentiate. Heterochromatin begins to appear. Brown (1984) has given us a model system in describing the transcription factors of the 5S RNA genes. Could heterochromatin be DNA without or with a reduced level of transcription factors? Some origins may be modified to respond to replication complexes of either early or late S-phase, perhaps others cannot. Before commenting further on these ideas, let us briefly review the way in which the late replicating X chromosome has developed.

The out-of-phase replication of the two X chromosomes, or the differential segments of these chromosomes, was the first example studied because the segments involved are so large and the differential gene expression had already been predicted by Lyon (1961) before she knew about the differential replication. She related the gene inactivation to the heterochromatic state of one of the X chromosomes that had been discovered by Barr and Bertram (1949). When the differential replication was revealed (Taylor 1960; German 1960) more attention was given to the fact that the differential segment continued replication longer than most other segments in the nucleus. For this reason the X could be identified in the human complement before the chromosome banding techniques were available. However, I was impressed by the completely out-of-phase replication in the Chinese hamster cells and suggested that Petersen (1964), a visiting scientist in my lab, study the timing of the X chromosome replication in human lymphocytes. Like the Chinese hamster, the study showed the complete asynchrony with one X chromosome replicated in the first 3 h and the other replicated in the last 3 h with no apparent overlap. We thought the asynchrony might have a connection with inactivation, but the mechanism was difficult to imagine at that time. The hope was that the nature of heterochromatin and its connection to gene inactivation might be revealed. Davidson et al. (1963) demonstrated that a gene, encoding glucose phosphate dehydrogenase (G6PD) is inactivated on the late replicating X, but many years passed before experimental evidence concerning the mechanism of inactivation became available. Riggs (1975) had proposed that the inactivation could be related to differential DNA methylation. The first break came with the demonstration by Mohandas et al. (1981) that the inactive genes on the late replicating X chromosome could be activated by treating the cells in culture with azacytidine. This drug is converted to deoxyazacytidylate and incorporated into DNA in place of deoxycytidylate. When DNA methylase reacts with this analog it not only fails to methylate that site, but is bound in an inactive complex perhaps long enough to prevent methylation at adjacent sites (for review see Cantoni and Razin 1985). The demonstration was made in a cell hybrid produced by fusing a human cell with a mouse cell and selecting a cell line without HPRT (hypoxanthine phosphoribosyl transferase), but with an inactive X chromosome marked by a translocation. When cells were selected for HPRT, two other X-linked genes, encoding G-6PD and PGK were activated at a frequency higher than when considered as independent events. However, the X chromosome did not become early replicating and the time of replication of the reactivated genes was not determined. In fact the change in methylation of the reactivated loci could not be determined. When Wolf and Migeon (1982) tried similar treatment on mouse cells in culture, they were unable to obtain reactivation of the same loci. Recently, Migeon et al. (1986) succeeded by producing a hybrid of mouse A9 cells ($HPRT^-$) with human chorionic villi cells. With these cells it was not necessary to treat with azacytidine, but only to select cells that were $HPRT^+$. Although the chorionic villi have a late replicating X chromosome, the inactivation is less stringent than in fibroblasts derived from the embryo. They found some of the genes in the chorionic villi cells to be expressed at a low level.

Shafer and Priest (1984) reported rather extensive changes in the time of replication of a number of chromosomal bands, as well as the late replicating X, in cells treated with azacytidine and examined a few divisions later. It is possible that in the experiments of Mohandas et al. (1981) and Wolf and Migeon (1982), the long period of selection necessary to isolate the reactivated cell lines resulted in the loss of cells with extensive changes in replication patterns. Certainly azacytidine is lethal to many of the treated cells. Recently, Jablonka et al. (1985) reported the advance in time of replication of the differential segment of late replicating chromosome of the vole, *Gerbillus gerbillus,* after treating the cells in culture with azacytidine. Although they had no mutant genes to check for reactivation, they found that the X-chromatin of the shifted chromosome had become sensitive to DNase I similar to other transcribable segments. Their success may also be due to the fact that they examined the cells a few cell cycles after the treatment rather than after many cell cycles required to isolate clones of cells from those originally treated.

These experiments indicate that azacytidine advances the time of replication of segments of the late replicating, inactive X. At the same time, certain suppressed genes are reactivated. Is the reactivation due to demethylation of the of specific sites in the genes or to the shift in time of replication? It could be both, but measurements by Jablonka et al. (1985) indicated an overall reduction of DNA methylation of only 15%. Activation of a gene encoding a trans-acting factor for replication could also be involved. No definitive answer to these intriguing questions is likely to be available until we are able to clone origins and find functional assays. However, there is evidence that trans-acting factors are involved in the control of time of replication.

Somssich et al. (1984) have demonstrated a shift in the time of replication of band E in chromosome 15 of the mouse and their observations suggest the action of trans-acting factors. The E-band contains about 10,000 kb DNA and normally replicates early in S-phase in most cells in culture. In certain T-cell tumors it has shifted to later replication. In hybrids produced by fusing tumor cells with non-tumor cells, the clones derived varied from highly tumorogenic to others of low tumorogenicity, measured by the ability to establish tumors in recipient mice. In a number of the cell hybrids the source of the chromosomes could be determined by having one marked with a translocation. Regardless of the source, the E-band on different chromosomes always replicated at the same time in the S-phase, either early or late in any one cell. This clearly indicates that the time of replication of this large segment (band E), consisting of many replicons at the molecular level, is regulated by trans-acting factors. The location of these factors could either be on chromosome 15 or possibly some other chromosomes of the complement. The significance of the shift in time of replication in terms of gene expression is not so clear. The shift to later replication might be expected to suppress some genes in the band, but the only known locus nearby appears to be the c-myc gene involved in the production of plasmacytomas and correlated with the 15D3/E translocation in which the distal part of the chromosome is moved to chromosome 6 or 12. According to Somssich et al. (1986), the shift in time of replication of the E-band is not from S_E to S_L, but represents a disappearance of the typical early replication band detected by BrdU incorporation in the last half of S-phase. These replication bands are different from either R-bands or G-bands (von Kiel et al. 1986). They examined ten chromosomal regions in man and the domestic cat with known genetic homology. In three of the regions replication banding and G banding are similar, but in seven chromosomal regions the homology seen with replication banding is greater than if G banding is compared. Likewise, R banding fails to reveal the same degree of homology as replication banding (replication banding refers to bands revealed by growing cells in BrdU during the last half of S-phase and staining by a procedure that makes dark bands in thymine-containing DNA replicated before BrdU was available). Two plasmacytoma cell lines from mice, J558 and MPC-11, were examined but did not have the variant replication pattern seen in the T-cell tumors. This situation may be comparable to the one to be described in more detail below by Brown et al. (1986) in which the cluster of immunoglobulin constant region genes are replicated very early in cells where they are functional, but in fibroblasts, hepatoma cells, MEL cells, and T-cells where the genes are not transcribed, they are replicated over a 3 h period from early to mid-S in the same sequence as their map position in the chromosome. Certainly the correlation between R bands and early replicating DNA and G-bands and late replicating DNA is not very precise. Whatever the explanation, shifts in time of replication of bands as large as 15E must be rare or these would have been discovered in comparing replication patterns of different tissues. Perhaps the shifts usually involve smaller units of the type described by Calza et al. (1984), in which small clusters, or rarely single large genes are involved in the shift.

With the above background, we can consider the mechanisms of regulation. The time of replication could be regulated by a combination of cis- and trans-acting factors analogous to the small BPV genome. The azacytidine could activate a gene, or genes, that produce the trans-acting factors necessary to make the shift in the time of replication in the X chromosome. This assumption still requires a difference between the two X chromosomes, perhaps at the level of the origins of each replicon, that could make the two types respond differently to trans-acting factors. However, it may be easier to imagine a mechanism that develops slowly to program an X before the switch becomes effective at the time trans-acting factors are produced. The switch from isocyclic to heterocyclic replication occurs during the course of one cell cycle at various times during early development. It should not be surprising that the reversal can occur in the same time interval whatever the mechanism. Like other events in development, there may be a determination stage followed by several to many divisions later by the differentiation which in this instance is a switch to late replication with consequent suppression of the genes. The switch in time of replication appears to be induced more readily in cell hybrids (Mohandas et al. 1981; Migeon et al. 1986), where both trans- and cis-acting factors may not operate as effectively as in the balanced differentiated diploid genome. The only cells besides the cell hybrids in which Somssich et al. (1984) found variants in the replication of 15E were embryonic liver cells and teratocarcinoma cells, both of which are capable of differentiation. The individual genes reactivated by azacytidine could be due to changes in methylation in their promoters, but to make this change effective a switch in the time of replication might also be required. If the time of replication turns out to have an important role in gene regulation, there are likely to be complex regulatory systems that can be disengaged at different levels to prevent irregular and spontaneous shifts. Producing cell hybrids between species might disengage one level and make the system more susceptible to manipulation. Certainly, the mechanism regulating the X chromosome is far from being solved, but I predict that it evolved in a genetic background in which the general mechanism of suppression in relation to time of replication was already established.

6 Possible Mechanisms of Suppression of Late Replicating DNA

Now we are ready to examine the clues which are available to indicate how the differential suppression of genes operates in connection with time of replication. There is no reason to think that cells utilize this system for the day to day operations of gene regulation. Only those loci that have to be used rarely in the life cycle or may not be needed again in the life of the organism would be relegated to a general suppression of the type visualized here. The expression of the 5S RNA genes in *Xenopus* is an example that has been reviewed by Brown (1984).

Xenopus has about 20,000 copies of the 5S RNA gene of the oocyte type and 400 of the somatic type. All are expressed during oogenesis when many copies of 5S RNA are required in the growth of the oocyte, but only the 400 somatic-type genes are expressed in most somatic cells. The genes have similar sequences, but are different enough so that the somatic type binds the transcription factor, designated TFIIIA, with a greater affinity than the oocyte type. The promoter for the 5S RNA genes is an internal one and to be transcribed by RNA polymerase III each gene must be complexed with TFIIIA and two other transcription factors, B and C. The TFIIIA is thought to bind first and the complex is completed by the addition of factors B and C. If the DNA is already complexed with all of the histones, the transcription factors do not compete effectively to make the genes transcribable. In the oocyte the transcription factors are very abundant, about 3×10^{10} molecules per cell or 5×10^{5} molecules per 5S gene. In somatic cells there is less than one molecule per 5S RNA gene. Because of the shortage of transcription factor molecules and their higher affinity for the somatic-type genes, only somatic-type 5S RNA is transcribed in most cells. Once bound to the genes the factors form a stable complex that is not displaced by histones. Many rounds of transcription can occur without disruption of the complex. However, in the absence of transcription factors in the medium a single round of replication disrupts the complex and the gene becomes inactive in the presence of histones (Bogenhagen et al. 1982; Brown and Schlissel 1985).

To make the system operate more effectively the cell has evolved an S-phase regulation in which the somatic-type genes are replicated in early S-phase, while the oocyte type are replicated later. Guinta and Korn (1986) used a fluorescence-activated cell sorter to separate growing cells of *Xenopus* into different fractions of the cell cycle. They measured the length of the cell cycle of cells in culture to be 40 h with an S-phase estimated to be 9 h. An exponentially growing population of cells was labeled with BrdU for 3 h and then fixed and stained with propidium iodide. The cells were sorted into G1, S1, S2, S3, S4, and finally G2 plus the mitotic group. The DNA was isolated from each fraction and separated by buoyant density in a cesium chloride gradient. Fractions were collected and applied to nitrocellulose filters in a slot blot apparatus. The DNA was then hybridized to a [^{32}P]-labeled probe specific for somatic 5S RNA genes. The fraction with highest buoyant density hybridized when it was obtained from cells in the first quarter of S-phase, but not from those at later stages of S-phase. The time of replication of the oocyte type is not as restricted; some heavy-light DNA hybridized from cells selected to be in the first quarter of S-phase. However, most of the oocyte-type 5S RNA genes replicated in the last quarter of S-phase with a smaller fraction in the third quarter. Replication could be restricted to the last half of S-phase considering that the sorting of cells on the basis of staining may not be perfect. In fact, another study of the same system by Gilbert (1986), who used similar techniques shows that most of the oocyte type genes replicate in the last quarter of S-phase, with very little if any replicated in the first half.

7 Are Time Compartments Very Significant in Gene Regulation?

There are many indications from genetic studies that heterochromatin may be inactive and much of the heterochromatin replicates late in S-phase (Lima-de-Faria and Jaworska 1968). Now that we distinguish so many types of DNA and control of the timing of replication may be at the level of replicons, it is probably not useful to think in terms of heterochromatin if we can be more precise in our definitions. I have suggested that it may be useful to consider two major compartments for replication. This does not imply that all DNA must fall into one of these classes. Some DNAs, particularly repetitive DNAs, may not operate under the same regulatory system. Certain replicons are switchable; others may not be. I would expect that to be a property of the origins. Assuming that there are origins, these sequences may operate in clusters, either responding to common trans-acting factors or possibly shifting independently of the cluster in some instances. If the time compartments prove to be of significance most of these questions will be answered experimentally in the future. Now there is only enough information to make them interesting for those who are willing to speculate.

There are several studies which indicate that the potential for expression of some genes is correlated with the time of replication as implied throughout this review. Considerable attention has been given to the globin genes (Furst et al. 1981; Goldman et al. 1984), but there are many unanswered questions about how the cluster operates. In erythroleukemic cells the cluster appears to be early replicating even though some are not inducible. In at least one cell, HeLa, the β-globin gene is replicated in late S. All of the constitutively active genes examined in several cell types are early replicating. In fibroblasts the genes that are expressed only in specialized tissues are late replicating. The time of replication in most tissues in which these genes are expressed has not been examined because appropriate cell lines are not available for synchronization of the division cycle. One exception is immunoglobulin heavy and light chains of the mouse studied by Calza et al. (1984) and Brown et al. (1986).

Brown et al. studied the time of replication of the murine immunoglobulin heavy chain constant region genes (IgC_H) in several cell lines. In MEL cells, fibroblasts, hepatoma cells, and T-cell lines, this cluster of genes replicate in the first half of S-phase. Furthermore, they replicate in the order of their sequence along the chromosome which is α, ε, γ2a, γ2b, γ1, γ3, δ, and μ in these cell lines, but in Ig-producing cells and pre-B lymphoid cells the same cluster replicates early with no detectable directionality.

Recently, we have approached the problem in a different way (Sturm and Taylor, unpublished). A retroviral vector, SNV plasmid (derived from avian spleen necrosis virus), has been constructed that will move the herpes simplex virus thymidine kinase (HSV TK) gene into mammalian cells (Watanabe and Temin 1983). The long terminal repeats in the vector are intact and downstream of the 5′ one is the encapsidation site. The HSV TK gene has been inserted intact except the polyadenylation region had to be removed to allow the full-sized modified genome to be transcribed and processed efficiently for encapsidation (Fig. 3). The other viral genes were removed and the modified viral genome with the TK gene

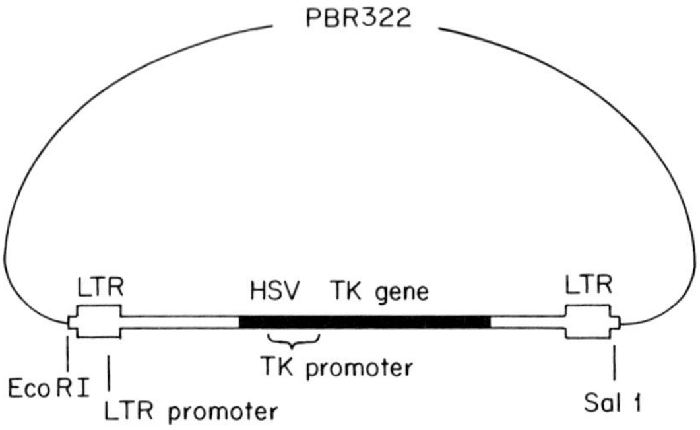

Fig. 3. Diagram of the genome of the retroviral vector (an avian spleen necrosis virus) for the HSV TK gene cloned into the bacterial plasmid, pBR322. (After Watanabe and Temin 1982)

was inserted into the bacterial plasmid pBR322 so that it could be amplified in *E. coli*. The plasmid was then transfected into chicken cells along with a plasmid containing a helper virus (a slow replicator) so that the modified viral genome with the TK gene was coated and released into the medium at a ratio of about 10:1 to helper virus. Buffalo rat cells (TK^-) can be infected with the viral mixture. Although these cells are nonpermissive for virus reproduction, the modified retroviral genome with the TK gene is efficiently integrated into the buffalo rat chromosomes. At low multiplicity of infection usually only one, or rarely two, copies of the TK gene are integrated into chromosomes of any one cell.

We selected transformed cells in HMT (hypoxanthine, methotrexate, thymidine) medium. If cells do not have thymidine kinase they quickly die in this medium. Once we had established that transformants were frequent, we isolated clones from nonselected cells that had been transfected and screened them by Southern blots to detect the HSV TK gene. Of 86 clones screened, 26 were found to have the gene. When the isolates were grown in HMT medium, only two were capable of growth. In the others the gene is defective or cryptic. Of 21 clones which tested positive for the sequences of the TK gene, eight were normal sequences on the basis of digestion with six different restriction endonucleases. We call these clones with a normal integrated TK gene, cryptic clones, because the apparently normal gene is hidden, i.e., not expressed. There is no reason to believe that all of these genes are defective in nucleotide sequence. One is certainly not since it reverts with a frequency greater than 10^{-6}, while most other cryptic clones revert at frequencies of less than 10^{-7}. The clone with the high spontaneous rate can be increased 1000-fold by treatment with azacytidine for one or two cell cycles. This observation suggests that this copy of the gene has become methylated and that the azacytidine leads to expression by inhibition of methylation of the new copies. We have also determined that this cryptic copy is replicated in early S-phase, both before and after activation by azacytosine. The time of replication was determined by a technique described by D'Andrea et al. (1983). Synchronous or asynchronous cells were pulsed labeled for 2 h with BrdU. If synchronized cells were used, replicate cultures were prepared and two were lysed immediately after the pulse and at each 2 h interval thereafter. The DNA was extracted and irradiated with long wavelength UV light after staining with Hoechst's dye. This treatment breaks the BrdU-containing DNA about 20 times as frequently as the thymine-containing DNA. Appropriate digestion with S_1 nuclease completes the break of both strands in the DNA labeled during the pulse. When Southern blots are prepared the bands are reduced in intensity, because of this breakage, in proportion to the amount of newly replicated DNA in each lane. Densitometer tracings indicate the fractions in which the TK gene was replicating most rapidly.

All of the functional genes examined are replicated in early S. So far, all of the cryptic genes are also early replicating genes. Although our working hypothesis had been that the viral vector would be integrated at random, it looks as if early replicating sequences are highly preferred. If that is correct, these observations indicate that only early replicating sequences can be expressed, and the virus has evolved integration mechanisms that favor transcribable sites. However, not all early replicating sites are equally available to RNA polymerases. Perhaps early replication is a prerequisite for gene expression, but a number of other levels of control are likely to operate that modulate or prevent expression of early replicating genes.

8 Transcription Factors for Polymerase II in Mammalian Cells

If we use the *Xenopus* 5S RNA genes as a model we might expect that the difference between early and late replicating DNA is that the former is complexed with transcription factors, while the latter is not. Maybe the factors are always in limited supply and are synthesized only in early S-phase. Of course, the situation could be much more complex. Jones et al. (1985) have found that the HSV TK gene requires at least three transcription factors which bind to specific regions of the promoter region (Fig. 4). Two Sp1 factors bind to regions that bracket a third segment binding a factor that operates with the TK gene, but not with several other promoters tried in an in vitro system. Sp1 was discovered to be essential for the transcription of the SV40 early genes where several binding sites are available. I know of no experimental evidence to indicate the affinity of these transcription factors relative to histones, but it is possible that once the sites are occupied, a stable complex is formed which makes the early replicating gene transcribable when the appropriate factors are available.

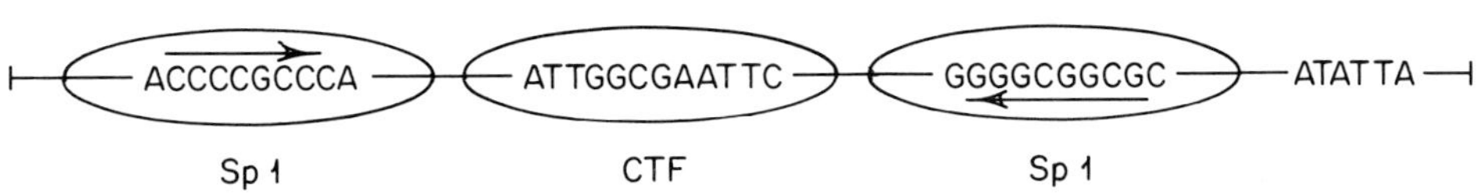

Fig. 4. Diagram of the region of the promoter of the HSV TK gene which binds transcription factors and makes the gene transcribable by RNA polymerase II. The *arrows* indicate reverse repeats that bind the same factor as the 21 bp repeats of SV40. The middle factor designated CTF binds at the CAAT box and all are 5′ to the TATTA box. (After Jones et al. 1985)

It is conceivable that transcription factors different from those produced in early S are also produced in late S-phase. These could bind a subset of the promoters of genes replicated in late S and make them available at certain stages in development without the necessity of switching their time of replication. The fact that evolution has preserved the late replicating genes rather than eliminating them as in chromosome diminution (W. Beerman 1953 and S. Beerman 1966) suggests that these late replicating genes may have retained some functions. We will require much more data on such regulatory mechanisms in order to make predictions about the potential for transcription of individual genes or gene clusters. Even when we know when they replicate and whether switches occur frequently or rarely during cell determination stages, we will have only a fraction of the necessary information. To predict transcription of genes in cell hybrids, for example, one needs to know the history of the gene, at least the time of replication in the parental cells and their potential for transcription in those cells. Does the time of replication change in a hybrid produced by cell fusion? For example, if a cell with a late replicating β-globin cluster is fused with an erythroleukemic cell with an early replicating β-globin cluster, what will happen? Will trans-acting factors switch one or the other of the clusters, or will both be stable? How will DNA methylation affect the genes? Perhaps methylation of CpG in a binding site for a transcription factor will change the binding affinity. In other cases the methylation may act at the level of replication timing, presumably by modifying the origin of replication and affecting the formation of a replication complex. We can only guess now, but at last we have possible molecular mechanisms for differentiation related to DNA replication that for so long was an imponderable mystery.

Acknowledgments. The preparation of this review was supported in part by a contract with the U.S. Department of Energy, Health Effects Research Division (DE-AS05-78 EV05854), and a grant from the National Institute on Aging (2R01AG01807-06).

References

Adegoke JA, Taylor JH (1977) Sequence programming of DNA replication over the S phase of Chinese hamster cells. Exp Cell Res 104:47–54

Amaldi F, Buongiorno-Nardelli F, Carnevalli F, Leoni L, Mariotti D, Pomponi M (1972) Replicon origins in Chinese hamster cell DNA. II Reproducibility. Exp Cell Res 80:79–87

Barr ML, Bertram E (1949) A morphological distinction between neurons of the male and female, and the behavior of the nucleolar satellite during accelerated nucleoprotein synthesis. Nature 163:676–677

Beerman S (1966) A quantitative study of chromatin diminution in embryonic mitoses of *Cyclops furcifer*. Genetics 54:567–576

Beerman W (1956) Nuclear differentiation and functional morphology of chromosomes. Cold Spring Harbor Symp Quant Biol 21:217–232

Berg L, Singh K, Botchan M (1986) Complementation of a BPV low copy mutant: evidence for a temporal requirement of the complementing gene. Mol Cell Biol 6:859–869

Bogenhagen DF, Wormington WM, Brown DD (1982) Stable transcription complexes of Xenopus 5S RNA genes: a means to maintain the differentiated state. Cell 28:413–421

Broach JR, Li Y-Y, Feldman J, Jayaram M, Abraham J, Nasmyth KA, Hick JB (1983) Localization and sequence analysis of yeast origins of DNA replication. Cold Spring Harbor Symp Quant Biol 47:1165–1173

Brown DD (1984) The role of stable complexes that repress and activate eucaryotic genes (a review). Cell 37:359–365
Brown DD, Schlissel MS (1985) A positive transcription factor controls the differential expression of two 5S RNA genes. Cell 42:759–767
Brown EE, Iqbal MA, Stuart S, Schildkraut CL (1986) Evidence that the murine immunoglobulin heavy chain constant region is part of an unusually large replicon. Mol Cel Biol 7:450–457
Burhans WC, Selegue JE, Heintz NH (1986) Isolation of the origin of replication associated with the amplified Chinese hamster dihydrofolate reductase domain. Proc Natl Acad Sci USA 83:7790–7794
Callan HG (1972) Replication of DNA in chromosomes of eukaryotes. Proc Roy Soc Lond B 181:19–41
Calza S, Adams JM, DelGindica T, Schildkraut CL (1984) Changes in gene position are accompanied by a change in time of replication. Cell 36:689–696
Cantoni GL, Razin A (1985) Biochemistry and biology of DNA methylation. Liss, New York
Chambers JC, Watanabe S, Taylor JH (1982) Dissection of a replication origin in Xenopus DNA. Proc Natl Acad Sci USA 79:5572–5576
Collins JM (1978) Rates of DNA synthesis during the S phase of HeLa cells. J Biol Chem 253:8570–8577
D'Andrea AD, Tantravahi V, Lalande M, Perle MA, Latt SA (1983) High resolution analysis of the timing of replication of specific DNA sequences during S phase of mammalian cells. Nucl Acids Res 11:4753–4774
Davidson RG, Nitowski HM, Childs B (1963) Demonstration of two populations of cells in a human female heterozygous for glucose-6-phosphate dehydrogenase variants. Proc Natl Acad Sci USA 50:481–485
Dutrillaux B, Courturier J, Richer C-L, Viegas-Pequignot E (1976) Sequence of DNA replication in 277 R- and Q-bands of human chromosomes using BrdU treatment. Chromosoma 58:51–61
Furst A, Brown EH, Braunstein JD, Schildkraut CL (1981) β-Globin sequences are located in a region of early-replicating DNA in murine erythroleukemia cells. Proc Natl Acad Sci USA 78:1023–1027
German J (1962) DNA synthesis in human chromosomes. Trans NY Acad Sci 24:395–407
Gilbert DM (1986) Temporal order of replication of *Xenopus laevis* 5S ribosomal RNA genes in somatic cells. Proc Natl Acad Sci USA 83:2924–2928
Goldman MA, Holmquist GP, Gray MC, Caston LA, Nag A (1984) Replication timing of genes and middle repetitive sequences. Science 224:686–692
Guinta DR, Korn LJ (1986) Differential order of replication of Xenopus 5S RNA genes. Mol Cell Biol 6:2536–2542
Hare JT, Taylor JH (1985) One role for DNA methylation in vertebrate cells is strand discrimination in mismatch repair. Proc Natl Acad Sci USA 82:7350–7354
Hiraga S, Suto T, Yoshida M, Kubota H, Ueyanna H (1982) In vitro replication of recombinant plasmids carrying chromosomal segments of *Xenopus laevis*. Proc Natl Acad Sci USA 79:3697–3701
Jablonka E, Goitein R, Marcus M, Cedar H (1985) DNA hypomethylation causes an increase in DNase-I sensitivity and an advance in the time of replication of the entire inactive X chromosome. Chromosoma 93:152–156
Jelinek WR, Toomey TP, Leinwand L, Duncan CH, Biro PA, Choudary PV, Weisman SM, Rubin CM, Houck CM, Deininger PL, Schmid CW (1980) Ubiquitous, interspersed repeated sequences in mammalian genomes. Proc Natl Acad Sci USA 77:1398–1402
Johnson EM, Jelinek WR (1986) Replication of a plasmid bearing a human Alu-family repeat in monkey Cos 7 cells. Proc Natl Acad Sci USA 83:4660–4664
Jones KA, Yamamoto KR, Tjian R (1985) Two distinct transcription factors bind to the HSV thymidine kinase promoter in vitro. Cell 42:559–572
Kiel K von, Hameister H, Somssich IE, Adolph S (1986) Early replication banding reveals a strongly conserved functional pattern in mammalian chromosomes. Chromosoma 93:69–76
Lima-de-Faria A, Jaworska H (1968) Late DNA synthesis in heterochromatin. Nature 217:138–142

Lusky M, Botchan M (1984) Characterization of the bovine papilloma virus plasmid maintenance sequences. Cell 36:391–401
Lusky M, Botchan M (1985) Genetic analysis of bovine papilloma plasmid type I trans-acting replication factors. J Virol 53:955–965
Lyon MF (1961) Gene action in the X chromosome of the mouse (*Mus musculus* L). Nature 190:372–373
Mechali M, Kearsey S (1984) Lack of specific sequence requirement for DNA replication in Xenopus eggs compared with high sequence specificity in yeast. Cell 38:55–64
Migeon BR, Schmidt M, Alexman J, Cullen CR (1986) Complete reactivation of X chromosomes from human chorionic villi with a switch to early DNA replication. Proc Natl Acad Sci USA 83:2182–2186
Mohandas T, Sparkes RS, Shapiro LJ (1981) Reactivation of an inactive human X chromosome: evidence for X inactivation by DNA methylation. Science 211:393–396
Morstyn G, Hsu SM, Kinsella T, Gratzner H, Russo A, Mitchell JB (1983) Bromodeoxyuridine in tumors and chromosomes detected with a monoclonal antibody. J Clin Invest 72:1844–1850
Perry P, Wolff S (1974) New Giemsa method for differential staining of sister chromatids. Nature 261:156–158
Petersen AJ (1964) DNA synthesis and chromosomal asynchrony. J Cell Biol 23:651–654
Pierron G, Durica DS, Sauer HW (1984) Invariant temporal order of replication of the four actin gene loci during the naturally synchronous mitotic cycles of *Physarum polycephalum*. Proc Natl Acad Sci USA 81:6393–6397
Riggs AD (1975) X inactivation, differentiation, and DNA methylation. Cytogenet Cell Genet 14:9–25
Roberts JM, Weintraub H (1986) Negative control of DNA replication in composite SV40-Bovine Papilloma Virus plasmids. Cell 46:741–752
Schempp W, Vogel W (1978) Decrease of DNA synthesis in aminotic fluid cells during the middle part of S phase, revealed by differential chromosome staining after incorporation of Brd U. Chromosoma 67:193–199
Schmidt M (1980) Two phases of DNA replication in human cells. Chromosoma 76:101–110
Shafer DA, Priest JH (1984) Reversal of DNA methylation with azacytidine alters chromosome replication patterns in human lymphocyte and fibroblast cultures. Am J Hum Genet 36:534–545
Somssich IE, Spira J, Hameister H, Klein G (1984) Correlation between tumorgenicity and banding pattern of chromosome 15 in murine T-cell leukemia cells and hybrids of normal and malignant cells. Chromosoma 91:39–45
Speit G, Vogel W (1986) Detection of bromodeoxyuridine incorporation in mammalian chromosomes by a bromodeoxyuridine antibody. II. Demonstration of sister chromatic exchanges. Chromosoma 94:103–106
Stubblefield ET, Klevecz R, Deaven L (1967) Synchronized mammalian cell cultures I. Cell replication cycle and macromolecular synthesis following brief colcemid arrest of mitosis. J Cell Physiol 69:345–354
Tice RR, Hollaender A (eds) (1984) Sister chromatid exchanges, Parts A & B. Plenum, New York London
Taylor JH (1958) Sister chromatid exchanges in tritium-labeled chromosomes. Genetics 43:515–529
Taylor JH (1960) Asynchronous duplication of chromosomes in cultured cells of Chinese hamster. J Biophys Biochem Cytol 7:455–464
Taylor JH (1978) Control of the initiation of DNA replication in mammalian cells. In: Molineux I, Kohiyama M (eds) DNA synthesis: Present and future. Plenum, New York, pp 143–159
Taylor JH (1984) Origins of replication and gene regulation. Cell Mol Biochem 61:99–109
Taylor JH, Woods PS, Hughes WL (1957) The organization and duplication of chromosomes as revealed by autoradiographic studies using tritium-labeled thymidine. Proc Natl Acad Sci USA 43:122–128
Vogel W, Autenrieth M, Speit G (1986) Detection of bromodeoxyuridine incorporation in mammalian chromosomes by a bromodeoxyuridine antibody. I. Demonstration of replication patterns. Hum Genet 264:1–4

Watanabe S, Taylor JH (1980) Cloning of an origin of DNA replication of *Xenopus laevis*. Proc Natl Acad Sci USA 77:5292–5296

Watanabe S, Temin HW (1983) Encapsidation sequences for spleen necrosis virus, an avian retrovirus, are between the 5′ long terminal repeat and the start of the gag gene. Proc Natl Acad Sci USA 79:5986–5990

Wolf SF, Migeon BR (1982) Studies of X-chromosome DNA methylation in normal human cells. Nature 295:667–671

Appendix

Wolfgang Beermann and I have shared a common interest in developmental biology of many years. The difference was that he made significant discoveries early in his research career that had great significance for our understanding of the changes in chromosomes related to development. I wandered through many byways before I came back to my early interest by studying DNA methylation. Along the way I became involved in the DNA replication story in a way that diverted my attention for many years, but never obscured my interest in and appreciation of the fundamental contributions of Beermann and his students. Another difference was in our training. I am 5 years older and therefore my Ph.D. training preceded the war at a time when the classical period of genetics was coming to a close, but cytochemistry and molecular biology had not yet become a part of the curriculum of most graduate schools, including the University of Virginia where I was trained. Some of my professors received their postdoctoral training in Germany in the 1890's, when what we call the classical period of genetics was just beginning. Beermann likewise must have been trained in classical cytogenetics, but at a time when cytochemistry and biochemistry was more a part of the training, I think. Yet both of us came into the field at an exciting time when great ideas were beginning to proliferate, but techniques for testing them were few. We had to learn and grow with the new biology as it developed in a way that is difficult for those who entered later to appreciate.

A Personal View of How the Importance of DNA was Discovered and the Making of a Molecular Biologist[2]

DNA was discovered in 1872 by Miescher in Tübingen, Germany. He isolated an acetic organic polymer from the nuclei of pus cells obtained from the wounds of soldiers who had fought in the Franco-Prussian War. Later, he obtained a similar material from trout sperm. The material was studied extensively in the succeeding years and during the early part of this century its components had been

[2] (Written in 1976 on the twentieth anniversary of our discovery of semi-conservative distribution of DNA in chromosomes by the use of tritiated thymidine.)

analyzed and shown to be phosphate, sugar, and four cyclic bases, two pyrimidines and two purines. It was thought to be a simple repeating polymer composed of tetranucleotides with the four bases, adenine, guanine, cytosine, and thymine in equal proportions. E. B. Wilson, the noted embryologist and cytogeneticist at Columbia University, had speculated on its properties as the hereditary material in the 1890's, but in 1925, when he published his final edition of the monumental text, *The Cell in Heredity and Development,* he had given up that youthful idea. Like his contempories, he had concluded that proteins were the stuff of which genes are made. DNA, or thymonucleic acid as it was called, was far too simple and evanescent, he thought, to be of much importance in the cell's life history. It was abundant in certain animal cells and a similar material, yeast nucleic acid, now called RNA, was known to be abundant in plant cells.

In 1944 O. T. Avery and his colleagues at the Rockefeller Institute for Medical Research located on the East River in Manhattan, demonstrated that DNA is the genetic coding material for bacteria. He showed that highly purified DNA obtained from virulent strains of *Pneumococcus* transformed avirulent types into deadly bacteria, which produced a mucopolysaccharide coating and killed mice in which the parental strain was harmless. Did the world take note and shout with joy, or even with anger that scientists were tampering with our genes? No, not really. The world was at war and many of its people had been in this state for 4 or 5 years. That year also happened to be the one that my Ph.D. in biology was awarded for research in genetics. I am not sure that I had ever heard of Avery, but my head had been filled with the concepts developed and published by T. H. Morgan, H. J. Muller, Calvin Bridges, C. D. Darlington and even John Belling, a cytologist who had spent some time in a then little-known university in Florida. When the degree was granted, I had already been in the South Pacific for sometime fighting the war of mosquitoes, malaria, and mud on lonely atolls and in islands surrounded by barrier reefs.

Soon after the publication of his monumental treatise, Avery retired to his native Tennessee for the remainder of his life. He was a modest, but perceptive man who had labored to find the transforming principle since about 1928 when Griffith discovered transformation of *Pneumococcus* in vivo. Griffith had injected mice with both dead virulent bacteria and live avirulent bacteria of the same species. He obtained a lethal infection. From the sick mice he isolated virulent bacteria similar to those dead ones injected with the live harmless ones. Few believed him, or paid much attention, but Avery and his colleagues labored through the years to find the active principle, the transforming material contributed by the dead bacteria.

Avery realized the significance of his discovery, as indicated by a letter that he wrote to Max Delbruck at that time. However, the long paper that documents the evidence that highly purified DNA is the transforming principle is difficult to read. Even Avery had been reluctant to accept the implications until he obtained the highly purified and specific enzyme, deoxyribonuclease, from his colleague at the Rockefeller Institute, Dr. Moses Kunitz. This enzyme was highly effective in destroying the purified, active transforming material, while various proteases were ineffective. In spite of the evidence, many geneticists considered this a special case applicable only to bacteria. Others thought that small amounts of residual protein in his preparations might be the active principle. Other experiments and a workable model for DNA were required to change the thinking about the nature of the gene.

A few years later when I went to the University of Oklahoma to teach genetics at that outpost of learning in a state made famous, or infamous, by Steinbeck's book *Grapes of Wrath,* and Rogers and Hammerstein's Broadway show *Oklahoma,* I was made aware of my own ignorance, as well as of that of the students who were to enroll in my genetics course that year. Toward the end of the first term one fiery, red-haired young lady, who had been delighted by my arrival from the eastern Halls of Learning, was now disappointed. She had read all of the textbooks I had assigned on genetics, but had not found the answer to the most important question: what is a gene? She wanted me to tell them of what it is composed and how it produces an identical copy of itself. None of the books had an answer and tears of anger and frustration came into her eyes when I also had to admit ignorance. She blamed me personally for the world's lack of knowledge, and felt sure that I was deceiving her to cover my own ignorance.

The next episode in this tale occurs about 10 years later, in late August of 1956, in a rather small open classroom in Storrs, Connecticut. The Genetics Society of America was holding its annual meeting at the University of Connecticut and I was scheduled for a 15-minute talk, one of the short papers on the program. Word had spread around that something dramatic was to be presented and the room was not only full, but others who could not get in stood outside the open windows and looked into the partially darkened room. I was presenting the results of some very recent, preliminary experiments of the labeling and distribution of chromosomes in dividing cells of the broad bean, *Vicia faba.* Walter L. (Pete) Hughes, Phillip S. Woods, and I had labeled thymidine with tritium, at that time a little-used isotope of hydrogen, and employed it to trace the behavior of DNA in the reproduction of chromosomes. DNA replication was known to occur preceding each division and our experiments were designed to test the hypothesis of Watson and Crick concerning the mechanism of that replication.

Among the audience was A. H. Sturtevant, the well-known and highly regarded *Drosophila* geneticist from CalTech. I did not know of his presence at the time, but he carried the word of our experiment back to CalTech, where he told George Beadle, Max Delbruck, and Linus Pauling, among others, of our results. Others left the meeting at Storrs, Connecticut, and went to Japan, where a world congress was in progress on the biochemistry of DNA. My colleague at Columbia University, Francis Ryan, heard about our experiments in Japan where he was attending the same meeting. Francis Crick, who was also in Japan, took the message back to Europe. In less than a week after we had reported our preliminary experiments, the results were known around the world, while 12 years earlier Avery's more fundamental and far-reaching, well-documented work had been known only to a few selected scientists, primarily biochemists who were interested in nucleic acids.

The alert ones among you will remember that 10 years have elapsed between this scene and the last one in which I was confronted by the impatient young lady at the mid-western university where I began my teaching career. I was now (1956) in what many regard as the center of the world, New York City, and worked daily down the hall from where Thomas Hunt Morgan, E. B. Wilson, A. H. Sturtevant, and Calvin Bridges a generation earlier had ushered in a new era in genetics. Their pictures were hanging on the walls of our seminar room, along with others such as H. J. Muller. He had worked somewhat isolated from the genetics group in a separate small laboratory on a different floor in the same building. That summer of 1956 I had plans to work at the Brookhaven National Laboratory as I had for the previous 3 years. However, this summer had more exciting prospects. I planned to prepare ^{3}H-thymidine and study its segregation in chromosomes at division. Where had I been for 10 years? I had been wandering and at times fasting in the wilderness to gain a better vision of the genetic material of cells. What wilderness? Oklahoma, Tennessee, and finally Manhattan and the catacombs of Columbia University's Schermerhorn Building.

During that period, I had one summer come to the pleasant shores of Lake Mendota, on which stands the University of Wisconsin, where the Genetics Society of America was celebrating the half century of progress in genetics, measured from the independent rediscovery of Mendel's work in 1900 by three biologists, Correns, Von Schermak, and de Vries in Germany, Austria, and Holland, respectively. I sat enthralled by the accomplishments of the noted geneticists that passed across the stage in those 2 days. However, I listened most carefully to Alfred E. Mirsky, a biochemist from the Rockefeller Institute for Medical Research, who told of his experiments with DNA and nucleohistones obtained from salmon or trout sperm as Miescher had years before. Mirsky and Pollister, later my colleague at Columbia University, had been able to isolate the material as a high polymer, i.e., a relatively undegraded highly viscous material, by a high salt extraction process. His topic was the role of nucleoproteins in inheritance. However, he mentioned Avery's work only in passing and used the occasion to downplay the significance of Avery's contribution and to restate the prevailing view among geneticists that the transforming material form *Pneumococcus* probably had a small contaminant of protein which was the real active component – the mysterious gene. There was no other mention of nucleic acids in the lectures which were later published as a book, *A Half Century of Progress in Genetics*. I was disappointed and had some of the same feelings as the fiery young lady who had confronted me 4 years before in Oklahoma. This was not only a celebration of 50 years of progress in what was becoming known as classical genetics; it marked the passing of an era. They had learned an amazing amount, but had failed to provide the answers to our most critical and fundamental questions in genetics. By that time I was becoming knowledgeable about the chemistry of macromolecules. I had finished my doctoral degree in biology with no credentials in biochemistry or physical chemistry. However, I was trying to become what a colleague of later years, Erwin Chargaff, was to define as a molecular biologist – a biologist, physicist, or any other brand of scientist who soon flocked to the new discipline and began to "practice biochemistry without a license".

I had reviewed organic chemistry while in the South Pacific and when the war was over and we were waiting to come home, I volunteered to teach the soldiers and received the only commendation of my short military career from the Commander of SOPAC for teaching physics in the University of the South Pacific – a school without a campus, with few textbooks, and war surplus materials for laboratory supplies. During my first year as an assistant professor at the University of Oklahoma, I finished a course in biochemistry which had not been required as part of my doctoral training at the University of Virginia. The next year, after I had moved to the University of Tennessee, I took a course in physical chemistry. This required a review and an extension of my rudimentary calculus, for even though as an undergraduate I had minored in math, I had somehow avoided a course in calculus.

I was now realizing that I was witnessing the passing of an era. Mendelian genetics and the cytology of E. B. Wilson, C. D. Darlington, A. J. Sharp, and John Belling could not provide the answers to the questions of the nature of the gene. In my cerebral wanderings I was already marching to the beat of a different drummer, for I had heard a voice crying in the wilderness, one Franz Schrader of Columbia University, whom I had met at scientific meetings. He was a classical cytologist and E. B. Wilson's successor, but his message was that the future of cytology and genetics would be in the chemist's laboratory. I was enthusiastically trying to follow his advice even if poorly equipped. I knew that Bernal and Astbury had made X-ray diffraction pictures of DNA fibers and found spacings about 3.5 Å, presumably the stacked bases. Proteins likewise had linkages with similar spacings, the peptide linkages. Were nucleoproteins copolymers of protein and nucleic acid chains? I proposed a model of a chromosome about that time (1950) in which I assumed that the chromatid was one long copolymer composed of a chain of DNA and a polypeptide chain bonded together and then coiled and supercoiled through seven levels to form the microscopically visible coils which many cytologists had studied by that time. The abstract in which I proposed the model is fortunately buried in some obscure issue of the abstracts for the annual meeting of the Genetics Society of America, I think. Not even I know where to find it and I trust no one will ever see it again.

I remember reading the paper by Linus Pauling in the Proceeding of The National Academy of Sciences, USA, describing his model for the α-helix and the pleated sheet structure for polypeptides. He based his model on the X-ray diffraction studies of crystalline peptide derivatives by Cory and Pauling and on intuition. I was very excited about this model and hoped it would be useful in understanding chromosome structure. In spite of the activities in the biochemical laboratories, as I indicated above, many who knew about the studies of Avery, McCloud, and McCarty indicating that the gene was pure DNA, were still temporizing and trying to ignore the evidence. I went uptown in Manhattan in the spring of 1952 to take Erwin Chargaff's course on nucleic acids. He was a great analytical chemist, but he depreciated any attempts to extrapolate from the then limited knowledge of the structure of DNA to the gene. He and his students had shown that the base composition of DNA varied from species to species and because the ratio of four bases was not 1 : 1 : 1 : 1 :, it could not be a simple repeating polymer of four nucleotides as had been assumed for many years. They had also identified a common characteristic of all DNA's which others were beginning to refer to as Chargaff's rule. The bases T+G were equal to C+A in all of the samples, or put in another way, 6-amino groups were equal to 6-keto groups. These facts were to me only puzzling and without meaning. Later when it became clear that in molar ratio A=T and C=G, the rule began to have an effect on model building, and Watson and Crick probably made use of it later in constructing their models. I remember when Pauling's triple-stranded model was published. Leonard Ornstein, a bright and promising student in Authur Pollister's lab at Columbia University, reviewed the paper for one of our seminars late in 1952. In the model Pauling had placed the phosphate groups inside and the bases outside along a triple-stranded helix. I did not fully understand the structure, but neither could I understand how it advanced our understanding of chromosomes or of gene structure. A few months later the Watson-Crick model was published in Nature, but I missed that paper. It was not until the Cold Spring Harbor Symposium was held late in June of 1953 that I realized anything of importance had happened. For reasons I no longer remember, I did not attend the Cold Spring Harbor meetings that year, but my colleague, Francis Ryan, returned from the Long Island meeting very excited and buoyed up by the happenings. Watson and Crick had presented their model and Wilkins had presented X-ray diffraction evidence supporting the model. It was soon the theme of a number of seminars and other scientific meetings, but I heard the story first hand when Francis Crick came to the US and gave a seminar a few months later at the Brooklyn Polytechnic Institute. He explained and demonstrated the model very clearly and concisely in the delightful manner for which he later became famous. He also explained the implications for a replicative mechanism by a template base pairing. At last I appreciated the implications and began to understand the enthusiasm that was pervading the embryonic molecular biologists of that time. Crick was truly surprised, I think, by all the attention his work was getting. It was after all a kind of side project that he had worked on at odd times with prodding by the enthusiastic Watson. His principal work on X-ray diffraction of helical polymers had attracted very little attention, even though he had spent much more time on those pursuits than on the DNA problem. As you know, he soon turned his full attention to DNA and its possible coding properties, for at last with a model that explained how a template guided replication could occur, DNA was now being accepted as the physical basis of heredity – the material of which genes were composed. Avery's work had been repeated and extended to other genetic loci by Hotchkiss, Harriet Taylor, and others at the Rockefeller Institute, where Avery and his students had labored so long and patiently in relative obscurity for so many years. Avery, unfortunately, did not live long enough to receive the Nobel Prize to which his work surely would have entitled him had it been appreciated sooner.

As with so many great discoveries, the scientific world was not ready for DNA in 1943 or 1944. We were on the verge of a great revolution in biology, and we wonder now why it took so long to see the Light. To illustrate the point more dramatically I go back to my early training and quote from the book of books that taught Lincoln how to use the English language. My father was a farmer for most of his adult life, as were his father and my great-grandfather on my mother's side. Both grandfathers were Baptist preachers in a time when such endeavors, in my part of the country – the Texas frontier – was not a full-time occupation; they expected to make their living by other means. However, a cousin of my age and the namesake of my grandfather, James Wesley Taylor, studied for the ministry and spent his career as pastor at different times to several of the largest Baptist churches in the south. I mention these facts to temper any criticism I may receive for sacrilegious use of holy writ. The Bible was a conspicuous book in our house and sometimes our bedtime stories, read by my father, were directly from the King James version of the Bible. By the time I was a mid-teenager I had carefully read both the old and the new testament and had memorized parts of it. I might mention that the two grandfathers were different kinds of Baptists. Great-grandfather Aaron Cardwell, as well as my mother and father, belonged to the Old School or Primitive Baptists who did not believe in having either Sunday schools or missions. However, when Grandfather Cardwell visited us, he and Dad would talk for hours about the Bible and its correct interpretation and the history of the division of the Baptists. I would listen. To them, if mentioned at all, Darwin and Scopes of Tennessee infamy were agents of the Devil. Yet they were amazingly liberal in their views in most other respects and never insisted that their children hold similar opinions. They were free thinkers. In college I joined the Missionary Baptist Church and once considered studying for the ministry, but finally my minister's narrow-mindedness and critical views of most biology, and especially the theory of evolution, alienated me and I finally stopped attending churches until many years later. When we lived in New City we occasionally attended the services at Riverside Church and when we moved to Weschester County in New York, we attended the Friends' Meeting in Scarsdale. When we moved to Tallahassee, we did not find the Friends, but joined the Unitarians. My most recent activity in the world of religion was to serve as their President during the period in which we acquired our first minister. I also designed and supervised the construction of the present Unitarian church. I mention these facts to explain why a bibical text came to my mind when considering the role of DNA in the nature of things and the acceptance by geneticists of Avery's findings.

I use the first 12 verses of the gospel according to St. John (I only substituted DNA for "WORD" and O. T. Avery for John; otherwise the following is a direct quote).

> In the beginning was the DNA, and the DNA was with God, and the DNA was God.
> 2. The same was in the beginning with God.
> 3. All things were made by him; and without him was not any thing made that was made.
> 4. In him was life; and the life was the light of men.
> 5. And the light shineth in darkness; and the darkness comprehended it not.
> 6. There was a man sent from God whose name was O. T. Avery.
> 7. The same came for a witness, to bear witness of the DNA, that all men through him might believe.
> 8. He was not the Light, but was sent to bear witness of that Light.
> 9. *That* was the true Light, which lighteth every man that cometh into the world.
> 10. He was in the world, and the world was made by him, and the world knew him not.
> 11. He came unto his own, and his own received him not.
> 12. But as many as received him, to them gave he power to become the sons of God, even to them that believe on his name.

Gene Amplification in Dipteran Chromosomes

A. Spradling [1]

1 Introduction

The giant polytene chromosomes characteristic of many Dipteran cells allow fundamental nuclear processes occurring during development to be directly visualized. Polytene chromosomes afford a detailed picture of "large-scale" structures within developmentally regulated interphase cells – features extending over more than about 5–10 kb of chromosomal DNA. Structures, such as chromomeres, constrictions, and puffs, must be disrupted for biochemical studies and cannot currently be reconstituted. However, following Wolfgang Beermann's recognition that polytene chromosome puffs were manifestations of developmentally regulated gene expression, much progress has been made in understanding the molecular basis of gene transcription. Genetic studies have identified small regulatory regions located near tissue-specific genes that mediate developmental control (see Rubin 1985). However, the functional significance of associated large-scale structures, such as puffs, remains poorly understood.

Studies of chromosome replication have progressed more slowly from the cytological to the molecular level. Polytene chromosomes provide one of the most detailed pictures available of how eukaryotic genomes replicate (reviewed in Rudkin 1972). Yet the fundamental units of chromosome reduplication remain to be defined in molecular terms, greatly limiting attempts to correlate the replication of specific DNA sequences with that of specific chromosome regions.

Instances of differential replication provide particularly attractive opportunities for analyzing replication. Differential replication may greatly magnify the small and temporary changes in the relative DNA concentration between different genomic regions occurring during a normal S-phase. Furthermore, alterations caused by disproportionate replication may persist for long periods of time. Study of polytene chromosomes revealed several examples of differential replication in Dipteran salivary gland cells. Heitz (1934) recognized that centromeric heterochromatin was underrepresented in *Drosophila* salivary gland polytene chromosomes. Differential replication occurring within the euchromatin of *Rhynchosciara angelae* was discovered by Breuer and Pavan (1955). In this review, progress in utilizing differential replication to analyze the control of chromosome replication will be surveyed. In particular, we discuss recent studies of *Drosophila* chorion genes where it has been possible to begin to analyze the molecular controls of DNA amplification within polyploid follicle cells.

[1] The Carnegie Institute of Washington, 115 W. University Parkway, Baltimore, Maryland 21210, USA.

Results and Problems in Cell Differentiation 14
Structure and Function of Eukaryotic Chromosomes
Edited by W. Hennig

2 Differential Replication of Heterochromatin

Following the report of Heitz (1934), many additional examples of incomplete heterochromatin replication have subsequently been described (see Rudkin 1972; Spear 1977; Spradling and Rubin 1981). Highly repetitive satellite DNA sequences are included within the underreplicated regions (Gall et al. 1971; Dickson et al. 1971). Underreplication of centromeric heterochromatin is not invariably associated with a polyploid cell cycle. Examples are known where only partial underreplication occurs during polyploidization (Mahowald et al. 1979; Hammond and Laird 1985), or where underreplication is limited to only certain chromosomes or heterochromatic elements. Furthermore, the extent of differential replication may be tissue-specific (Endow and Gall 1975; Ribbert 1979; Redfern 1981).

The difficulties in isolating and working with genomic DNAs containing highly repetitive sequences have limited progress in understanding the regulation or the mechanism of differential heterochromatin replication. Methods for defining specific chromosomal replicons within heterochromatin used during normal and differential replication would seem to be a prerequisite to further progress. Currently, the ability to recover heterochromatic genomic regions intact from genomic libraries is uncertain, and contiguous clones cannot usually be recognized. This has prevented the mapping of specific heterochromatic regions larger than about 10–20 kb. Furthermore, the absence of any known function for the underreplicated sequences in the affected tissues has discouraged genetic analysis of the problem.

Ribosomal genes replicate differentially in polytene salivary gland cells (Hennig and Meer 1971; Spear and Gall 1973) and in the DNA of adult flies (Tartof 1971; Renkawitz and Kunz 1975). Here, both molecular and genetic analysis of the problem have proceeded further. Differences in the nontranscribed spacer sequences separating tandemly repeated rDNA genes and in the structure of insertions located in some 28S coding sequences allow rDNA genes to be separated into subfamilies (reviewed in Beckingham 1982). During polytenization of the salivary gland, rDNA subfamilies containing functional noninserted genes replicate selectively (Endow and Glover 1979; Endow 1980; Kunz et al. 1982; Bellikoff and Beckingham 1985). The particular rDNA genes which replicate may vary between individual polytene nurse cell nuclei (Belikoff and Beckingham 1985). Some genes from within the rDNA clusters on both the X and Y chromosomes may replicate during polytenization, however genes within a single nucleolus organizer are often utilized selectively. Frequently, the relative proportions of different gene subfamilies in salivary gland DNA resembles the distribution in one of the parental nucleolus organizers. It is therefore likely that a sufficient number of genes from within a cluster replicate during polyploidization to provide a representative sample of subfamilies.

The absence of knowledge concerning the large-scale arrangement of different rDNA genes within the nucleolus organizing regions has limited analysis of differential rDNA polytenization. The replicon structure of rDNA is unknown during normal or differential replication, although replicating rDNA molecules can be recognized in the electron microscope (McKnight et al. 1978; Saffer and Miller

1986). Two major classes of models need to be distinguished. The first postulates that each rDNA repeat is an independent replicon. Differential replication would, therefore, result from the selective activation of certain repeat types which might enjoy an advantage in the competition for limiting replication factors. Alternatively, rDNA replication might be controlled in larger units containing many rDNA repeats. Differentially replicating repeat classes would derive from a continuous subregion of the nucleolus organizer which remained capable of replication during polytenization. The size of the rDNA clusters (4×10^3 kb) virtually insures that they contain multiple replicons, because a single replicon can duplicate at most about 10^3 kb in a 600 min S-phase at measured elongation rates, which do not exceed 3 kb min^{-1} (Blumenthal et al. 1973) and may be closer to 0.3 kb min^{-1} in polyploid cells (Steinemann 1981). It remains unknown whether replication origins are present in some or all individual rDNA repeats or whether differential replication is mediated by non-rDNA sequences interrupting the array of tandemly repeated genes. The observation that individual polytene nuclei may largely replicate rDNA repeats characteristic of one nucleolus organizer or the other (Belikoff and Beckingham 1985) is difficult to reconcile with models postulating independent competition for replication of each individual rDNA repeat. Recently, the development of methods for separating very large DNA molecules using pulsed-gradient electrophoresis (Schwartz and Cantor 1983) holds promise that it may be possible to map large sections of nucleolus organizer regions.

3 Differential Replication of Euchromatin

3.1 DNA Puffs in Sciarids

In 1955 Breuer and Pavan reported that three polytene chromosome regions in the salivary gland of *Rhyncosciara angelae* underwent an unusual process of puff formation (Fig. 1). After the regression of several large puffs in prepupae, polytene chromosome bands reappearing within the formerly puffed regions were observed to stain much more intensely relative to neighboring bands than prior to puffing. Pavan proposed that the increased staining reflected the local synthesis of DNA; these sites were termed "DNA puffs" to distinguish them from the normal "RNA puffs" where additional staining is not observed. Detailed analysis of puffing within the salivary gland, intestine, and Malpighian tube chromosomes of *Rhynchosciara* larvae revealed a total of ten DNA puffs (Guevara and Basile 1973). All formed only in the salivary gland of 4th instar larvae. DNA puff formation was subsequently observed in the salivary glands of several other Sciarid species (Swift 1962; Gabrusewycz-Garcia 1964; Crouse and Keyl 1968).

The timing of DNA puff formation in *Rhynchosciara* suggested that they represented a process of specific gene amplification. At the end of the 4th larval instar, synchronously developing *Rhynchosciara* larvae weave a communal cocoon composed of protein fibers secreted from the salivary glands. The DNA puffs correspond to chromosome regions that are highly decondensed during at least part

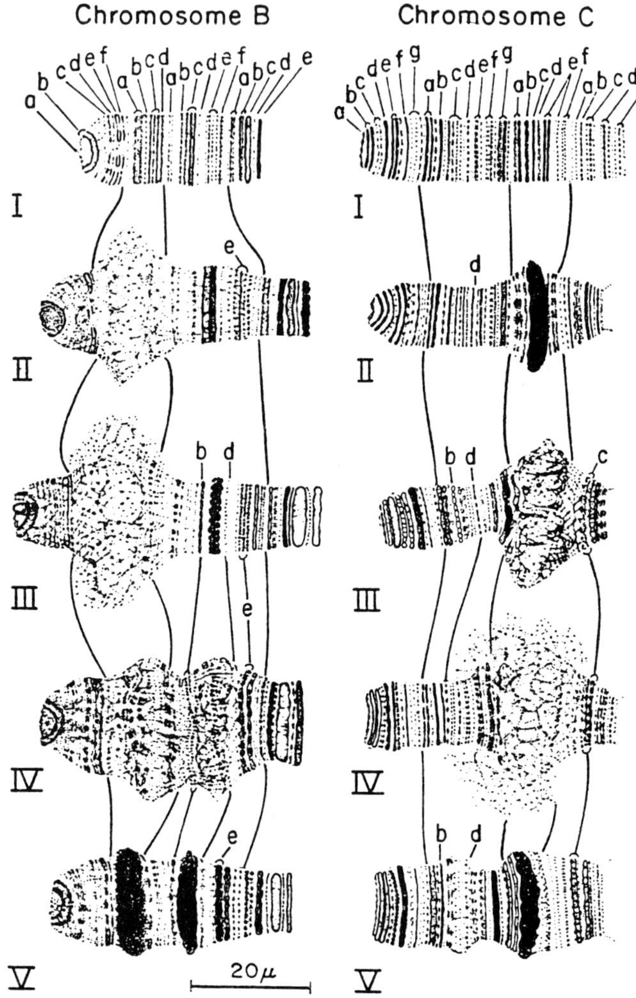

Fig. 1. DNA puff formation in salivary gland chromosomes of *Rhynchosciara angelae*. *Camera lucida* drawings by Mrs. M. E. Breuer of the first four regions at the distal ends of *chromosomes B* and *C* during the 4th instar of *Rhynchosciara angelae* are shown. The *numerals I–V* refer to puffing stages at increasingly later times of larval development. Although not indicated, each chromosome is subdivided into numerical regions beginning at the tip. Each numerical region is further divided into subdivisions denoted by *lower case letters*. (Pavan and Da Cunha 1969)

of the process of cocoon formation in late 4th instar larvae. The puffs then regress during the prepupal stage. This sequence of events suggested that genes specifically required in the salivary gland during late larval development to encode salivary gland secretory proteins were amplified in the DNA puffs. Early cytophotometric (Rudkin and Corlette 1957) and autoradiographic (Ficq and Pavan 1957) studies verified that differential DNA synthesis occurs at the sites of DNA puffs. Work by Lara's group associated specific salivary gland polypeptides (Winter et al. 1980) and messenger RNAs (Bonaldo et al. 1979) with specific puff sites. The isolation of specific cDNA and genomic clones from the sites of the C3 and C8 puffs allowed the direct demonstration that approximately 16-fold amplification of the coding sequences at these sites occurred during DNA puff formation (Glover et al. 1982; Millar et al. 1985).

3.2 *Drosophila* Chorion Gene Amplification

3.2.1 Mechanism of Amplification

Drosophila ovarian follicle cells (for review see Mahowald and Kambysellis 1980) cease division at about stage 5 of oogenesis and subsequently become polyploid (Mahowald et al. 1979; Hammond and Laird 1985), but never display polytene chromosomes. Beginning at stage 8 they sequentially produce a series of different products which are secreted in large amounts. These include three yolk proteins during stages 8–9 (Brennan et al. 1982), at least five vitelline membrane proteins during stages 9–10 (Petri et al. 1976; Fargnoli and Waring 1982), and at least

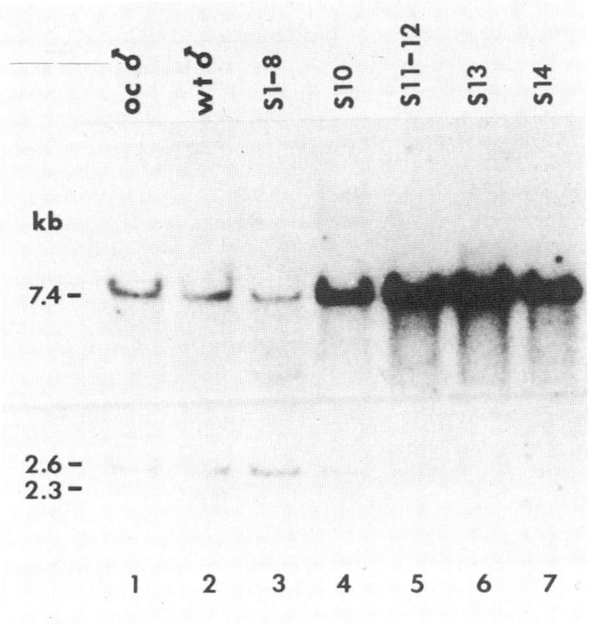

Fig. 2. *Drosophila* chorion gene amplification. DNA was prepared from adult males (*lanes 1* and *2*) and from egg chambers of stages 1–8 (*lane 3*), stage 10 (*lane 4*), stages 11–13 (*lane 5*), stage 13 (*lane 6*), and stage 14 (*lane 7*). After digestion with EcoRI and transfer to nitrocellulose, the blot was hybridized with probes specific for a 7.7 kb fragment derived from the third chromosome chorion gene cluster, and a control probe not undergoing amplification that labeled bands of 2.6 kb and 2.3 kb. The large increase in the relative intensity of the 7.7 kb and the 2.6 kb bands in late-stage egg chamber DNA compared to stage 1–8 or male DNA is a reflection of chorion gene amplification. (After Spradling and Mahowald 1980)

15 chorion structural proteins during stages 11–14 (Petri et al. 1976; Waring and Mahowald 1979). Following the completion of the eggshell during stage 14, the follicle cells degenerate.

Isolation of cDNA and genomic clones encoding four major chorion proteins (s38, s36, s18, and s15) revealed that single copies of each gene were present in the germ line genome, arranged in clusters on the X chromosome and on chromosome 3. However, quantitation of the chorion gene copy number in DNA from the egg chambers of developing ovarian follicles (Fig. 2) demonstrated that genes in both clusters undergo amplification beginning in stage 8–9 (Spradling and Mahowald 1980). The s36 and s38 genes within the X chromosome cluster amplify about 16-fold, while s18 and s15 genes on chromosome 3 amplify 60-fold (Fig. 2). In contrast, genes encoding other follicle cell secretory proteins, including the yolk proteins and several vitelline membrane proteins, have not been observed to amplify.

Many of the cytological features of DNA puff formation in Rhynchosciara correspond closely to the characteristics of chorion gene amplification within the polyploid follicle cells. DNA puffs all initiate amplification at about the same time, corresponding to approximately day 62 of larval development. However, the time of actual puffing varies. For example, puff B2 is first observed in period II, while puff C3 opens in period III and puff B3 in period IV (Fig. 1). Likewise, both chorion gene clusters begin amplification in stages 8–9, but genes within the X cluster are transcribed during stages 11–13, while the chorion genes on chromosome 3 are not activated until stage 13 or 14. The staining of bands giving rise to puff C3 increases rapidly compared to the changes in progenitor bands of puff B3. This suggests that in Rhynchosciara the rate of DNA synthesis may not be identical at each site. In *Drosophila* the third chromosome cluster initiates an average of six rounds of replication during the time the X chromosome cluster initiates only four rounds.

Because the chromosomal regions surrounding the chorion gene clusters consist almost entirely of unique DNA sequences, it was possible to isolate them by the process of chromosome walking (Spradling 1981). Amplification extends 40–50 kb along the chromosome on either side of each gene cluster. However, the magnitude of amplification decreased with distance along the chromosome from the chorion genes (Fig. 3 B, C). Despite the changes in copy number, no alterations in restriction fragments between amplified and unamplified DNA within the regions were found. These observations suggested that amplification occurred by disproportionate chromosomal replication from a site near or within the chorion genes themselves (Fig. 3 A). Termination of replication forks progressing in opposite directions from the gene clusters would give rise to the observed amplification gradients. Direct visualization of chromatin from *Drosophila* egg chambers allowed Osheim and Miller (1983) to identify transcriptionally active chorion genes. Replication forks were frequently observed proceeding away from active s36 and s38 gene pairs in stage 11 chromatin. Nested replication forks with dimensions very close to those expected were detected only in chromatin from postamplification egg chambers. These observations strongly support the disproportionate replication model.

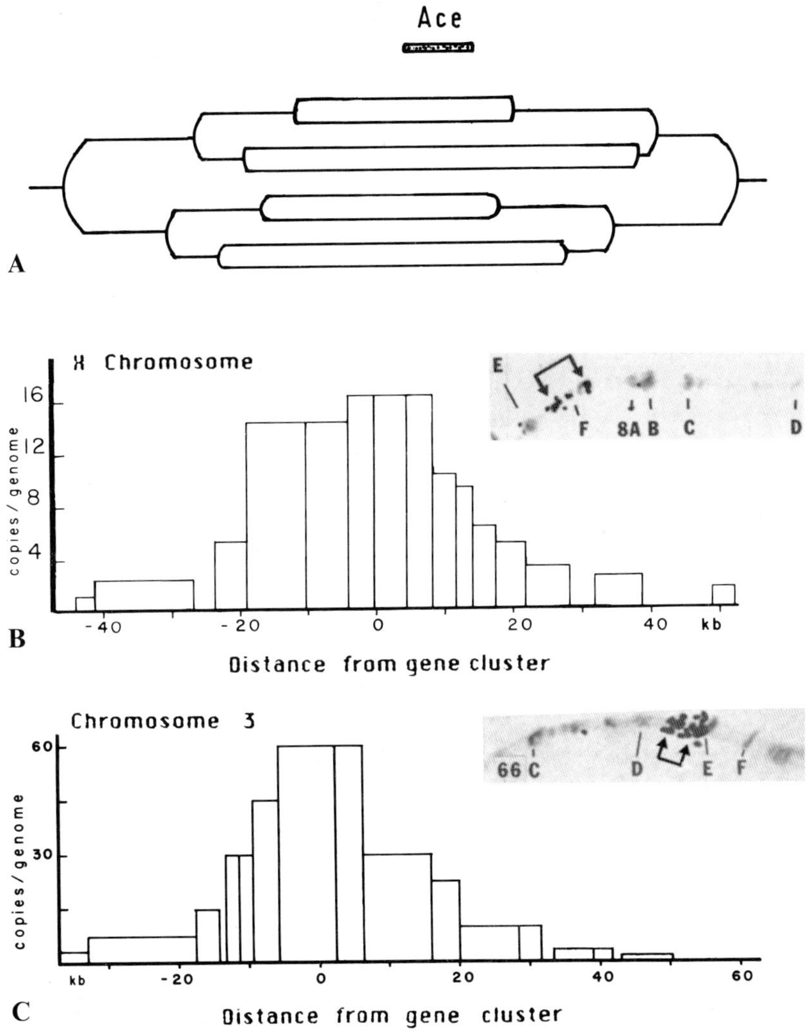

Fig. 3A–C. Structure of chromosomal domains of chorion gene amplification. **A** The disproportionate replication model of chorion gene amplification. The approximate location of an amplification control element (*ACE*) is indicated. **B** The amplification of chromosomal DNA sequences surrounding the X chromosome chorion gene cluster is indicated. The zero point on the ordinate is taken as the EcoRI site between s36 and s38 (see Fig. 4). The *inset* shows the cytogenetic extent of the amplified region as determined by in situ hybridization with probes taken from each end. **C** The amplification of chromosomal DNA sequences surrounding the chorion gene cluster on chromosome 3 is shown. The zero point on the ordinate represents the Sal I site near the s15 gene (see Fig. 4). The *inset* shows the cytogenetic extent of the amplified region as determined by in situ hybridization using probes from each end. (After Spradling 1981)

3.2.2 Chromosomal Units of Amplification

Do the 80–100 kb amplified domains correspond to specific polytene chromosome bands? To investigate this question, probes from each end of the X chromosome and third chromosome amplified regions were hybridized in situ to salivary gland polytene chromosomes (Spradling 1981). Figure 3 shows that both domains include more than one salivary gland polytene band. The X-linked replicon spans bands from about the middle of region 7E to band 7F3-4, while bands between 66D11 and 66D15 are amplified within chromosome 3.

It is interesting to compare these results with the cytological picture in *Rhynchosciara* DNA puffs. Although each DNA puff may be mapped to a single band in its early stages, more than one band generally becomes involved. For example (Fig. 1), puff B2 derives from bands in sections a–d, while puff B3 involves the three bands in section c. Band d initially produces the C3 puff, but other adjacent bands are eventually included. Thus, in both *Drosophila* and Rhynchosciara, amplification involves sequences within a small number of adjacent polytene bands. These results are consistent with the concept that each amplified region represents a single normal chromosomal replicon. Replicon lengths in *D. virilis* salivary gland chromosomes varied between 15–600 kb (Steinemann 1981), a distribution similar to the DNA content of individual *D. melanogaster* polytene bands (Rudkin 1972). However, the mean replicon size of 140 kb was equivalent to 3.4 average chromomeres (Steinemann 1981), similar to the 80–100 kb chorion gene replicons.

If we assume that Rhynchosciara DNA puffs, like *Drosophila* chorion genes, amplify by disproportionate replication, then recondensation of the amplification gradient into normal polytene chromosome bands following puff regression would be expected to produce a strongly amplified central band flanked by bands showing smaller specific increases in staining intensity. However, chromomeric structure does not appear to survive amplification; only a single, very dark band appears at the site of each DNA puff. The appearance of these bands differs from normal salivary gland chromomeres and has been described as heterochromatic. The total amount of DNA replicated per strand during amplification of the two *Drosophila* clusters can be calculated from Fig. 3 as about 1600 kb (chromosome 3) and 600 kb (X chromosome). Upon recondensation into a single band these amounts of DNA would produce very dark bands such as those observed at the site of DNA puffs.

3.2.3 Sequences Regulating Amplification

The disproportionate replication model suggests that amplification is controlled by mechanisms which regulate the initiation of rounds of replication at specific sites near or within the chorion gene clusters (see Fig. 3 A). Such cis-regulatory sites have been termed "amplification control elements" (*ACE*). The existence of amplification control elements was demonstrated by the ability of specific subregions from within the two gene clusters to undergo normal amplification following chromosomal rearrangement (Spradling and Mahowald 1981) or P element-mediated transformation (de Cicco and Spradling 1984).

Transposons bearing a 7.7 kb EcoRI fragment containing genes s18, s15, and s19 from the third chromosome gene cluster underwent follicle cell-specific amplification following introduction at diverse chromosome locations (de Cicco and Spradling 1984; Orr-Weaver T, Spradling AC, unpublished). Transposons containing other DNA regions from the cluster never amplified. However, the level of amplification induced by transposons containing the critical sequences (*ACE3*) was affected by the chromosomal site of integration. It varied between 2- and 40-fold, in contrast to normal 60-fold amplification. At some sites no amplification was observed. Furthermore, as the size of the chorion DNA within the transposons was reduced, the frequency of position effects increased (Orr-Weaver T, Spradling AC, unpublished; Kafatos et al. 1985). Similar results were obtained using transformants containing sequences derived from the X chromosome cluster to map an amplification control element (*ACE1*) (see Kalfayan et al. 1985).

The problem of position effects limited our ability to define elements sufficient for amplification to regions of about 3–4 kb within each cluster. However, sequences necessary for amplification were mapped in greater detail by constructing a series of small deletions throughout the sufficient region, and testing the remaining sequences for their ability to induce amplification (Orr-Weaver T, Spradling AC, unpublished; Kalfayan et al. 1985). In the case of both gene clusters, only a very small region appeared to be necessary for amplification. These regions are shown in Fig. 4. Both essential regions are located upstream from the initiation site of a major chorion gene. *ACE1* is located between 9 and 488 nucleotides upstream from the s38 gene, while *ACE3* resides between 310 and 630 nucleotides

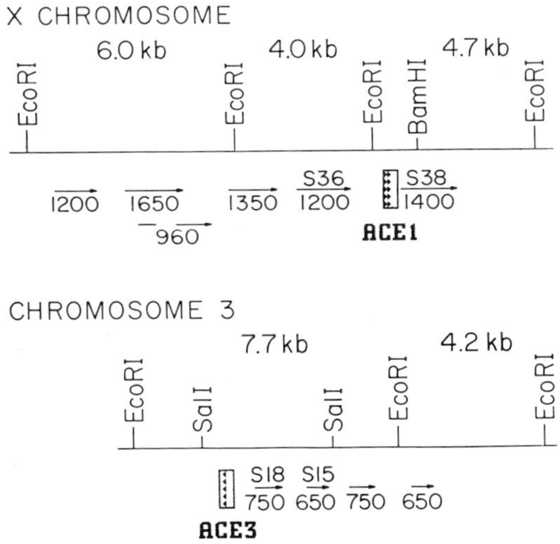

Fig. 4. Localization of sequences essential for chorion gene amplification. Restriction maps of the X chromosome and chromosome 3 chorion gene clusters are shown. Regions of each cluster shown to be essential for amplification in transformation experiments are indicated by a *box*. The approximate location and orientation of a conserved A/T-rich sequence described in the text is indicated by small *triangles* within the essential regions

upstream from the s18 start site. The *ACE3* region represents the only essential segment for amplification in the third chromosome cluster, however, a small portion of the X cluster within the 3′ portion of the s38 gene has yet to be tested. Sequence comparison between the two elements revealed only one small segment of significant homology. It lies either 464 or 492 base pairs upstream from the initiation sites, and includes the consensus sequence TT(T/C)TATTGTA(T/A)T. In both regions the conserved sequence lies within or near short tandemly repeated sequences. In *ACE1*, a variant of the consensus sequence differing by only one nucleotide is itself repeated, since it occurs within an eightfold repeat of the pentanucleotide sequence AATAC located just upstream of the consensus. The location of these elements are indicated by triangles in Fig. 4.

3.2.4 Model of Amplification Control

Many previous observations in a wide variety of systems have suggested interactions between the regulation of transcription and DNA replication. Primer RNA transcription is required for replication initiation at the ColE1 origin (Ito and Tomizawa 1980) and bacteriophage lambda replication requires "transcriptional activation" (Furth et al. 1982). A role for chorion RNA transcription in the induction of amplification was suggested by the observation of transient s15 RNA production during stage 9 (Thireos et al. 1980) and by the presence of s18 and s15 within the 3.8 kb Sal I fragment which induced developmentally regulated amplification and transcription (Wakimoto et al. 1986). However, the deletion experiments described above ruled out a role for either s18 or s15 transcription in chromosome 3 amplification. Deleted transposons lacking any detectable transcription from these genes were capable of inducing normal amplification. Likewise, elimination of s38 transcription by promoter deletion did not eliminate amplification under the control of *ACE1* sequences (Spradling AC et al., unpublished).

These results and the mapping of cis-regulatory sequences described above suggested a new model for the regulation of chorion gene amplification (Orr-Weaver T, Spradling AC, unpublished). In the case of both amplified domains, sequences essential for amplification coincided with sequences required for the transcription of a nearby major chorion gene. Thus, on chromosome 3, the deletion of the essential sequences upstream of s18 eliminated s18 transcription as well as amplification. Sequences within the *ACE1* element were required for s38 transcription (Kalfayan et al. 1985) as well as for amplification. These results suggested that a common element might regulate the developmental specificity of both chorion gene amplification and the transcription of the linked chorion gene.

One simple version of this model is diagramed in Fig. 5. Each amplification control element is postulated to contain an origin of DNA replication used during amplification (ori) as well as an "enhancer-like" transcriptional control element (enh). The transcriptional control element would act both to activate the origin and to stimulate transcription from the associated chorion gene promoter. Amplification begins earlier than the onset of s18 or s38 transcription. According to the model, activation of the enhancer, due to a change in a trans-acting factor in

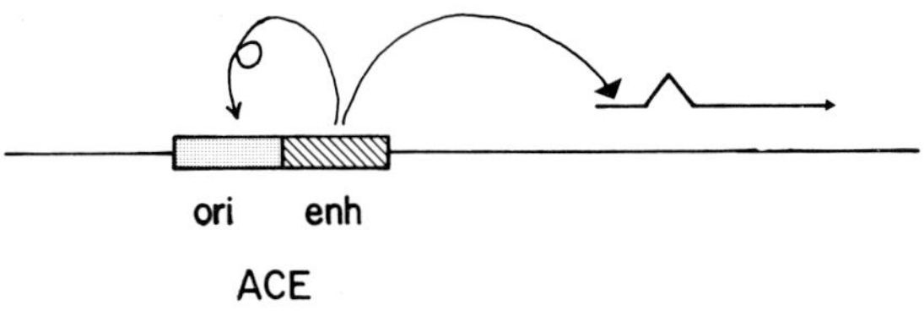

Fig. 5. Model for the control of chorion gene amplification, depicting the proposed structure of the essential sequences mapped in Fig. 4. Each essential region contains an origin of DNA replication (*ori*) and a transcriptional control element (*enh*). The transcriptional control element is proposed to interact with the linked chorion gene promoter to regulate chorion mRNA transcription and also to regulate the initiation of rounds of replication at the origin during amplification

stage 8–9, would be sufficient to activate amplification. Specific chorion gene induction would require additional specific factors associated with the specific temporal programs of s18 and s38 synthesis during stages 11–14.

The proposed model is similar to mechanisms which may regulate the replication of several eukaryotic viruses, including polyoma (de Villiers et al. 1984), bovine papilloma virus (Lusky and Botchan 1986), and Epstein-Barr virus (Reisman et al. 1985). In these cases a transcriptional regulatory sequence located near the origin has also been implicated in the control of viral DNA replication.

The structure and location of the control regions is consistent with our model. The A/T-rich consensus sequence conserved between the two clusters is similar to the core sequence of yeast *ARS* elements that have been implicated as replication origins (Broach et al. 1983). This conserved sequence may form part of the origin sequences used during amplification. Several *Drosophila* genes have been shown previously to contain enhancer-like transcriptional regulatory elements in their 5′ upstream regions (see Rubin 1985). This model predicts that the tissue specificity of amplification could be altered by inserting other tissue-specific enhancers near the control elements.

4 Conclusion

Chromosome replication is usually envisioned as a passive part of development. In most cells, the entire genome replicates during the S-phase; the developmental significance of any given region is believed to depend on the transcriptional activity or inactivity of its component genes. Only in rare instances of differential replication, such as those described here in Dipteran cells, does replication regulate gene expression. Even in these cases the effect is thought to be purely quantitative – by increasing gene dosage when products must be produced at extremely high rates.

This conventional view of a generally passive role played by chromosome replication may prove to be oversimplified. Genomic replication is an ordered and tissue-specific process (reviewed in Hand 1978). Genes active in a given tissue are generally early replicating, while inactive genes may replicate at any time during S-phase and frequently do so late (Goldman et al. 1984). Following chromosome rearrangement, the time of replication and the activity of a transcriptional unit may be altered in parallel (Calza et al. 1984). Replication has been proposed to turn on gene expression in *Physarum* (Pierron et al. 1982) and to specifically turn off Neurospora and yeast histone gene transcription.

The model proposed here for the regulation of chorion gene amplification and previously for the regulation of polyoma virus provides an interesting perspective on these observations. The association between transcriptional control elements and replication origins postulated for the chorion gene control regions may occur frequently. In this view, the early replication of genes in tissues in which they are expressed would result from the fact that the associated transcriptional control element is active and simultaneously stimulates a linked replication origin inactive in other tissues.

It will be interesting to determine if these ideas are useful in understanding other examples of differential replication in polytene cells. However, progress in answering these questions can only come when additional replicons are defined at the molecular level and their cis-regulatory sequences analyzed.

References

Beckingham K (1982) Insect rDNA. In: Busch H, Rothblum L (eds) The cell nucleus, vol X. Academic Press, New York, pp 206–269

Belikoff EJ, Beckingham K (1985) A stochastic mechanism controls the relative replication of equally competent ribosomal RNA gene sets in individual dipteran polyploid nuclei. Proc Natl Acad Sci USA 82:5045–5049

Blumenthal A, Kriegstein H, Hogness D (1973) The units of DNA replication in *Drosophila melanogaster* chromosomes. Cold Spring Harbor Symp Quant Biol 38:205–223

Bonaldo MF, Santelli RV, Lara FJS (1979) The transcript from a DNA puff of Rhynchosciara and its migration to the cytoplasm. Cell 17:827–833

Brennan MD, Weiner AJ, Goralski TJ, Mahowald AP (1982) The follicle cells are a major site of vitellogenin synthesis in *Drosophila melanogaster*. Dev Biol 89:225–236

Breuer ME, Pavan C (1955) Behavior of polytene chromosomes of *Rhynchosciara angelae* at different stages of larval development. Chromosoma 7:371–386

Broach JR, Li YY, Feldman J, Jayaram M, Abraham J, Nasmyth KA, Hicks JB (1983) Localization and sequence analysis of yeast origins of DNA replication. Cold Spring Harbor Symp Quant Biol 47:1165–1173

Calza RE, Eckhardt LA, DelGiudice T, Schildkraut CL (1984) Changes in gene position are accompanied by a change in time of replication. Cell 36:689–696

Crouse HV, Keyl HG (1968) Extra replication in the DNA puffs of *Sciara coprophila*. Chromosoma 25:357–364

de Cicco D, Spradling A (1984) Localization of a cis-acting element responsible for the developmentally regulated amplification of Drosophila chorion genes. Cell 38:45–54

de Villiers J, Schaffner W, Tyndall C, Lupton S, Kamen R (1984) Polyomavirus DNA replication requires an enhancer. Nature 312:242–246

Dickson E, Boyd I, Laird CH (1971) Sequence diversity of polytene chromosome DNA from *Drosophila hydei*. J Mol Biol 61:615–627

Endow SA (1980) On ribosomal gene compensation in Drosophila. Cell 22:149–155

Endow SA, Gall J (1975) Differential replication of satellite DNA in polyploid tissues of *Drosophila virilis*. Chromosoma 50:175–192

Endow SA, Glover DM (1979) Differential replication of ribosomal gene repeats in polytene nuclei of Drosophila. Cell 17:597–605

Fargnoli J, Waring GL (1982) Identification of vitelline membrane proteins in *Drosophila melanogaster*. Dev Biol 92:306–314

Ficq A, Pavan C (1957) Autoradiography of polytene chromosomes of *Rhynchosciara angelae* at different stages of larval development. Nature 180:983–984

Furth M, Dove M, Meyer B (1982) Specificity determinants for bacteriophage λ DNA replication. III. Activation of replication in λri^c mutants by transcription outside of ori. J Mol Biol 154:65–83

Gabrusewycz-Garcia N (1964) Cytological and autoradiographic studies in *Sciara coprophila* salivary gland chromosomes. Chromosoma 15:312–344

Gall J, Cohen E, Polan M (1971) Repetitive DNA sequences in Drosophila. Chromosoma 33:319–344

Glover DM, Zaha A, Stocker AJ, Santelli RV, Pueyo MT, De Toledo SM, Lara FJS (1982) Gene amplification in Rhynchosciara salivary gland chromosomes. Proc Natl Acad Sci USA 79:2947–2951

Goldman MA, Holmquist GP, Gray MC, Caston LA, Nag A (1984) Replication timing of genes and middle repetitive sequences. Science 224:686–692

Guevara M, Basile R (1973) DNA and RNA puffs in Rhynchosciara. Caryologia 26:275–295

Hammond MP, Laird CD (1985) Chromosome structure and DNA replication in nurse and follicle cells of *Drosophila melanogaster*. Chromosoma 91:267–278

Hand R (1978) Eucaryotic DNA: organization of the genome for replication. Cell 15:317–325

Hennig W, Meer B (1971) Reduced polyteny of ribosomal RNA cistrons in giant chromosomes of *Drosophila hydei*. Nat New Biol 233:70–72

Heitz E (1934) Über alpha- und beta-Heterochromatin sowie Konstanz und Bau der Chromomeren bei Drosophila. Biol Zentralbl 54:588–609

Itoh T, Tomizawa J (1980) Formation of an RNA primer for initiation of replication of Col E1 DNA by ribonuclease H. Proc Natl Acad Sci USA 77:2450–2454

Kafatos FC, Mitsialis SA, Spoerel N, Mariani B, Lingappa JR, Delidakis C (1985) Studies on the developmentally regulated expression and amplification of insect chorion genes. Cold Spring Harbor Symp Quant Biol 50:537–548

Kalfayan L, Levine J, Orr-Weaver T, Parks S, Wakimoto B, de Cicco D, Spradling AC (1985) Localization of sequences regulating Drosophila chorion gene amplification and expression. Cold Spring Harbor Symp Quant Biol 50:527–535

Kunz W, Grimm C, Franz G (1982) Amplification and synthesis of rDNA: Drosophila. In: Busch H, Rothblum L (eds) The cell nucleus, vol XII. Academic Press, New York, pp 156–184

Lusky M, Botchan M (1986) Transient replication of bovine papilloma virus type plasmids: cis and trans requirements. Proc Natl Acad Sci USA 83:3609–3613

Mahowald AP, Kambysellis MP (1980) Oogenesis. In: Ashburner M, Wright TRF (eds) The genetics and biology of Drosophila, vol 2D. Academic Press, New York, pp 141–224

Mahowald AP, Caulton JH, Edwards MK, Floyd A (1979) Loss centrioles and polyploidization in follicles of *Drosophila melanogaster*. Exp Cell Res 118:404–410

McKnight SL, Bustin M, Miller OL (1978) Electron microscopic analysis of chromosome metabolism in the *Drosophila melanogaster* embryo. Cold Spring Harbor Symp Quant Biol 42:741–754

Millar S, Hayward DC, Read CA, Browne MJ, Santelli RV, Vallejo FG, Pueyyo MT, Zaha A, Glover DM, Lara FJS (1985) Segments of chromosomal DNA from *Rhynchosciara americana* that undergo additional rounds of DNA replication in the salivary gland DNA puffs have only weak ARS activity in yeast. Gene 34:81–86

Osheim Y, Miller O (1983) Novel amplification and transcriptional activity of chorion genes in *Drosophila melanogaster* follicle cells. Cell 33:543–553

Pavan C, da Cunha AB (1969) Chromosomal activities in Rhynchosciara and other Sciaridae. Annu Rev Genet 7:425–450

Petri WH, Wyman AR, Kafatos FC (1976) Specific protein synthesis in cellular differentiation. III. The eggshell proteins of *Drosophila melanogaster* and their program of synthesis. Dev Biol 49:185–199

Pierron G, Sauer HW, Toublan B, Jalouzount R (1982) Physical relationship between replicons and transcription units in *Physarum polycephalum*. Eur J Cell Biol 29:104–113

Redfern CPF (1981) Satellite DNA of *Anopheles stephensi* Liston (Diptera: Culcidae) Chromosomal location and underreplication in polytene nuclei. Chromosoma 82:561–581

Reisman D, Yates J, Sugden B (1985) A putative origin of replication of plasmids derived from Epstein-Barr virus is composed of two cis-acting components. Mol Cell Biol 5:1822–1832

Renkawitz R, Kunz W (1975) Independent replication of the ribosomal RNA genes in the polytrophic-meroistic ovaries of *Calliphora erythrocephala, Drosophila hydei* and *Sarcophaga barbata*. Chromosoma 53:131–140

Ribbert D (1979) Chromomeres and puffing in experimentally induced polytene chromosomes of *Calliphora erythrocephala*. Chromosoma 74:269–298

Rubin G (1985) Molecular biology of development. Cold Spring Harbor Symp Quant Biol 50:905–908

Rudkin GT (1972) Replication in polytene chromosomes. In: Beermann W, Reinert J, Ursprung H (eds) Results and problems in cell differentiation, vol 4. Springer, Berlin Heidelberg New York, pp 59–85

Rudkin GT, Corlette SL (1957) Disproportionate synthesis of DNA in a polytene chromosome region. Proc Natl Acad Sci USA 43:964–968

Saffer LD, Miller OL (1986) Electron microscopic study of *Saccharomyces cerevisiae* rDNA chromatin replication. Mol Cell Biol 6:1148–1157

Schwartz DC, Cantor CR (1984) Separation of yeast chromosome-sized DNAs by pulsed field gradient gel electrophoresis. Cell 37:67–75

Spear B (1977) Differential replication of DNA sequences in Drosophila chromosomes. Am Zool 17:695–706

Spear B, Gall J (1973) Independent control of ribosomal gene replication in polytene chromosomes of *Drosophila melanogaster*. Proc Natl Acad Sci USA 70:1359–1363

Spradling AC (1981) The organization and amplification of two clusters of Drosophila chorion genes. Cell 27:193–202

Spradling AC, Mahowald A (1980) Amplification of chorion genes during oogenesis in *Drosophila melanogaster*. Proc Natl Acad Sci USA 77:1096–1100

Spradling AC, Mahowald A (1981) A chromosome inversion alters the pattern of specific DNA replication in Drosophila follicle cells. Cell 27:203–210

Spradling AC, Rubin GM (1981) Drosophila genome organization: conserved and dynamic aspects. Annu Rev Genet 15:219–264

Steinemann M (1981) Chromosomal replication in *Drosophila virilis* III. Organization of active origins in the highly polytene salivary gland cells. Chromosoma 82:289–307

Swift H (1972) Nucleic acid and cell morphology in Dipteran salivary glands. In: Allen JM (ed) The molecular control of cellular activity. McGraw-Hill, New York, pp 73–125

Tartof KD (1971) Increasing the multiplicity of ribosomal RNA genes in *Drosophila melanogaster*. Science 171:294–297

Thireos G, Griffin-Shea R, Kafatos F (1980) Untranslated mRNA for a chorion protein of Drosophila melanogaster accumulates transiently at the onset of specific gene amplification. Proc Natl Acad Sci USA 77:5789–5793

Wakimoto B, Kalfayan L, Spradling A (1986) Developmentally regulated expression of Drosophila chorion genes introduced at diverse chromosomal positions. J Mol Biol 187:33–45

Waring GL, Mahowald AP (1979) Identification and time of synthesis of chorion proteins in *Drosophila melanogaster*. Cell 16:599–607

Winter CE, de Bianchi AG, Terra WR, Lara FJS (1980) Protein synthesis in the salivary glands of *Rhynchosciara americana*. Dev Biol 75:1–12

Note added in proof

Orr-Weaver T, Spradling AC (unpublished) is now:

Orr-Weaver T, Spradling AC (1986) *Drosophila* chorion gene amplification requires on upstream region regulating s18 transcription. Mol Cell Biol 6:4624–4633

Spradling AC et al. (unpublished) is now:

Spradling AC, de Cicco DV, Wakimoto BT, Levine JF, Kalfayan LJ, Cooley L (1987) Amplification of the X-linked *Drosophila* chorion gene cluster requires a region upstream from the s38 chorion gene. EMBO J 6:1045–1053

The Significance of Plant Transposable Elements in Biological Processes

Zs. Schwarz-Sommer [1]

1 Introduction

Development and differentiation of living organisms as individuals and evolution of populations leading to new species have been and still are two major problems fascinating scientists.

The process of the organized development of an individual is puzzling because it requires a program which is inherent in the living system itself. At the DNA level at which this program is realized a mechanism may exist that dictates at what time and to what extent genes are expressed or shut off. Flexibility at this level is only required for the ability to respond to specific signals in a concerted way.

The process of evolution requires flexibility of individual genomes and has no program. Inherent to all living systems, however, are mechanisms to realize flexibility which at the DNA level means to generate diversity. This diversity then can serve as a playground for subsequent selection.

The complexity of the processes necessary for development of an organism as well as for its evolution prohibits asking in either cases for a single principle explaining all phenomena. It is the purpose of this paper to point out one possible tool of nature, namely transposable elements, contributing to diversification of genomes and genes as a consequence of evolution as well as their potential as pacemakers during development. The long history of their genetics and the recent, rapid accumulation of molecular data on plant transposable elements is the main reason for concentrating on plants in this report.

2 A Brief Introduction into the Structure and Function of Plant Transposable Elements

Many recent reviews deal with the versatility, structure and function of plant transposable elements (Fedoroff 1983; Nevers et al. 1986; Döring and Starlinger 1986). Here, I only wish to summarize those characteristics which make the understanding of the following parts easier.

1. The presence of a transposable element within a gene is manifested by mutational instability (Fig. 1). Phenotypically this is revealed by somatic variegation.

[1] Max-Planck-Institut für Züchtungsforschung, Egelspfad, 5000 Köln 30, FRG.

Results and Problems in Cell Differentiation 14
Structure and Function of Eukaryotic Chromosomes
Edited by W. Hennig

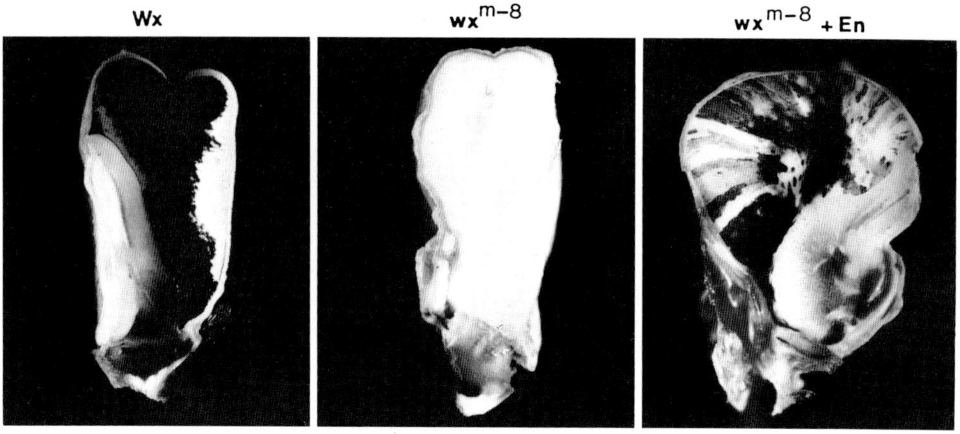

Fig. 1. Somatic instability at the waxy (*Wx*) locus resulting in a variegated phenotype in the endosperm of the wx^{m-8} mutant of *Zea mays*. Wild-type *Wx* endosperm (*left*) stains dark with iodine, whereas insertion of a 2.2-kb-long receptor element termed inhibitor or I in the mutant wx^{m-8} (McClintock 1962; Schwarz-Sommer et al. 1984) decreases gene expression and abolishes staining (*middle*). Due to somatic excision events triggered by the presence of the autonomous Enhancer (*En*, see Peterson 1953, 1965) element stained sectors appear in the wx^{m-8} endosperm. (Schwarz-Sommer et al. 1984)

Integration of an element alters or abolishes the expression of the gene involved in the manifestation of a given phenotype. However, somatically the locus can be restored if the element excises. Since this may not occur in every cell of a developing tissue, a mosaic phenotype will result.

2. The process of excision may be governed by the element residing at the mutable locus (autonomous or regulatory element). Receptor or nonautonomous elements are transposition-deficient: they can only excise if an autonomous element belonging to the same family of elements is also present in the genome, thus providing receptor elements with functions necessary for the excision event. In molecular terms receptor elements are deletion derivatives of structurally related autonomous elements (Fedoroff et al. 1983; Pereira et al. 1985).

3. The majority of plant transposable elements possess terminal inverted repeats the size and the composition of which are specific for every element system. Exceptions to this rule exist in which either long, direct repeats at the termini are present or no prevalent structure can be detected (Fig. 2). We recently discovered by chance a retroposon-like structure in maize because it was integrated into an expressed gene. By these structural criteria every type of plant transposable element finds parallels in other organisms including bacteria, yeast, flies and mammals.

4. Upon integration of an element a specific number of base pairs at the target site of the insertion becomes duplicated (Fig. 2). This property of transposable elements is also universal.

5. Excision of an element from a gene may or may not restore the wild-type gene activity, and in the majority of cases it is accompanied by more or less complex rearrangements at the insertion site (for review, see Saedler and Nevers

ELEMENT	TERMINI	STRUCTURE	DUPLICATION	REFERENCE
Ac-Ds	IR(11)		8	*Fedoroff et al., 1983*
Tam3	IR(12)		8	*Sommer et al., 1985*
En(Spm)-I	IR(13)		3	*Schwarz-Sommer et al., 1984* *Pereira et al., 1985*
Tam1	IR(13)		3	*Bonas et al., 1984*
Mu1	IR(215)		9	*Barker et al., 1984*
cin3	IR(624)		9	*Blumberg, 1985*
Bs1	DR(304)		6	*Johns et al., 1985*
cin2	DR(142-146)		3	*Blumberg, 1985*
cin1	IR(6)		5	*Shepherd et al., 1984*
Tz86	no		10	*Dellaporta et al., 1984*
cin4	no		7(6)	*Schwarz-Sommer et al. 1986**

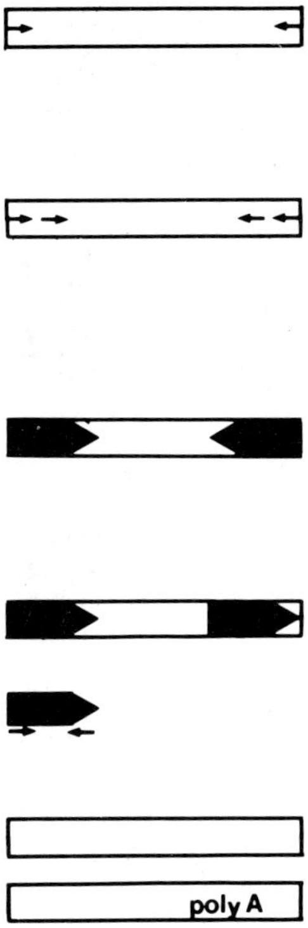

Fig. 2. Compilation of structures identified in plant insertion elements. The general architecture of each type of element is presented. For a more detailed compilation, see Doring and Starlinger (1986). * Indicates that the sequence of cin4 is not yet published. *Small arrows* indicate short inverted sequence duplications. *Thick arrows* represent long regions at the termini of an element which may be in direct or inverted orientation. The homology of cin1 with a single copy of long terminal repeats of other eukaryotic elements is indicated by a *single arrow* containing small inverted repeats

1985). Imprecise excision of elements restoring wild-type gene activity, but not the wild-type gene structure, has only been found in plants.

3 How Transposable Elements May be Involved in Development

The term-controlling element as emphasized by McClintock (1956a, b, 1967) was supposed to indicate the ability of a piece of (transposable) genetic information, foreign to the locus, to alter expression of unrelated genes. In the following section I want to illustrate how this can be utilized for developmental control.

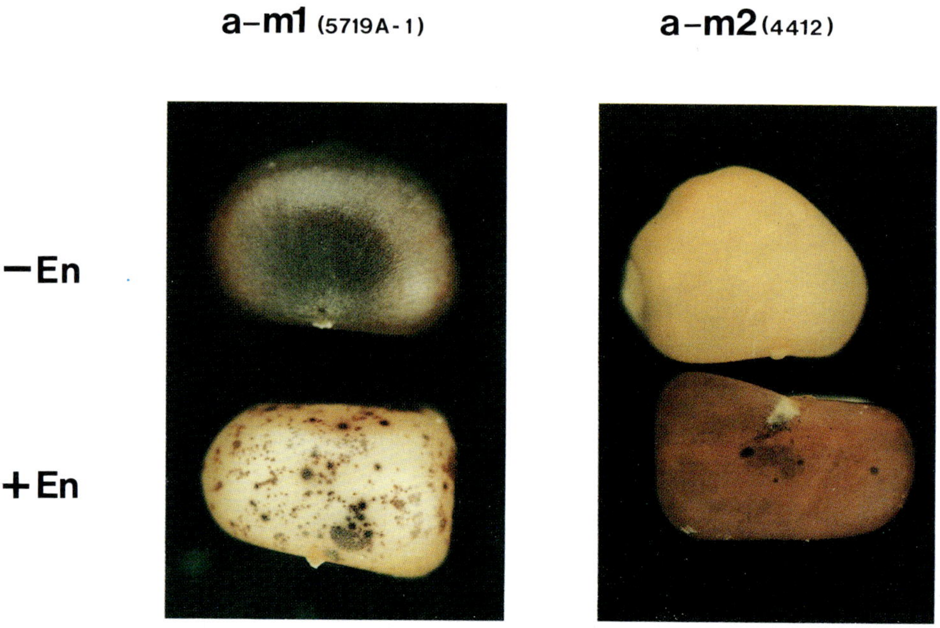

Fig. 3. Extreme effects of transposable elements on gene expression. The allelic mutants *a-m1 5719A-1* (McClintock 1965 and Schwarz-Sommer et al. 1985 b) and *a-m2 4412* (Reddy and Peterson 1985) were chosen to demonstrate two different, controlling-element-induced phenotypes. Both alleles are independent mutations caused by the insertion of an inhibitor element at the *A1* locus of *Zea mays*. In the absence of an autonomous element (En or Spm) the aleurone of kernels carrying the *5719A-1* allele is colored (*top left*), whereas the aleurone of those kernels carrying the *4412* allele (*top right*) is colorless. The presence of an autonomous element *suppresses* color in the *5719A-1* allele (*bottom left;* the dark spots arise by excision events restoring wild-type gene activity). In contrast, the autonomous element *induces A1* gene activity in the *4412* allele (*bottom right*). Molecular analysis of these alleles is described by Schwarz-Sommer et al. 1985 b and 1987

Alteration of gene expression by insertion mutations may depend on the site of insertion within a gene and on the structure of the inserted element itself. As illustrated in Fig. 3 in the absence of transpositional activity insertions may decrease gene expression (in this case aleurone color) or may hardly affect it. Molecular analysis of such mutants revealed differences in the structure of the corresponding elements indicating the role of element-specific sequences on the transcription of the affected genes (Gierl et al. 1985; Schiefelbein et al. 1985; Schwarz-Sommer et al. 1985 b; Tacke et al. 1986). Introduction of active elements into the same plants carrying the mutant alleles described above, again, either can suppress residual gene activity or can increase gene expression. In addition, somatic excision events can restore gene activity (colored spots) in a somatic cell lineage. Depending on the time scale of activity such restoration may occur early in development (large spots) or late (small spots) and can be more or less frequent. The common source of such varying effects is the autonomous element residing in the genome and dictating, by its own activity, when and where excisions may occur. If the autonomous element is expressed in a tissue-specific manner, its control over its receptor becomes tissue-specific as well (McClintock 1965; Peterson 1966; Fowler and Peterson 1978). Furthermore, since elements (receptors or autonomous ones) can change chromosomal locations, stocks can be generated in which several, otherwise independent, genes can become controlled simultaneously. Thus, transposable elements possess the potential to regulate developmental processes because they may exist as a module (the receptor element) affecting more than one gene which can be controlled in a concerted way by a common "pacemaker" (the autonomous element). The fact that the plant transposable elements studied so far do not seem to be involved in the control of developmentally relevant genes may solely indicate that the elements of importance have not yet been mobilized to give rise to phenotypically detectable mutations. One also can suspect that mobilization of such elements is deleterious to the organism; thus, they cannot become available by classical methods for genetic and molecular analysis.

4 Transposable Elements are Mobilized by Genomic Stress

All transposable elements detected in plants so far are natural components of the genome and they can become activated by genomic stress (for reviews, see Fedoroff 1983; Dellaporta and Chomet 1985; Nevers et al. 1986). The nature of the genomic stress seems to play a secondary role for this activation process and the effects may reflect a general response of the organism to "shock" or "danger" (McClintock 1984). Thus, transposable elements in maize became active in material exposed to chromosome breakage, X-ray exposure, by virus infection or by crosses between unrelated lines (for review, see Nevers et al. 1986). As outlined by McClintock (1978, 1984) most of the consequences of genomic shock or stress, like large-scale reorganization of the genome, but also more subtile differences, can easily be governed by transposable elements. Once activated mobile elements transpose and thus can occupy different chromosomal positions. In a freely breeding population this event generates the substrate for the inversion and deletion of large and small chromosome segments. All these rearrangements observed to be caused by genetically identified transposable elements (for reviews see Dellaporta and Chomet 1985; Nevers and Saedler 1977) can serve as a source of diversity for subsequent selection during evolution. The observation that structural properties of various middle repetitive and highly repetitive DNA segments of the plant genome resemble that of transposable elements (Flavell et al. 1981; Shepherd et al. 1984; Gupta et al. 1983, 1984) provides some support for the idea that transposition is one of the mechanisms spreading DNA to dispersed chromosomal locations.

5 Plant Transposable Elements as Generators of Sequence Diversity for Molecular Evolution

Excision of transposable elements from a locus, if it occurred in germinal cells, leads to a stable (nonmutable) progeny. Such an excision event is often seen as a phenotypic reversion, however, not always to the wild-type phenotype. In addition, biochemical analysis of proteins in wild-type revertants revealed a surprise: the product of genes affected by the mutation in the presence of an element was altered in the revertant progeny (Dooner and Nelson 1979; Echt and Schwartz 1981; Tuschall and Hannah 1982). Excision of the element therefore did not restore wild-type functions, but rather generated mutations. DNA sequence analysis of revertant genes shed light on the origin of these mutations, i.e. the result of imprecise excision of plant transposable elements (Sachs et al. 1983; Bonas et al. 1984; Pohlman et al. 1984; Weck et al. 1984; Schwarz-Sommer et al. 1985a; Sommer et al. 1985; Chen et al. 1986). This type of mutation termed footprints of elements (Schwarz-Sommer et al. 1985a), ranges from small rearrangements, such as inversions, duplications and small deletions affecting few nucleotides around the insertion site of the element, to more complex events resulting in large deletions (Sommer et al. 1985). All events analyzed so far can be explained by a molecular model proposed for excision of plant transposable elements (Saedler and Nevers 1985).

Insertion of elements can affect virtually every region of a transcription unit. Thus, the footprints generated after excision of the elements may have manifold consequences, affecting promoters as well as the primary structure of proteins. In contrast to classical point mutations (which also can be the result of excisions), more severe alterations can occur in one step after visitation of a gene by an element leading even to change in the size of the gene product.

In the search for evidence of such transposon-induced alterations during evolution it seemed promising to compare different alleles of a gene for the presence

```
LC          ATATGGAAAGC  CGACCT GGCGGAGG  AAGGCAGCTTCCACGA-     -AAGACCCTGAG GTAGATCAGT  CT..   ...T  GGCTGCACAA
a-m1        ATATGGAAAGC  CGACCT GGCGGAGG  AAGGCAGCTTCCACGA-     -AAGACCCTGAG GTAGATCAGT  CTCTAG CTCT  GGCTGCACAA
a-m(papu)   ATATGGAAAGC  CGA... ......GG  AAGGCAGCTTCCACGA-     -AAGACCCTGAG GTAGATCAGT  CT..   ...T  GGCTGCACAA

PROGENITOR?                 CGAGG                                                         CTCT
```

Fig. 4. Footprints of putative transposable elements within the *A1* gene of *Zea mays*. The DNA sequence of three alleles (each indicated at the *left*) of the wild-type *A1* gene have been determined (Schwarz-Sommer et al., manuscript in prep). The exon within the presented portion of the transcription unit is *boxed in*. Only that region of the exon is shown by sequence data which is relevant for the presentation of allelic differences. The putative progenitor sequence at the *bottom* indicates that the insertion of a putative transposable element might have occurred at this site generating the duplication of this small region. Upon excision the altered sequences (*shadowed boxes*) may be generated in the different alleles. Nucleotides unaffected by the excision event are indicated by *heavy letters*. The mechanism by which such alterations may be generated is described in detail by Saedler and Nevers (1985) of imprecise duplications present in one allele, but absent in the other. These then could be interpreted as an indication of visitation of the gene by a transposable element. As one of several examples a section of the *A1* gene of *Zea mays* is depicted in Fig. 4. This example shows that such footprints are not restricted to intron sequences (although their frequency there is much higher than in exons, see Schwarz-Sommer et al. 1985a; Werr et al. 1985; Zack et al. 1986). The putative excision event documented in this case leads to deletion (or insertion) of three amino acids within an expressed gene present in both variants in wild-type populations. The frequency in which this type of allelic alteration can be detected is limited because size differences in a functional protein may not be tolerated at every site.

Insertions and excisions of plant transposable elements are frequent events. Visitation by elements thus may lead to a rapid diversification of genes which in a population may become important for subsequent selection. Induction of mobility of transposable elements as a consequence of genomic stress could thus allow not only large-scale chromosomal rearrangements, but also could be part of the mechanisms involved in molecular evolution.

6 Conclusions

The present report attempts to discuss the role that plant transposable elements may play in developmental and evolutionary processes. The ability to transpose and to influence gene expression in a concerted manner suggests that such genetic entities can become significant in the developmental control of individuals. Transposition, furthermore, is one of the mechanisms required in large-scale chromosome rearrangements which are essential for the flexibility of genomes. More subtile rearrangements within expressed genes generated by excisions of transposable elements can become the source of molecular evolution of proteins by producing mutations which affect more than one amino acid. Since transposable elements are natural components of genomes and since they can be mobilized by genomic stress, they may be part of the mechanisms which generate sequence divergency essential for subsequent selection and adaptation of living systems to changing environmental conditions.

References

Barker RF, Thompson DV, Talbot DR, Swanson J, Bennetzen JL (1984) Nucleotide sequence of the maize transposable element Mul. Nucleic Acids Res 12:5955–5967

Blumberg vel Spalve J (1985) Die Cin2- und Cin3-DNA-Elemente aus *Zea mays*. PhD Thesis, University of Cologne

Bonas U, Sommer H, Saedler H (1984) The 17 kb Tam1 element of *Antirrhinum majus* induces a 3 bp duplication upon integration into the chalcone synthase gene. EMBO J 3:1015–1019

Chen CH, Freeling M, Merckelbach A (1986) Enzymatic and morphological consequences of Ds excisions from maize *Adh1*. Maydica XXXI:93–108

Dellaporta SL, Chomet PS (1985) The activation of maize controlling elements. In: Hohn B, Dennis ES (eds) Genetic flux in plants. Springer, Wien New York, pp 169–216

Dellaporta SL, Chomet PS, Mottinger JP, Wood J, Yu, SM et al. (1984) Endogenous transposable elements associated with virus infection in maize. Cola Spring Harbor Symp Quant Biol 49:321–328

Dooner HK, Nelson OE (1979) Heterogenous flavonoid glucosyltransferases in purpule derivatives from a controlling element-suppressed *bronze* mutant in maize. Proc Natl Acad Sci USA 76:2369–3371

Döring HP, Starlinger P (1986) Molecular genetics of transposable elements in plants. Annu Rev Genet 20:175–200

Echt CS, Schwartz D (1981) Evidence for the inclusion of controlling elements within the structural gene at the waxy locus in maize. Genetics 99:275–284

Fedoroff NV (1983) Controlling elements in maize. In: Shapiro JA (ed) Mobile genetic elements. Academic Press, London New York, pp 1–63

Fedoroff N, Wessler S, Shure M (1983) Isolation of the transposable maize controlling elements Ac and Ds. Cell 35:235–242

Flavell RB, O'Dell M, Hutchinson J (1981) Nucleotide sequence organization in plant chromosomes and evidence for sequence translocation during evolution. Cold Spring Harbor Symp Quant Biol 45:501–508

Fowler RG, Peterson PA (1978) An altered state of a specific En regulatory element induced in a maize tiller. Genetics 90:761–782

Gierl A, Schwarz-Sommer Zs, Saedler H (1985) Molecular interactions between the components of the En-I transposable element system of *Zea mays*. EMBO J 3:579–583

Gupta M, Bertram I, Shepherd NS, Saedler H (1983) Cin1, a family of dispersed repetitive elements in *Zea mays*. Mol Gen Genet 192:373–377

Gupta M, Bertram I, Shepherd NS, Saedler H (1984) Repetitive sequences and their organization on genomic clones. EMBO J 1:133–139

Johns MA, Mottinger J, Freeling M (1985) A low copy number, *copia*-like transposon in maize. EMBO J 4:1093–1102

McClintock B (1956 a) Intranuclear systems controlling gene action and mutation. Brookhaven Symp Biol 8:58–74

McClintock B (1956 b) Controlling elements and the gene. Cold Spring Harbor Symp Quant Biol 21:197–216

McClintock B (1962) Topographical relations between elements of control systems in maize. Carnegie Inst Wash Year Book 61:448–461

McClintock B (1965) The control of gene action in maize. Brookhaven Symp Biol 18:162–184

McClintock B (1967) Genetic systems regulating gene expression during development. Dev Biol Suppl 1:84–112

McClintock B (1978) Mechanisms that rapidly reorganize the genome. Stadler Genet Symp 10:25–48

McClintock B (1984) The significance of responses of the genome to challenge. Science 226:792–801

Nevers P, Saedler H (1977) Transposable genetic elements as agents of gene instability and chromosomal rearrangements. Nature 268:109–115

Nevers P, Shepherd NS, Saedler H (1986) Plant transposable elements. Adv Bot Res 12:103–203

Pereira A, Schwarz-Sommer Zs, Gierl A, Bertram I, Peterson PA, Saedler H (1985) Genetic and molecular analysis of the enhancer (En) transposable element system of *Zea mays*. EMBO J 4:17–23

Peterson PA (1953) A mutable pale green locus in maize. Genetics 38:682–683

Peterson PA (1965) A relationship between the Spm and En control systems in maize. Am Nat 99:391–398

Peterson PA (1966) Phase variation of regulatory elements in maize. Genetics 54:249–266

Pohlman RF, Fedoroff N, Messing J (1984) The nucleotide sequence of the maize controlling element activator. Cell 37:635–643

Reddy LV, Peterson PA (1985) Spm and I element changes with the *a-m2 8004* allele in maize. Mol Gen Genet 200:211–219

Sachs MM, Peacock WJ, Dennis ES, Gerlach WL (1983) Maize Ac/Ds controlling elements – A molecular viewpoint. Maydica 28:189–303

Saedler H, Nevers P (1985) Transposition in plants: a molecular model. EMBO J 4:585–590

Schiefelbein JW, Raboy V, Fedoroff NV, Nelson OE (1985) Deletions within a defective Suppressor-Mutator element in maize affect the frequency and developmental timing of its excision from the *bronze* locus. Proc Natl Acad Sci USA 82:4783–4787

Schwarz-Sommer Zs, Gierl A, Klösgen RB, Wienand U, Peterson P, Saedler H (1984) The Spm (En) transposable element controls the excision of a 2-kb DNA insert at the *wx-m8* allele of *Zea mays*. EMBO J 3:1021–1028

Schwarz-Sommer Zs, Gierl A, Cuypers H, Peterson PA, Saedler H (1985a) Plant transposable elements generate the DNA sequence diversity needed in evolution. EMBO J 4:591–597

Schwarz-Sommer Zs, Gierl A, Berndtgen R, Saedler H (1985b) Sequence comparison of "states" of *a-m1* suggests a model of Spm (En) action. EMBO J 4:2439–2443

Schwarz-Sommer Zs, Shepherd N, Tacke E, Gierl A, Rohde W, Leclercq L, Mattes M, Berndtgen R, Peterson PA, Saedler H (1987) Influence of transposable elements on the structure and function of the *A1* gene of *Zea mays*. EMBO J 6:287–294

Shepherd NS, Schwarz-Sommer Zs, Blumberg vel Spalve J, Gupta M, Wienand U, Saedler H (1984) Similarity of the cin1 repetitive familiy of *Zea mays* to eukaryotic transposable elements. Nature 307:185–187

Sommer H, Carpenter R, Harrison BJ, Saedler H (1985) The transposable element Tam3 of *Antirrhinum majus* generates a novel type of sequence alterations upon excision. Mol Gen Genet 199:225–231

Tacke E, Schwarz-Sommer Zs, Peterson PA, Saedler H (1986) Molecular analysis of states of the *A1* locus of *Zea mays*. Maydica XXXI:83–92

Tuschall DM, Hannah LC (1982) Altered maize endosperm ADP-glucose pyrophosphorylases from revertants of *shrunken-2-dissociation* allele. Genetics 100:105–111

Weck E, Courage U, Döring HP, Fedoroff N, Starlinger P (1984) Analysis of *sh-m6233*, a mutation induced by the transposable element Ds in the sucrose synthase gene of *Zea mays*. EMBO J 3:1713–1716

Werr W, Frommer WB, Maas C, Starlinger P (1985) Structure of the sucrose synthase gene on chromosome 9 of *Zea mays*. EMBO J 4:1373–1380

Zack CD, Ferl RJ, Hannah LC (1986) DNA sequence of a *shrunken* allele of maize: evidence for visitation by insertional sequences. Maydica XXXI:5–16

Arrangement of Chromosomes in Interphase Cell Nuclei

H. SAUMWEBER [1]

1 Introduction

Understanding of the architecture of the eukaryotic cell nucleus is an absolute requirement to explain the precisely coordinated processes of gene expression, genome replication and nuclear division.

Understanding of this architecture has been propagated at a different pace at different levels of organization. Much has been learned at the level of greatest resolution, however, surprisingly little is known about higher order organization. The nucleosome fiber has been generally accepted as the basic structural motif (Klug et al. 1980), however, different views already exist on its next level of coiling (Finch and Klug 1976; Jorcano et al. 1980). The looped domain organization has been outlined by several investigators (Benyajati and Worcel 1976; Cook and Brazell 1977; Hancock and Hughes 1982) and lacking information on organizational motifs in between both radial loop and sequential coiling models have been suggested for the highest level of chromatin organization in mitotic chromosomes (Marsden and Laemmli 1979; Sedat and Manuelidis 1977; Bak et al. 1977). The arrangement of a chromosome in an interphase nucleus and its interaction with other chromosomes is even less well understood.

It is this last aspect of chromosomal arrangement during interphase I would like to address in the present talk. At the beginning I will briefly summarize what has been concluded for interphase chromosomes from studies of mitotic and meiotic cells in different systems. Most of this topic has been covered by excellent reviews of Commings (1980) and Avivi and Feldman (1980). In a second part I will discuss what we have learned by the study of polytene cell nuclei.

2 Polar Chromosome Organization

The first reports on chromosomal order date back to 1885 when Rabl, studying prophase plant nuclei, suggested a model which predicts that chromosomes maintain an anaphase-like configuration and remain in distinct domains throughout the cell cycle (Rabl 1885). In his report the centromeres are grouped

[1] Institut für Entwicklungsphysiologie, Universität Köln, Gyrhofstr. 3, 5000 Köln, FRG.

Results and Problems in Cell Differentiation 14
Structure and Function of Eukaryotic Chromosomes
Edited by W. Hennig

together next to the centrioles with the chromosome arms arranged in parallel and attached at the opposite side of the nucleus. Later, Boveri (1888, 1909) provided evidence for this model and more recently, the so-called Rabl orientation has been described in a variety of plant and animal cell nuclei (Comings 1980).

Centromeric and telomeric elements have been considered as responsible for this kind of organization and also as indicative for its persistence in interphase (Fussell 1975). Consequently, their arrangement in interphase nuclei has been tested using a variety of techniques. By autoradiography of late labelling centromeric and telomeric regions Fussell (1975) demonstrated a Rabl-like organization in onion root tips and observations made by Ellison and Howard (1981) using a fluorescent DNA-binding dye on *Drosophila* blastoderm embryos suggested a similar organization, which in addition showed orientation with respect to the axes of the embryo. However, Hsu et al. (1971) using a staining technique specific to constitutive heterochromatin suggested a cell type-specific pattern for the arrangement of centromeres in mouse cells. A scattered centromere arrangement was also demonstrated by Moroi et al. (1981) in mammalian cell lines using antibodies specific to centromeres.

It has been discussed whether this Rabl orientation is a passive relic of previous mitoses or whether it has some functional meaning. One of the suggestions was that it may be a prerequisite to orderly meiotic pairing. The so-called bouquet formation in meiotic prophase represents a similar arrangement (Hughes-Schrader 1943; Moens 1969; see Fig. 1 a, b) and Hughes-Schrader has shown that it is brought about by active movement of the chromosomes (Hughes-Schrader 1943). She also argued that this process might be independent of a passive telophasic orientation.

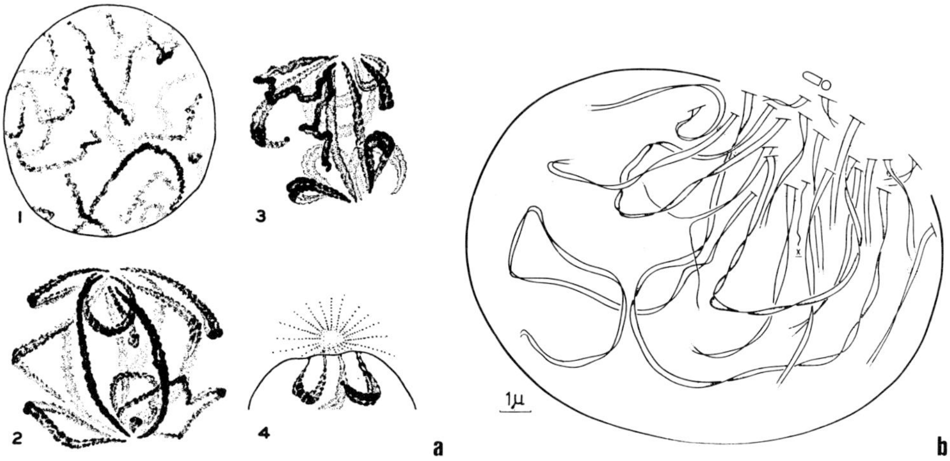

Fig. 1 a, b. Bouquet formation at meiosis. **a** Observed in mantids by Hughes-Schrader using light microscopy (*1* nonpolarized early pachytene; *2* second polarization in mid-pachytene; *3* later stage in second polarization; *4* detail of same after different staining.) **b** After reconstruction from EM photographs of serially sectioned spermatocyte nuclei of locusts by Moens. All autosomal bivalents form a complete loop. Note the attachment of the chromosome ends to a confined region on the nuclear envelope. (**a** Hughes-Schrader 1943 and **b** Moens 1969)

Another consequence of the Rabl model is that the chromosomes occupy distinct nuclear domains throughout the cell cycle. Boveri demonstrated 1888 that this was true in the case of *Ascaris* egg nuclei (Boveri 1888) and more recently this has also been observed in other species. Curiously, in some species chromosomes are packed in discrete vesicles during certain stages of the cell cycle (see Ito et al. 1981 and references therein). Chromosomes have been visualized as discrete entities during the cell cycle by Stack and colleagues (1977) using specific staining methods and after reconstruction from serial sections of amphibian interphase nuclei by Murray and Davies (1979). However, due to limitations of preservative and analytical methods, any further conclusions on higher order structure are precluded in both cases. By UV laser irradiation of distinct nuclear areas with the subsequent recording of unscheduled DNA synthesis or by localization of the distribution of irradiated chromatin with specific antibodies, Cremer and coworkers could also demonstrate a territorial organization of interphase chromosomes (Zorn et al. 1979; Hens et al. 1983).

In conclusion, there is substantial evidence for a Rabl orientation of interphase chromosomes in a variety of species and cell types. However, how generalizable it is, and whether it is a passive relic of mitosis (stable and undisrupted by preparation methods in some cell types) or whether it has a functional meaning (at least in some cell lineages) remains open. In addition, this section illustrates the incompleteness and uncertainty in our knowledge concerning interphase chromosome structure which, at present, is mainly due to the lack of suitable methods for detailed observation and analyses without disruption of the existing organization.

3 Chromosomal Interaction

Evidence for a nonrandom placement of interphase chromosomes in cell nuclei has been suggested from the observation of their nonrandom arrangement on the metaphase plate and from a nonrandomness of exchanges following treatment with chromosome-breaking agents. In principle, three types of interaction can be taken into consideration:

1. Interaction between nonhomologous chromosomes within a haploid set;
2. Interaction between homologues within a set; and
3. Interaction between different sets of chromosomes as in interspecies hybrids.

1. Nonrandom interaction between nonhomologous chromosomes was first reported by Hughes-Schrader (1946) and subsequently reported in many plant cells (for a review see Avivi and Feldman 1980). Ashley (1979) analyzed haploid generative nuclei and prophase nuclei from root tips of *Ornithogalum virens*. She concluded that the same organization persisted in both cell types, resulting in the diploid case in a ring of chromosomes where the homologues were opposite one another (see Fig. 2). Costello (1970) reported an identical order of the chromosome set in both gametes of *Polychoerus carmelensis* in a sectioned egg at first cleavage metaphase.

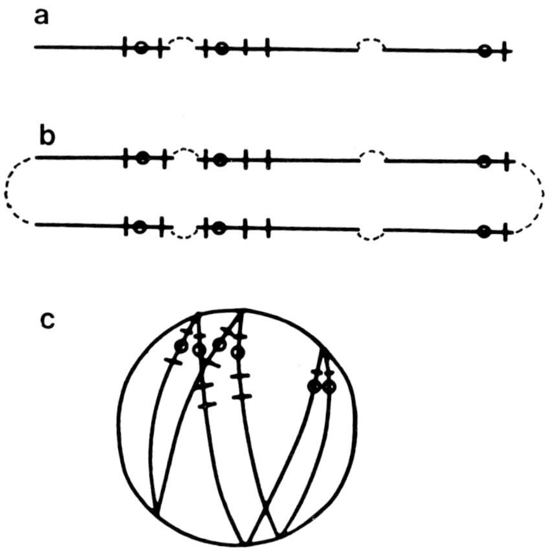

Fig. 2. Conserved chromosome arrangement in the diploid nucleus of *Ornithogalum virens* as proposed by Ashley. **A** The arrangement of the three nonhomologous chromosomes in a generative pollen nucleus. **B** Proposed arrangement at time of fertilization. **C** The diploid nucleus showing **B** folded up. *Circles* represent centromeres; *short bars* the location of C-bands; the end to end connections are shown by *dashed lines*. Note the Rabl orientation. The model has been deduced from conserved telomeric contacts in haploid generative nuclei and from diploid root tip nuclei during prophase. (Ashley 1979)

Additional evidence for a specific spatial relationship within one chromosomal set resulted from studies of the distribution of chromosomal exchanges. Studies in plants (Sax 1940; Evans and Bigger 1961; Kumar and Natarajan 1966) indicated a nonrandom arrangement and Werry et al. (1977) correlated a specific pattern inferred from the frequency of induced interchanges following irradiation with a nonrandom position observed at metaphase.

An ordered arrangement of nonhomologues was noted by Avivi et al. (1982a) measuring mean distances of metaphase chromosomes in squashed root tip cells of common wheat. Similarly, Heslop-Harrison and Bennett (1983) suggested a highly nonrandom arrangement of chromosomes after evaluating the distances between centromeres as determined from models of serial sectioned root tip cells from *Aegilops umbellulata*. Bennett (1983) suggested a model in which chromosomal arrangement should be predictable. According to this model, chromosomes in a haploid genome are uniquely ordered according to corresponding arm length. However, although Coates and Smith (1984) found evidence for a nonrandom association of chromosomes in the grasshopper, there was no evidence for an order predicted by the Bennett model.

2. Somatic association of homologues during the mitotic cycle has been clearly demonstrated in Diptera by Metz (1916) and was also reported by classical plant cytologists early this century (see Avivi and Feldman 1980). It has been suggested that homologue pairing, which is required during meiosis, is maintained

throughout the life cycle of an organism. In plants this tendency of pairing was corroborated by more recent investigations of squashed mitotic cells (cf. Avivi et al. 1982a) and it can be concluded from analyses of radiation- or chemically-induced interchanges that this arrangement persists during interphase (Evans and Bigger 1961; Kumar and Natarajan 1966; Werry et al. 1977). Conflicting results reported by Bennett (1983) can be reconciled with the common idea if one assumes that this association is expressed spatially along the chromosome and not necessarily confined to the centromere region used in Bennett's measurements.

However, there appear to be conflicting results with respect to the occurrence of somatic pairing in mammals. Most of the data have been obtained by analyzing squashed metaphase chromosomes and it may be that the chromosomal arrangement is fragile to this treatment. This may be particularly true during metaphase when pairing is potentially disrupted by the spindle as has been demonstrated by Becker (1969) in *Drosophila* heterozygotes for the normal X and the sc^8 inversion. In addition, statistical treatment of the data may be inadequate (Coll et al. 1980; Lacadena and Ferrer 1978). Although more indirectly, the data of Zorn et al. (1979) on Chinese hamster ovary cells suggest that somatic pairing may not be very tight. After UV laser irradiation of interphase cells the authors found no significantly higher damage in homologous compared to heterologous chromosomes in the following metaphase.

3. Avivi et al. (1982b) in analyzing mean distances of chromosomes in common wheat found significantly greater distances between chromosomes of different genomes than between nonhomologues of the same genome. Bennett (1983) suggested a similar genome separation in barley and concluded: "... whatever its basis, it follows that the simple haploid genome is a basic structural unit in nuclear architecture. But what of ordered arrangements within this unit?"

In conclusion, the data suggest a significant degree of chromosomal interrelationship. It remains to be established whether there is a strict order between the heterologous chromosomes or whether this arrangement is more relaxed and loosely determined as discussed by Comings (1980). Homologue pairing as a prerequisite to orderly meiotic pairing might be maintained in a more or less tight fashion throughout the life cycle of an organism and association of haploid genomes may be taken as an indication of an existing general nuclear architecture.

4 A Direct Approach to Interphase Chromosome Arrangement: The Polytene Cell

For a direct approach to the problem of chromosome arrangement in interphase, the nuclei of intact polytene cells of Diptera would appear to provide a convenient model system. Data obtained by studying the phenomenon of ectopic pairing (Kauffmann and Iddless 1963) and the observation of nonrandom telomere contacts (Hinton and Atwood 1941; Berendes and Meyer 1968; Kauffmann and Gay 1969) already suggested a certain degree of order. In 1975 Skaer and Whytock published their interpretation of the three-dimensional structure of living nuclei from *Drosophila, Simulium,* and *Chironomus*. Their data were obtained

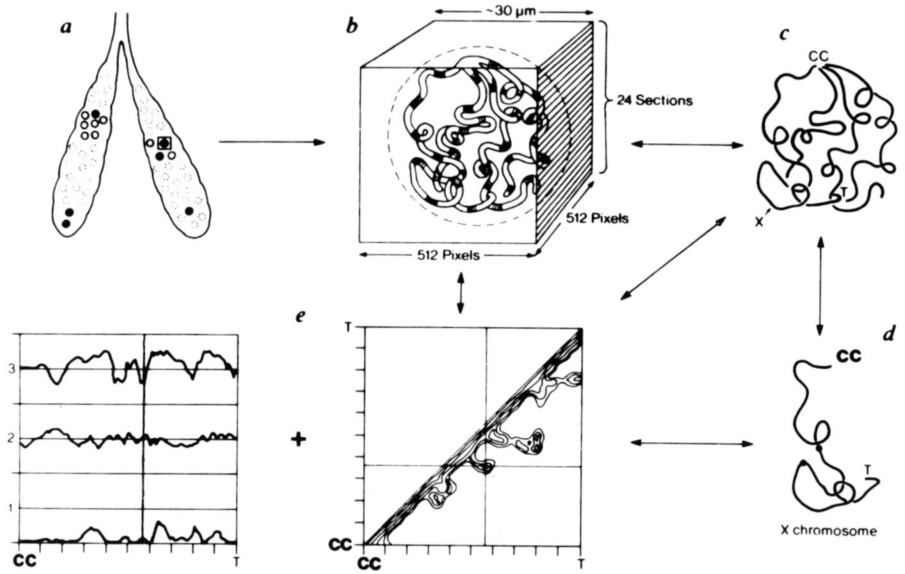

Fig. 3 a–e. Data recording, model building and quantitative evaluation processes. **a** Nuclei of a sample gland which was used for analyses. **b** Stack of images from which the chromosome paths were traced using an interactive modelling program. **c** Resulting stick figure model which can be correlated to the cytological map position by reference to the image stack. **e** The chromosome pathways can be analyzed for several parameters and after normalization of the graph axes to the cytological map positions nuclei from different glands can be compared. The crosshair in **e** corresponds to the position of the loci in the model, which are shown in **d** by the *two black dots*. (For details see Mathog et al. 1984; Hochstrasser et al. 1986) (Mathog et al. 1984)

by optical sectioning of nuclei in living glands using Nomarski optics. However, their conclusions were rather general mainly due to the complexity of the observed structures with a concomitant lack of adequate means for analyzing and interpreting their data.

Sedat and co-workers (Agard and Sedat 1983; Mathog et al. 1984; Gruenbaum et al. 1984; Hochstrasser et al. 1986) using a similar approach studied the three-dimensional arrangement of polytene chromosomes in the *Drosophila* salivary gland. I would like to summarize their findings in this section.

Nuclei in intact glands were stained with a DNA-specific fluorescent dye and recorded in a series of images of optical sections, which were stored in a computer. We followed the path of the chromosome arms in the image stack using an interactive modelling program. Subsequently, quantitative properties could be derived from the resulting stick figure models and could be used for a detailed description of the chromosome folding (see Fig. 3). By using this approach the models of 24 nuclei from three glands have been evaluated.

By a direct inspection of the stereo-pair models of the nuclei (see Fig. 4) several organizational features can be demonstrated.

1. The chromosome arms are always maintained in separate spatial domains, respecting each other's domain, although they are highly contorted and closely packed.

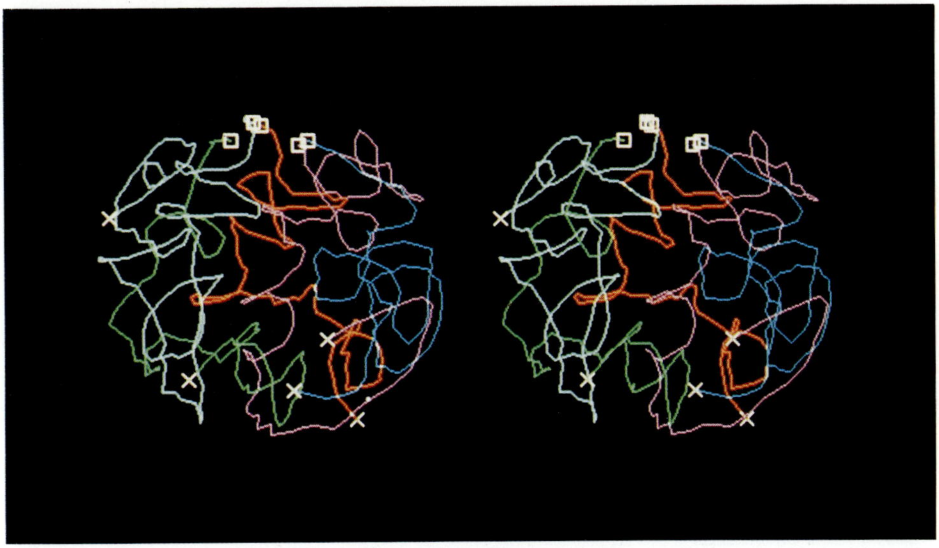

Fig. 4. Stereo-pair of a stick figure model of a salivary gland nucleus. The major chromosome arms are given in different colors. *Squares* refer to the centomere; *crosses* to the telomere ends. Note the Rabl orientation. (Hochstrasser et al. 1986)

Figure 5 see page 231

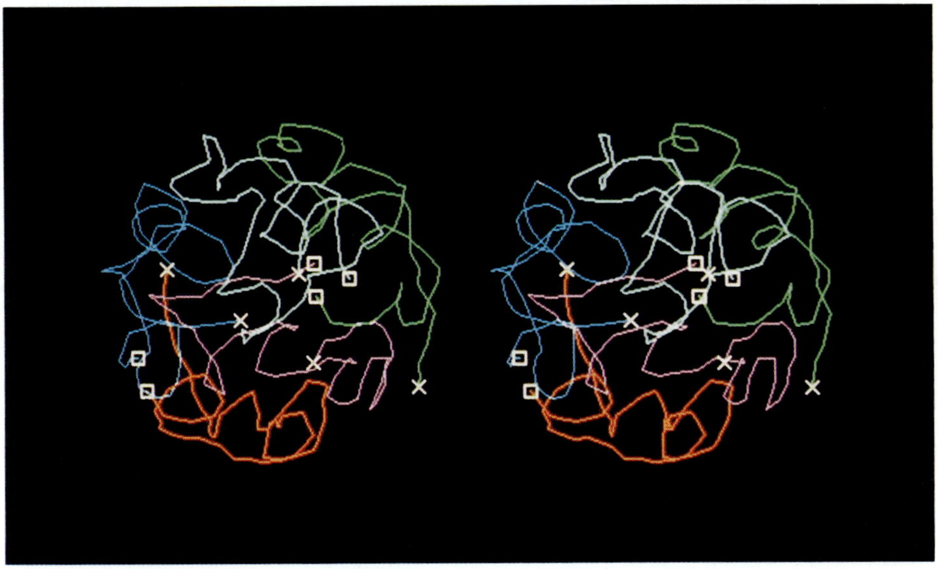

Fig. 6. Stereo-pair of the stick figure model of a nucleus with split chromosome center. *Symbols* as in Fig. 4. Note that the chromosomes are still in separate domains. (Hochstrasser et al. 1986)

2. The aggregated centromere regions are always positioned against the nuclear envelope as has been reported for *Drosophila* embryos (Ellison and Howard 1981). However, there is no preferred arrangement of the chromosome center with respect to the axes of the gland.

3. The telomeres of the chromosomes are in 80% of the cases in the hemisphere opposite the chromosome center.

4. The chromosome arms are almost always next to each other. The X chromosome, although almost always between the two autosomes, is not preferentially positioned next to a particular arm.

So far the data are consistent with a Rabl orientation. If this is a structural relic of the final mitosis in the embryo, then the data imply a high degree of positional stability, since the nuclei have increased in volume several hundredfold and there have been nine to ten rounds of internal DNA replication. Another striking feature is that the chromosomes strictly respect their domains. It can be inferred that the same may be true for the diploid interphase chromosome with the effect that the chromosomes do not interfere with each other during mitosis.

Further information on chromosome organization has been extracted by quantitative analyses of the models.

5. The chromosome coils, in general, appear to be right-handed as can be shown using a standard vector operation. This was not immediately apparent by visual inspection and this may account for the fact that no such chirality has been reported previously.

6. A set of 15 loci regularly contacts the nuclear envelope with a high frequency (see Fig. 5). All but one of these loci coincide with positions of intercallary heterochromatin and in physical sections they often appear to be attached to the nuclear surface. There were additional loci which often appeared to be in contact with the nuclear envelope and which corresponded to known positions of intercallary heterochromatin. However, a larger number of nuclei is required to prove the significance of this observation. The heterochromatic nature of these loci may allow them to bind more firmly to the reforming nucleus following mitosis, thus accounting for the observed selectivity. Quick (1980) showed the adhesion of ectopic fibers to the nuclear envelope and heterochromatic material is also known to bind to nuclear envelopes in diploid nuclei (cf Comings 1980). We also observed a number of sites which are not constantly at the nuclear periphery. There are no particular chromatin structures correlated to these sites, so their biological significance remains to be established.

7. Five nuclei have been found with the chromosome center region split (either a chromosome 2 or chromosome 3 was split off; see Fig. 6 on page 229). Apart from centromere-telomere polarization all the structural features outlined in the previous sections still exist in these nuclei. This implies that (1) the structures at the nuclear envelope to which the chromosomes attach are not uniquely configured relative to these sites and it is also unlikely that the nuclear envelope per se defines the geometry of a chromosome. (2) The chromosomal domains are apparently established and maintained by a chromosomal impedance rather than by a general nuclear scaffolding scheme. (3) Obligatory pairwise interactions between loci on different chromosomes might be improbable. However, it does not preclude that the chromosome arm in its domain may interact with itself.

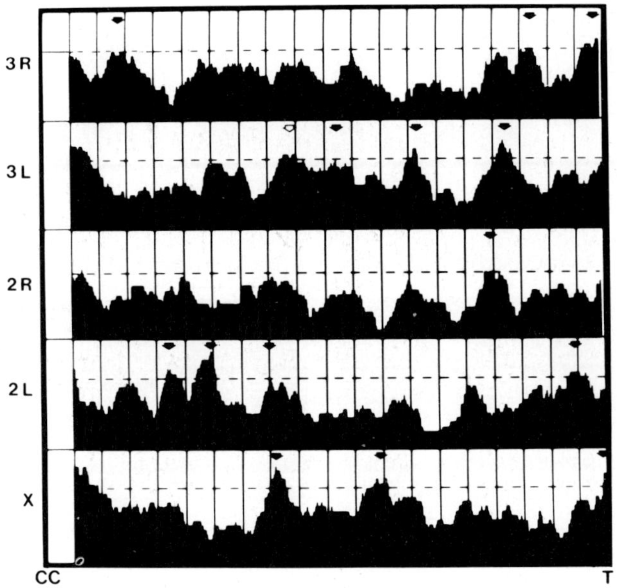

Fig. 5. Nuclear surface contact frequencies for the five major chromosome arms. Recorded were loci within 1 μ of the surface. The cytological position is plotted along the abscissa with the chromosome center (*CC*) to the *left* and the telomere (*T*) on the *right*. The ordinate range is 0–24 (24 nuclei analyzed). A cutoff frequency was determined for each chromosome arm to exclude the recording of contact sites occurring by chance. For details see Hochstrasser et al. (1986). The cutoff is shown by the *dashed line*. *Arrows* mark the position of the contact sites; *dark arrows* indicate an overlap with the position of intercallary heterochromatin, which is not the case for the *white arrow*. (Hochstrasser et al. 1986)

8. In order to determine whether the single chromosome arms assume distinct configurations, rank ordering of intradistance plots has been performed (Hochstrasser et al. 1986). In summary, the results show that there is no long-range interaction of loci more than two cytological subdivisions apart, which can be observed in a large fraction of the nuclei examined. Either no long-range interactions exist or there are several subsets of such interactions depending on the cell position within the gland, or a particular configuration is only important in early diploid stages of development.

9. In addition, we performed a preliminary analysis of the spatial arrangement of active genes on chromosomes in fixed polytene nuclei. Comparison of the distribution of chromosomal loci marked by antibodies against RNA-binding proteins in spread chromosomes and in intact optical sectioned nuclei suggested a nonrandom clustering of these sites in the nuclear space. Data obtained from reconstructed nuclei of physically sectioned glands which had been labelled by a pulse of tritiated uridine, a DNA-specific fluorescence dye, and antibodies against RNA-associated proteins again led to the observation that active sites might be restricted in their spatial distribution to certain nuclear regions (clustering). Spatially close sites, although probably apart on linearly arranged chromosomes, ap-

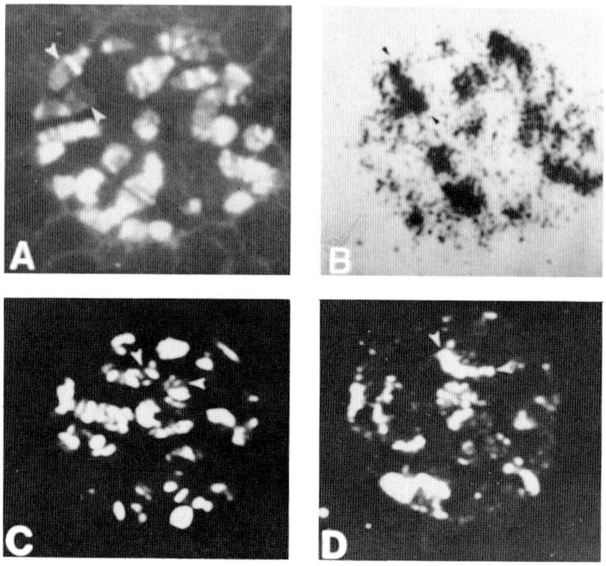

Fig. 7 A–D. Physically sectioned salivary gland nuclei. In **A, B** the glands were pulse-labelled. **A** DNA staining by a fluorescence dye; **B** autoradiography of the same section. **C, D** The distribution of an RNA-binding protein. See DNA staining as above. **D** Indirect immunofluorescence with an antibody against an RNA-binding protein. Some of the connections between active sites are indicated by *arrows*. (Gruenbaum et al. 1984)

pear connected to each other as was apparent both from the distribution of uridine label and antibody (see Fig. 7).

5 Concluding Remarks

The existence of a nonrandom spatial organization within the interphase nucleus appears to be beyond any doubt. Questions pertain to their flexibility and importance for cellular processes. At the one extreme there is the view of a highly ordered structure enabling the concerted regulation of gene batteries and directing the export of their products into a structured cytoplasm, as has been hypothesized recently (Blobel 1985). At the other end is the conservative statement that the observed organization merely reflects the packing constraints for the chromosome arms inside the nuclei. In summarizing that which is presently known, I have tried to elucidate that reality might be somewhere in between and I also have tried to demonstrate that we now have the tools to determine where it is.

References

Agard DA, Sedat JW (1983) Three-dimensional architecture of a polytene nucleus. Nature 302:676–681

Ashley T (1979) Specific end-to-end attachment of chromosomes in *ornithogalum virens*. J Cell Sci 38:357–367

Avivi L, Feldman M (1980) Arrangement of chromosomes in the interphase nucleus of plant cells. Hum Genet 55:281–295

Avivi L, Feldman M, Brown M (1982a) An ordered arrangement of chromosomes in the somatic nucleus of common wheat, *Triticum aestivum* L. Chromosoma 86:1–16

Avivi L, Feldman M, Brown M (1982b) An ordered arrangement of chromosomes in the somatic nucleus of common wheat, *Triticum aestivum* L. Chromosoma 86:17–26

Bak AL, Zeuthen J, Crick FHC (1977) Higher-order structure of human mitotic chromosomes. Proc Natl Acad Sci USA 74:1595–1599

Becker HJ (1969) The influence of heterochromatin, inversion-heterozygosity and somatic pairing on X-ray induced mitotic recombination in *Drosophila melanogaster*. Mol Gen Genet 105:203–218

Bennett MD (1983) The spatial distribution of chromosomes. In: Brandham P, Bennett MD (eds) Proc 2nd Kew Chromosome Conf. Allen and Unwin, London, pp 71–79

Benyajati C, Worcel A (1976) Isolation, characterization, and structure of the folded interphase genome of *Drosophila melanogaster*. Cell 9:393–407

Berendes HD, Meyer GF (1968) A specific chromosome element, the telomere of *Drosophila* polytene chromosomes. Chromosoma 25:184–197

Blobel G (1985) Gene gating: a hypothesis. Proc Natl Acad Sci USA 82:8527–8529

Boveri Th (1888) Zellen Studien. Jenaische Z Naturwissenschaft 22:685–882

Boveri T (1888) Die Befruchtung und Teilung des Eies von *Ascaris megalocephala*. Zellen-Studien 2:1–189, Fischer, Jena

Coates DJ, Smith D (1984) The spatial distribution of chromosomes in metaphase neuroblast cells from subspecific F_1 hybrids of the grasshopper *Caledia captiva*. Chromosoma 90:338–348

Coll MD, Cuadras CM, Egozcue J (1980) Distribution of human chromosomes on the metaphase plate. Symmetrical arrangement in human male cells. Genet Res 36:219–234

Comings DE (1980) Arrangement of chromatin in the nucleus. Hum Genet 53:131–143

Cook PR, Brazell IA (1977) The superhelical density of nuclear DNA from human cells. Eur J Biochem 74:527–531

Costello DP (1970) Identical linear order of chromosomes in both gametes of the acoel turbellarian *Polychoerus carmelensis:* a preliminary note. Proc Natl Acad Sci USA 67:1951–1958

Ellison JR, Howard GC (1981) Non-random position of the A-T rich DNA sequences in the early embryos of *Drosophila virilis*. Chromosoma 83:555–561

Evans HJ, Bigger TRL (1961) Chromatid aberrations induced by gamma irradiation. II. Non randomness in the distribution of chromatid aberrations in relation to chromosome length in *vicia faba* root-tip cells. Genetics 46:277–289

Finch JT, Klug A (1976) Solenoidal model for superstructure in chromatin. Proc Natl Acad Sci USA 73:1897–1901

Fussell CP (1975) The position of interphase chromosomes and late replicating DNA in the centromere and telomere regions of *Allium cepa* L. Chromosoma 50:201–210

Gruenbaum Y, Hochstrasser M, Mathog D, Saumweber H, Agard DA, Sedat JW (1984) Spatial organization of the Drosophila nucleus: A three-dimensional cytogenetic study. J Cell Sci Suppl 1:223–234

Hancock R, Hughes ME (1982) Organisation of DNA in the interphase nucleus. Biol Cell 44:201–212

Hens L, Baumann H, Cremer T, Sutter A, Cornelis JJ, Cremer C (1983) Immunocytochemical localization of chromatin regions UV-microirradiated in S-phase or Anaphase. Exp Cell Res 149:257–269

Heslop-Harrison JS, Bennett MD (1983) The spatial order of chromosomes in root-tip metaphases of *Aegilops umbellulata*. Proc R Soc Lond B 218:225–239

Hinton T, Atwood KC (1941) Terminal adhesions of salivary gland chromosomes in Drosophila. Proc Natl Acad Sci USA 27:491–496

Hochstrasser M, Mathog D, Gruenbaum Y, Saumweber H, Sedat JW (1986) Spatial organization of chromosomes in the salivary gland nuclei of *Drosophila melanogaster*. J Cell Biol 102:112–123

Hsu TC, Cooper JEK, Mace ML Jr, Brinkley BR (1971) Arrangement of centromeres in mouse cells. Chromosoma 34:73–87

Hughes-Schrader S (1943) Polarization, kinetochore movements, and bivalent structure in the meiosis of male mantids. Biol Bull 85:265–300

Hughes-Schrader S (1946) A new type of spermatogenesis in Icerine coccids with a linear alignment of chromosomes in the sperm. J Morphol 78:48–87

Ito S, Dan K, Goodenough D (1981) Ultrastructure and ^{3}H-Thymidine incorporation by chromosome vesicles in sea urchin embryos. Chromosoma 33:441–453

Jorcano JL, Meyer G, Day LA, Renz M (1980) Aggregation of small oligonucleosomal chains into 300 Å globular particles. Proc Natl Acad Sci USA 77:6443–6447

Kaufmann BP, Iddles MK (1963) Ectopic pairing in salivary-gland chromosomes of *Drosophila melanogaster*. Port Acta Biol Ser A 7:225–248

Kaufmann BP, Gay H (1969) The capacity of the fourth chromosome of *Drosophila melanogaster* to establish end-to-end contacts with the other chromosomes in salivary-gland cells. Chromosoma 26:395–409

Klug A, Rhodes D, Smith J, Finch JT, Thomas JO (1980) A low resolution structure for the histone core of the nucleosome. Nature 287:509–516

Kumar S, Natarajan AT (1966) Kinetics of two-break chromosome exchanges and the spatial arrangement of chromosome strands in interphase nucleus. Nature 209:796–797

Lacadena JR, Ferrer E (1978) Homologous somatic association in *Crepis* species: a critical comparison of several statistical approaches. Chromosoma 67:77–86

Marsden MPF, Laemmli UK (1979) Metaphase chromosome structure: Evidence for a radial loop model. Cell 17:849–858

Mathog D, Hochstrasser M, Gruenbaum Y, Saumweber H, Sedat J (1984) Characteristic folding pattern of the polytene chromosomes in Drosophila salivary gland nuclei. Nature 308:414–421

Metz CW (1916) Chromosome studies in Diptera. II. The paired association of chromosomes in the Diptera and its significance. J Exp Zool 21:213–279

Moens PB (1969) The fine structure of meiotic chromosome polarization and pairing in *Locusta migratoria* spermatocytes. Chromosoma 28:1–25

Moroi Y, Hartman AL, Nakane PK, Tan EM (1981) Distribution of kinetochore (centromere) antigen in mammalian cell nuclei. J Cell Biol 90:254–259

Murray AB, Davies HG (1979) Three-dimensional reconstruction of the chromatin bodies in the nuclei of mature erythrocytes from the newt *Triturus cristatus:* The number of nuclear envelope attachment sites. J Cell Sci 35:59–66

Quick P (1980) Junctions of polytene chromosomes and the inner nuclear membrane. Experientia 36:456–457

Rabl C (1885) Über Zelltheilung. Morph Jahrb 10:214–330

Sax K (1940) An analysis of X-ray induced chromosomal aberrations in Tradescantia. Genetics 25:41–68

Sedat J, Manuelidis L (1977) A direct approach to the structure of eucaryotic chromosomes. Cold Spring Harbor Symp Quant Biol 42:331–349

Skaer RJ, Whytock S (1975) Interpretation of the three-dimensional structure of living nuclei by specimen tilt. J Cell Sci 19:1–10

Stack S, Brown D, Dewey W (1977) Visualization of interphase chromosomes. J Cell Sci 26:281–291

Werry PAThJ, Stoffelsen K, Engels FM, Van der Laan F, Spanjers AW (1977) The relative arrangement of chromosomes in mitotic interphase and metaphase in *Haplopappus gracilis*. Chromosoma 62:93–101

Zorn C, Cremer C, Cremer T, Zimmer J (1979) Unscheduled DNA synthesis after partial UV irradiation of the cell nucleus. Exp Cell Res 124:111–119

Heterochromatin and the Phenomenon of Chromosome Banding

D. Schweizer, J. Loidl, B. Hamilton [1]

1 Mammalian G-/R-Bands

1.1 DNA Composition in G- and R-Bands

1.1.1 DNA Base Pair-Specific Fluorescent Dyes as Labels for dA+dT-Rich G-Bands and dG+dC-Rich R-Bands

DNA-binding, base pair-specific fluorescent dyes impart a longitudinal differentiation in mammalian metaphase chromosomes consisting of specific nonperiodical patterns of alternating bright and dull bands of characteristic width. Human prometaphase chromosomes exhibit more than 2000 light and dark bands. A very small band at the limit of light microscope resolution contains about 1250 kb (Holmquist 1986a).

AT-specific fluorescent dyes such as quinacrine, 33258 Hoechst and DAPI illuminate the so-called G-bands (see Schweizer 1981). The GC-specific dyes chromomycin A_3, mithramycin, and olivomycin impart a reciprocal fluorescent pattern to the chromosomes (Schweizer 1976; van de Sande et al. 1977). Thus, fluorochromes with complementary DNA base specificity stain R-bands and G-bands in a reverse manner as if the two kinds of bands differ in overall base composition. DNA biochemical data support this conclusion (Table 1). The overall difference visualized by DNA affinity labels is likely to reflect the compositional compartmentalization of the mammalian genome into a mosaic of isochores (Bernardi et al. 1985), very long DNA segments predominantly AT- or GC-rich.

Earlier, the use of pairs of selected DNA ligands provided additional indirect evidence that G- and R-bands have reciprocal base pair bias (Schweizer 1976, 1980, 1981). The staining difference of a fluorescent primary stain exhibiting a given base pair binding and/or fluorescence specificity is enhanced when an appropriate nonfluorescent counterstain with complementary binding affinity is employed (Fig. 1). The enhancement mechanisms have been studied in some detail (Sahar and Latt 1978, 1980; Schweizer 1981). Either binding competition or energy transfer are predominantly responsible for the observed contrast enhancement. Energy transfer between dyes bound to DNA in proximity is efficient over approximately 35 Å (about 10 bp) (Sahar and Latt 1980). This result can be reconciled with a model where the enhanced chromosome bands that are resistant

[1] Institut für Botanik der Universität Wien, Rennweg 14, 1030 Wien, Austria.

Results and Problems in Cell Differentiation 14
Structure and Function of Eukaryotic Chromosomes
Edited by W. Hennig

Table 1. Cytochemical and molecular properties of R-band DNA and G-band DNA

R-bands	G-bands	
Quinacrine dull	Quinacrine bright	[1]
DAPI/actinomycin D dull	DAPI/actinomycin D bright	[2]
Chromomycin/distamycin bright	Chromomycin/distamycin dull	[2]
Potential zDNA-forming sites	–	[3]
GC-rich	AT-rich (e.g., in Chinese hamster 3.2% richer in AT than R-bands)	[4]
GGCGGG-rich promoters	CAAT, TATA box promoters	[5]
CCGG-rich	CCGG poor	[5, 6]
Alu sequences	Hind III repeat sequences	[7]
Constitutive genes	Tissue-specific genes	[8]
Early replicating	Late replicating	[9]

[1] Caspersson et al. (1972)
[2] Schweizer (1981)
[3] Viegas-Péquignot et al. (1983)
[4] Holmquist et al. (1982)
[5] Holmquist (1986a)
[6] Holmquist and Caston (1986)
[7] Manuelidis and Ward (1984)
[8] Goldman et al. (1984)
[9] Ganner and Evans (1971)

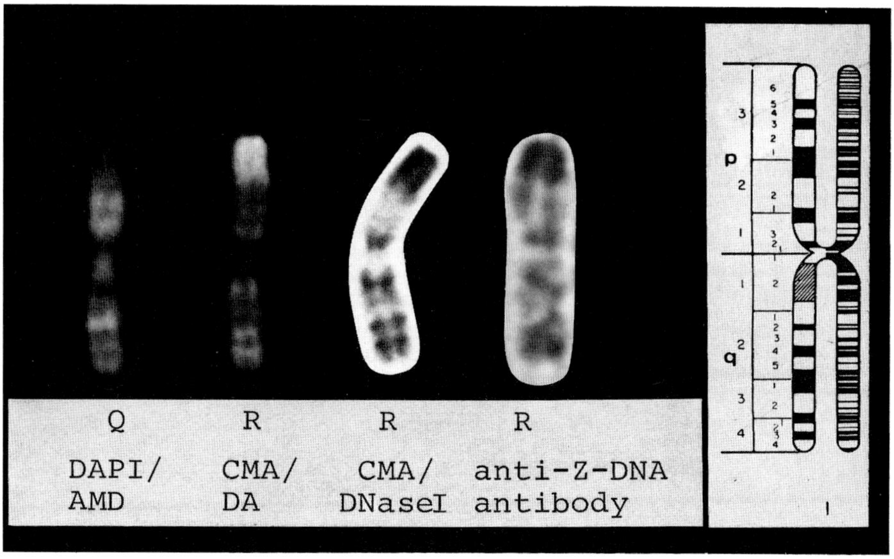

Fig. 1. DNA base pair-specific banding as exemplified on human chromosome # 1. *From left to right* counterstain-enhanced banding: the two dye pairs DAPI/actinomycin D (AMD) and chromomycin A_3 (CMA)/distamycin A (DA) impart reciprocal patterns. R-bands are revealed by CMA fluorescence, and also by DNase I digestion of fixed CMA stained chromosomes followed by Giemsa-staining (Schweizer 1977), and after staining with antibodies to Z-DNA (kindly provided by Dr. D.J. Arndt-Jovin, MPI, Göttingen). *Right* diagram of a banded human chromosome # 1 (Paris Conference 1971) with one chromatid schematically exhibiting "high resolution" prophase differentiation (Yunis 1976)

to quenching are likely to harbor many DNA clusters 20 bp or longer predominantly of the base pair composition favored by the primary stain.

1.1.2 Fractionation and Characterization of Mammalian G- and R-Band DNA

Gerald Holmquist and his coworkers were the first to link a cytological knowledge of chromosome bands with molecular data. They developed means for fractionating mammalian genomic (euchromatic) DNA into two approximately equal major components of R- and G-band DNA (Holmquist et al. 1982; Goldman et al. 1984). They separated early replicating DNA from late replicating DNA. R-bands normally complete their replication at about mid S-phase and usually before G-bands initiate replication. Bromodeoxyuridine (BrdUrd)-substitution of either early or late replicating DNA allowed a fractionation of the two components by equilibrium density gradient centrifugation. By this approach, a determination of the overall G+C content in R-band DNA and G-band DNA was possible (Table 1). To determine what genes and what defined interspersed repeat sequences can be attributed to either G- or R-band DNA or to both components, digested and electrophoretically separated mammalian DNA was probed with defined sequences.

Base composition fractionation using equilibrium density gradient centrifugation in conjunction with DNA ligands allowed the fractionation of mammalian main band DNA into light (AT-rich) and heavy (GC-rich) components (Bernardi et al. 1985). It was suggested that this kind of heterogeneity in base composition is associated with chromosomal G- and R-banding. As outlined above, the genome distribution of defined sequences to the different "isochores" was probed using the Southern transfer method. The results obtained, together with data from other laboratories, are summarized in Table 1 and in the following paragraphs. The matter has been extensively reviewed by Holmquist (1986a, b, c).

Southern blot analysis, in conjunction with R- and G-genome fractionation or base composition fractionation of mammalian DNA showed that most families of interspersed repetitive sequences are not "randomly" dispersed but are predominantly confined to either R-band or G-band DNA. The mammalian chromosome banding patterns of R- and G-bands reflect a differential arrangement of long and short interspersed repeated sequences (LINEs and SINEs) (Holmquist and Caston 1986). The biochemical data using the transfer method are substantiated by results obtained by in situ hybridization to human metaphase chromosomes using biotinylated probes and a nonradioactive detection system. Alu sequences belonging to the short interspersed repeat SINE family preferentially "stain" R-bands, while Hind III probes of KpnI belonging to the long interspersed repeat LINE family preferentially hybridized to G-bands (Manuelidis and Ward 1984). As almost all the interspersed repetitive sequences in the mammalian genome are or were mobile elements, one may ask why there is apparently no or little interchange of family members of retroposon sequences between types of bands (see Sect. 1.3). The problem of mobile genetic elements as they relate to mammalian chromosome structure has been reviewed recently in some detail (Holmquist 1986c).

1.2 Mammalian Chromosome Bands and DNA Replication

Each mammalian chromosome band represents a replicon cluster containing about 20 replicons 50 to 350 kb long (Goldman et al. 1984). DNA replication in mammals is temporally biphasic: there are two classes of replicon clusters corresponding to early replicating R-bands and late replicating G-bands, respectively (Holmquist et al. 1982). Bands visualized by differential BrdUrd incorporation during S-phase followed by BrdUrd-sensitive histochemical staining are called replication bands (Fig. 2) (Latt 1973; Schempp 1980; Holmquist et al. 1982). Bands visualized by trypsin Giemsa or quinacrine fluorescent staining (Fig. 1) are called structural bands. There is in general good overlap between structural R-bands and early replicating R-type bands and between structural G-bands and late replicating G-type bands (see Kroisel et al. 1985). However, the coincidence of replication patterns and the structural-banded prepattern is not absolute. The two female X-chromosomes have identical structural band patterns but different replication band patterns (Fig. 2). Additionally, the correlation in autosomes may not be constant during ontogeny nor at the level of gene resolution (Goldman et al. 1984; Jablonka et al. 1985).

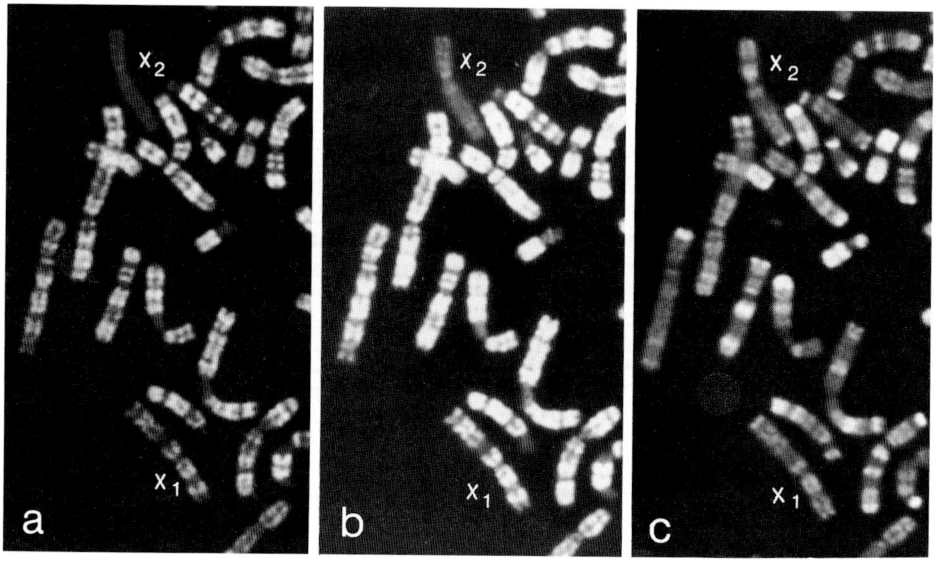

Fig. 2 a–c. Fluorescent staining of human chromosomes labelled with 100 μg ml^{-1} BrdUrd 8 h before harvest to show on the same metaphase cell either replication R-bands (**a, b**) or structural R-bands (**c**). **a** and **b** are dark and light prints of the same photographic negative to show that the late and the early replicating X-chromosomes are different both with respect to overall brightness and band pattern. By contrast, the structural R-bands are identical in both X-chromosomes (**c**). Replication bands (**a, b**) were visualized by DAPI-pH 11-staining. Structural R-bands (**c**) were obtained by chromomycin/distamycin. (Kindly provided by S. Strehl, Vienna)

1.2.1 Replication Timing and Transcriptional Activity of G-Bands and R-Bands

Late replication in the S-phase is usually equated with genetic inertness. Most C-bands, as well as the inactive ("Lyonized") X-chromosomes in mammalian females, replicate late and are genetically inert. Goldman et al. (1984), using their R-/G-replication band fractionation and the Southern transfer method, presented evidence that mammalian genes which are potentially active in a given cell type replicate early and that genes which are permanently inactive replicate late. Constitutively active "housekeeping" genes always reside in early replication R-bands, whereas tissue-specific "ontogenetic" genes usually reside in G-bands. DNA and chromatin properties considered to be related to gene activity, such as DNase I-nicking sensitivity (Kerem et al. 1984) or DNA methylation (Gartler and Riggs 1983) have been discussed in relation to the above finding (Goldman et al. 1984; Holmquist 1986a).

In the case of ontogenetic genes it was postulated that the correlation between replication of tissue-specific genes and the late replicating G-bands is not absolute: a tissue-specific gene located in a late replicating G-band environment is likely to replicate itself early in those cell types that express that particular gene (Holmquist 1986b).

1.2.2 Constancy and Conservation of Mammalian Chromosome Band Patterns

Structural G-/R-band patterns in mammals are species-specific and their ontogenetic constancy is documented (Burkholder and Comings 1972; Caspersson et al. 1972). Chromosome R- and G-band patterns are also highly conserved in evolution. Genetically homologous regions in mammalian chromosomes of different closely related and more distantly related species, such as man and the great apes (chromosomes # 1), man and mouse (human 1p and mouse # 4, same autosomal associations), or man and the cat (human 1p and chromosome # 1 in *Felis catus*), show the same kind of banding patterns. Recently, a close scrutiny of this relationship revealed that it is the *R-type early replication patterns* that are most strongly conserved during evolution (von Kiel et al. 1985). Structural G-band patterns are less strongly conserved. In this context one may speculate that the 10% difference in nuclear DNA contents between ox and goat, two related species with strongly homologous chromosome band patterns (Sumner and Buckland 1976; Mayr et al. 1985) resides in structural G-bands, i.e., in the "G-genome" rather than in the "R-genome" (as defined by Goldman et al. 1984).

1.3 Evolution of the "Banded" Mammalian Genome

From what has been said above, there must exist strong selective pressure for the conservation of units with different replication times, particularly early replication in the case of functional genetic material coding for housekeeping properties. As a consequence, the ancestral genome diverged into the "early" and "late", i.e., the R- and the G-genomes found in extant mammals. The sequence divergence rate in R-bands has been found to be slow compared to that in G-bands (Holmquist 1986a; Table 1). Sequence comparisons also suggest that the rate of sequence homogenization *within* the two genomes is higher than *between* the R- and G-genomes. For instance, G-bands on different chromosomes are much more similar in DNA composition than adjacent G- and R-bands on the same chromosome. In other words, there is coevolution of sequences in R-bands, and in G-bands, but divergence between the R- and G-band DNA.

In summary it appears that the partition of the mammalian genome into temporal clusters of replicons is more ancient than the cytologically detectable differentiation into structural G-/R-band patterns (Holmquist et al. 1982).

2 Heterochromatin and C-Banding

The C-band method is a rapid means of detecting constitutive heterochromatin in both the dividing and interphase nucleus. C-bands are heritable chromosome elements that harbor repetitive DNA sequences. "Heterochromatic DNA" (Brutlag 1980) is generally equated with tandemly arranged highly repeated sequences or satellite DNAs (see Singer 1982). The sensitivity of the Giemsa C-band histochemical method lies in the range of 1×10^{-2} pg of DNA, i.e., about 10^7 bp (Schweizer and Ehrendorfer 1983).

In only a few known cases are C-bands apparently absent from a species, i.e., not detectable by cytological means. When present they are preferentially located at the telomeres, the centromeres, and surrounding the nucleolus organizing region. In a variety of organisms intercalary C-bands do occur.

The mechanisms of C-banding have been analysed in some detail (Comings 1978; Holmquist 1979) and structural aspects of heterochromatin and satellite DNA have been reviewed extensively (Appels and Peacock 1978; John and Miklos 1979; Miklos and John 1979; Appels 1981).

In the following section we turn to the discussion of a model for the evolution of C-band patterns and its implications on hypotheses on C-heterochromatin function.

2.1 Evolution of C-Band Patterns

An interpretation of the C-band pattern of an extant species is only possible when we have independent data concerning the evolutionary history of the karyotype. Because this is usually not the case, our understanding of the modes of evolution of C-band patterns has to depend principally on comparative studies within natural populations. These involve one of three approaches:

a) Analysis of C-band variation in species pairs where the direction of karyotypic changes can be reasonably inferred.

b) Investigation of C-band variation in and between natural populations of a single species, i.e., cases of polymorphism and polytypism.

c) Comparisons of a large number of related, and even unrelated, species to detect similarities and differences in C-band patterns.

The observed principles and rules should provide clues to the mechanisms of cytological and molecular evolution of C-band patterns.

2.1.1 Regularities in C-Band Distribution Patterns

Mammals tend to have a simple centromeric/telomeric C-band distribution with the notable exception of cetaceous whales, where intercalary bands are the rule (Árnason 1974). However, the banding phenomenon has to be interpreted on a larger scale and findings in other animal groups as well as plants have to be included. There are many observations of C-band patterns from many species, which make it possible to discern the generalities as to the karyotype position of C-bands (King and John 1980; Greilhuber and Loidl 1983; Loidl 1983; Schweizer and Ehrendorfer 1983; John et al. 1985). The subject has recently been reviewed (Schweizer and Loidl 1986) and, therefore, the major regularities and rules are here presented without substantiating data (for details see Schweizer and Loidl 1986).

a) *C-Band Distribution and Chromosome Size.* Within a given complement, the shorter chromosomes and chromosome arms tend to have telomeric bands, while the longer chromosomes or chromosome arms tend to have intercalary bands. The shorter chromosome arms tend to have more C-band material than longer arms (Schweizer and Loidl 1986). The gain of terminal heterochromatin may be such that the total length of an initially short arm approaches the length of the longest arm in a given karyotype. In karyotypes with interstitial bands only, these are more likely to occur in the longer chromosome arms (Fig. 3).

b) *„Äquilokale Heterochromatie“ (Equilocal Heterochromatin Distribution)* (Heitz 1933). Similar C-bands in nonhomologous members of the same diploid set tend to be located at similar sites, i.e., at the centromere and/or interstitially at roughly similar distance from the centromere.

c) *Interstitial Bands.* The distribution of interstitial bands is nonrandom [see (b) "equilocality"], and the positions and amounts are not simply a function of arm size. The positions of interstitial bands show a relationship to the arm lengths of the remaining chromosomes within a karyotype. For each interstitial band, there exists a chromosome arm in the population, the telomere of which is as far from the centromere as the proximal border of that interstitial band is from the centromere (Loidl 1983). The telomeres of the shorter chromosome arms thus define the positions of the interstitial bands in the longer arms (Fig. 3).

2.1.2 The Role of Chromosome Disposition in the Three-Dimensional Space of the Mitotic Interphase Nucleus

On the basis of the generalities outlined above, it is conceptually simple and attractive to think that the localization of C-bands in a given karyotype is constrained by the spatial organisation of the chromosomes in the mitotic nucleus (Barnes et al. 1985; Schweizer and Loidl 1986). The so-called Rabl polarization (see Cremer et al. 1982; Saumweber, this vol.) of chromosomes refers to the fact that the anaphase-like disposition of the chromosomes (with the centromeres directed to the spindle pole of the preceding anaphase) is maintained through inter-

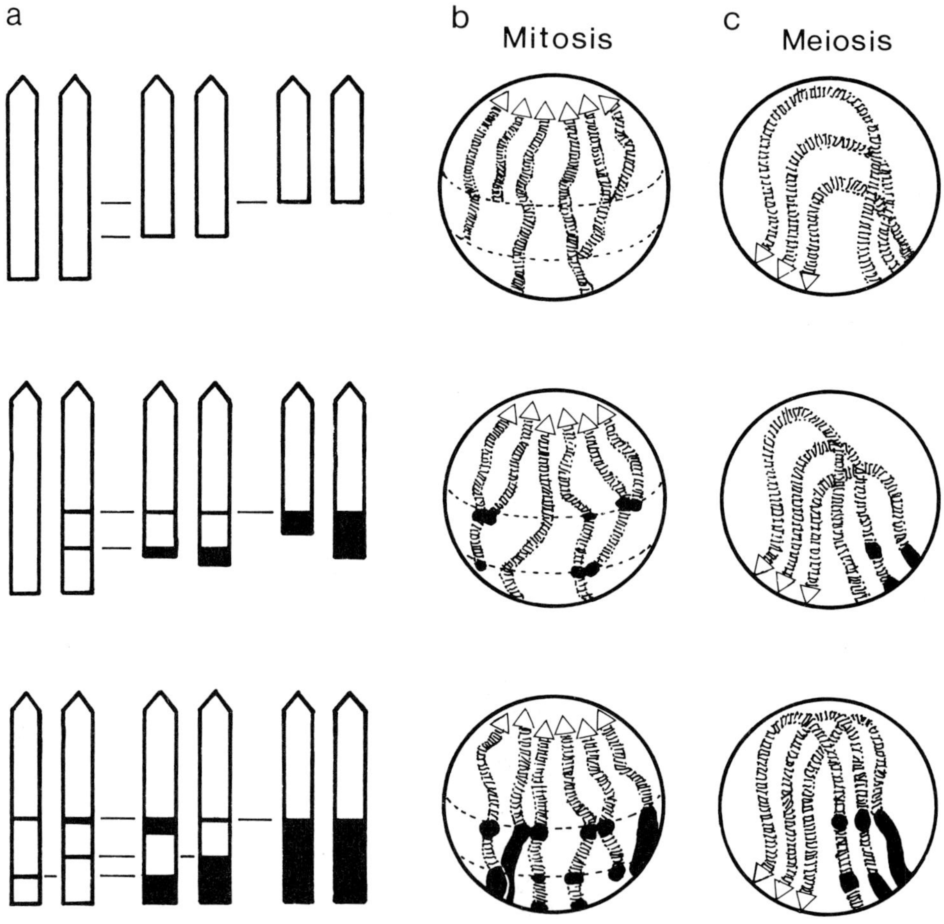

Fig. 3 a–c. Evolution of intercalary and telomeric C-bands. Hypothetical sequence of the formation of band patterns (lane **a**): Intercalary bands in longer chromosome arms are formed at positions which have the same centromere distance as the telomeres of shorter chromosome arms. Shorter arms tend to accumulate terminal heterochromatin. Lane **b** shows the corresponding situation in mitotic nuclei with Rabl polarization of chromosomes. Telomeres of shorter chromosome arms come into contact with equidistant intercalary regions in longer chromosomes. At these contact zones transfer of heterochromatic DNA sequences occurs. In lane **c** the situation in meiotic prophase is depicted. Band patterns do not reflect this kind of spatial chromosome disposition

phase and into the succeeding mitosis. The Rabl polarization brings centromeres together while chromosome ends are attached to the nuclear membrane. The site of attachment is largely a function of the length of a given chromosome arm. Clearly, centromeres will be close enough to come into contact, but also chromosome ends (telomeres), although probably much less frequently, will come into contact with those of similar arms and with equilocal interstitial regions of longer nonhomologous chromosome arms.

2.1.3 A Model for C-Band Patterns

The model we have proposed (Schweizer and Loidl 1986) assumes that all potential contact zones between a given telomere and those intercalary chromosome regions which lie at a similar distance from the centromere are potential sites of *heterochromatin transfer*. The model assumes that, over evolutionary time, this kind of restricted heterochromatin transfer is effective in mitotically dividing cells within the germ line. The incipient evolution of C-band patterns may be largely constrained by nucleotypic characters (chromosome arm size and volume, nuclear volume) and the polarized disposition of the chromosomes at mitotic interphase.

A possible constraint of nuclear organization at first meiotic prophase (zygotene/pachytene) can be ruled out in this situation. It is known from studies on C-band distribution and variation carried out on a large number of grasshopper species (see John et al. 1985) and liliaceous plants (see Greilhuber 1984) that incipient distribution patterns of intercalary C-bands are a reflection of Rabl-polarization rather than bouquet polarisation (Fig. 3). It is likely, however, that meiotic constraints represent major determinants of the independent evolution of specific chromosomes or groups of chromosomes such as the sex chromosomes, chromosome 1 in crested and marbled newts (Sims et al. 1984), and the megameric chromosome pair in certain grasshoppers (White 1973).

The operation at meiotic prophase of processes responsible for the maintenance of satellite DNA sequence homogeneity is a second important feature of our model (see below).

Predictions of the Model. The model implies that metacentric chromosomes will tend to have similar (symmetric) band patterns in their two arms. It further predicts that a karyotype consisting of chromosomes with approximately similar arm lengths will tend to have telomeric rather than intercalary bands. A karyotype consisting of members with different sizes and arm ratio, on the other hand, will tend to possess intercalary bands in addition to telomeric bands. Concerted evolution of C-band patterns is to be expected in nonhomologous chromosome arms of similar lengths of a given species.

If, as proposed, karyotype architecture and chromosome arm lengths determine C-band patterns, then parallelism, or convergence, is to be anticipated for similarly shaped karyotypes. Therefore, similar C-band patterns may be found in similar chromosomes in more distant or even in unrelated species with similar karyomorphology.

2.1.4 The Role of Meiosis in the Evolution of C-Band Patterns

Meiotic recombination between different structural morphs of C-banded chromosomes can be a source of heterochromatin diversity and of new structural types as defined by C-band patterns.

Recombination and information transfer, including nonreciprocal exchanges such as gene conversion, operating between heterochromatic DNA of homologous and also of nonhomologous chromosomes, have to be invoked to explain the reported cases of concerted evolution of constitutive heterochromatin (Dover

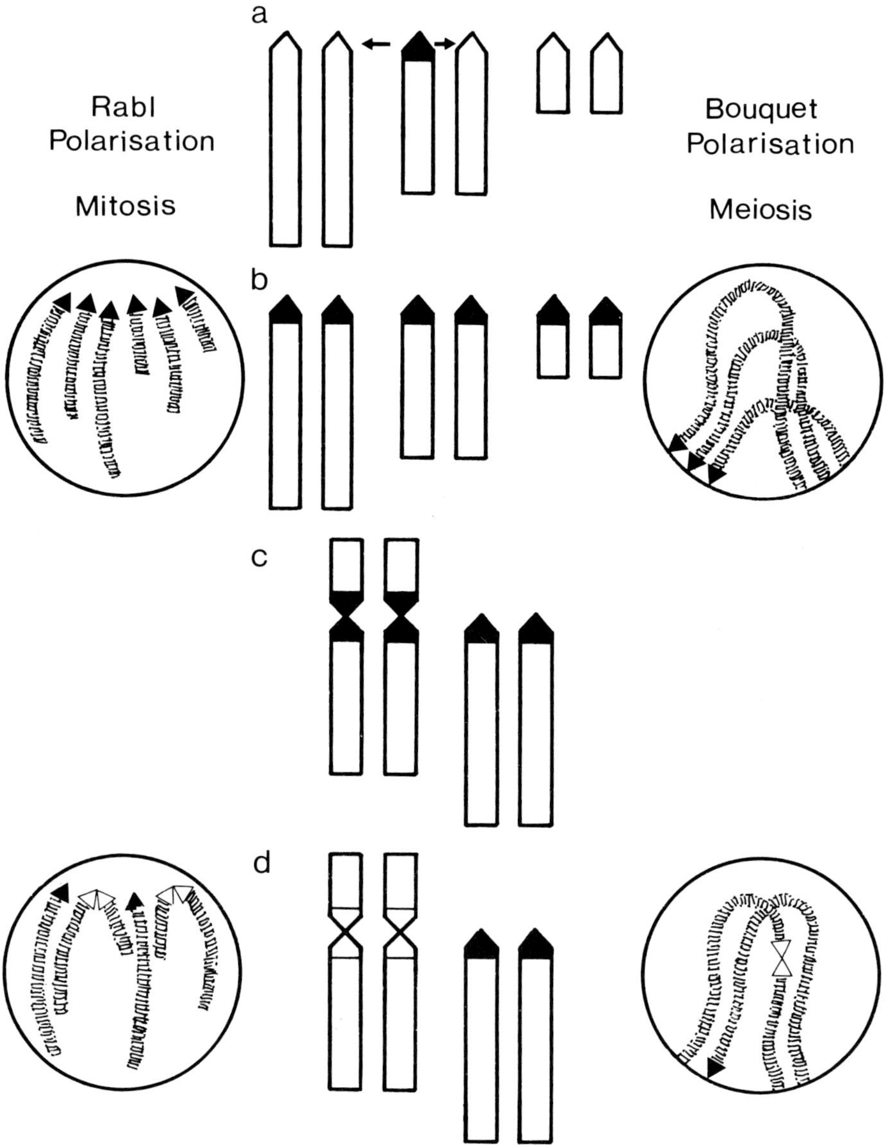

Fig. 4 a–d. Hypothesis for the evolution of centric C-bands. **a** Accumulation of heterochromatic sequences and their transfer to homologous and nonhomologous chromosomes (*arrows*). In **b** the close contact in mitotic and meiotic nuclei and sequence homogeneity of centric C-bands in telocentric chromosomes is shown. **c** a Robertsonian fusion creates (sub-) metacentrics out of telocentrics. **d** situation as encountered in *Sus scrofa* where centromeric heterochromatin in metacentrics and telocentrics is of different sequence composition (drawn as open and filled centromeres, respectively). C-band sequences in all telocentrics are similar, those in metacentrics have diverged. While in mitotic nuclei the relative positions of centromeres are not altered substantially by the Robertsonian event, centromeres of metacentrics are spatially isolated in meiotic pachytene. This kind of nuclear constraint is suggested to promote the independent satellite DNA evolution between groups of chromosomes et al. 1982). Recombination requires proximity and interchromosomal contact. Proximity of C-bands is determined by the spatial disposition of chromosomes. In order to assess these nuclear constraints, a comparison of the meiotic vs. mitotic spatial nuclear organization (bouquet polarization vs. Rabl polarization) was carried out in Robertsonian systems, where the importance of chromosome arrangement is especially clear.

Figure 4 demonstrates that the disposition of acro- or telocentric chromosomes is not significantly changed at mitotic ana-/telophase by a Robertsonian rearrangement (fusion or dissociation) and hence also during interphase (mitotic Rabl polarization). By contrast, the disposition of bivalents at meiotic prophase (zygotene/pachytene) is significantly altered. Centromeric regions, which in telocentrics are clustered at the membrane of the zygotene/pachytene nucleus, are spaced out when involved in a fusion event. This difference in spatial arrangement of the two chromosome types (telocentric and biarmed) may provide a rationale for the reported differences in chemical composition of centric heterochromatin in biarmed and telocentric chromosomes in stable Robertsonian systems (Kurnit et al. 1978; Schwarzacher and Schweizer 1982). On the other hand, the proximity of heterochromatic DNA in the zygotene and pachytene nucleus, i.e., at stages when the enzymatic machinery for both reciprocal and nonreciprocal recombination is available, may provide a cytological basis for those molecular processes which are responsible for the maintenance of satellite DNA homogeneity between groups of C-bands on both homologous and nonhomologous chromosomes (for discussion see Arnheim et al. 1980; Dover et al. 1982; Dover and Flavell 1984).

Thus, in polymorphic Robertsonian systems, at an incipient stage of evolution of a bimodal karyotype by centric fusion of telocentrics, the biochemical structure of the heterochromatic centric regions in the fusion submetacentrics should be similar to centromeric heterochromatin in telocentrics and, additionally, it should not be very different between different fusion submetacentrics. This situation is indeed found in the European wild pig, *Sus scrofa scrofa* (Mayr et al. 1984), and in the mouse fusions recently analyzed by Redi et al. (1986). On the other hand, in stable Robertsonian systems, where biarmed chromosomes and telocentrics have been separated over longer evolutionary time, the compositions of centromeric heterochromatin in the two chromosome groups of a given karyotype are expected to have diverged.

2.1.5 *Sus scrofa* as a Model System for the Study of Centric C-Band Composition

Male *Sus scrofa domestica* (2n = 38, XY) has 24 biarmed autosomes and an XY pair which is also biarmed, together with 12 acrocentric chromosomes. Counterstain-enhanced fluorochrome staining of mitotic metaphase chromosomes revealed a qualitative difference of centric C-bands in the biarmed autosomes, Nos. 1–12, and in the telocentric chromosomes Nos. 13–18 (Schnedl et al. 1981; Schwarzacher et al. 1984). The centromeres of members of the latter group are positively stained by distamycin/DAPI, whereas the centric regions of biarmed autosomes Nos. 1–12 are highlighted by the GC-specific fluorescent dye chromomycin A_3. The autosomal heterochromatin differentiation in *Sus scrofa domestica*

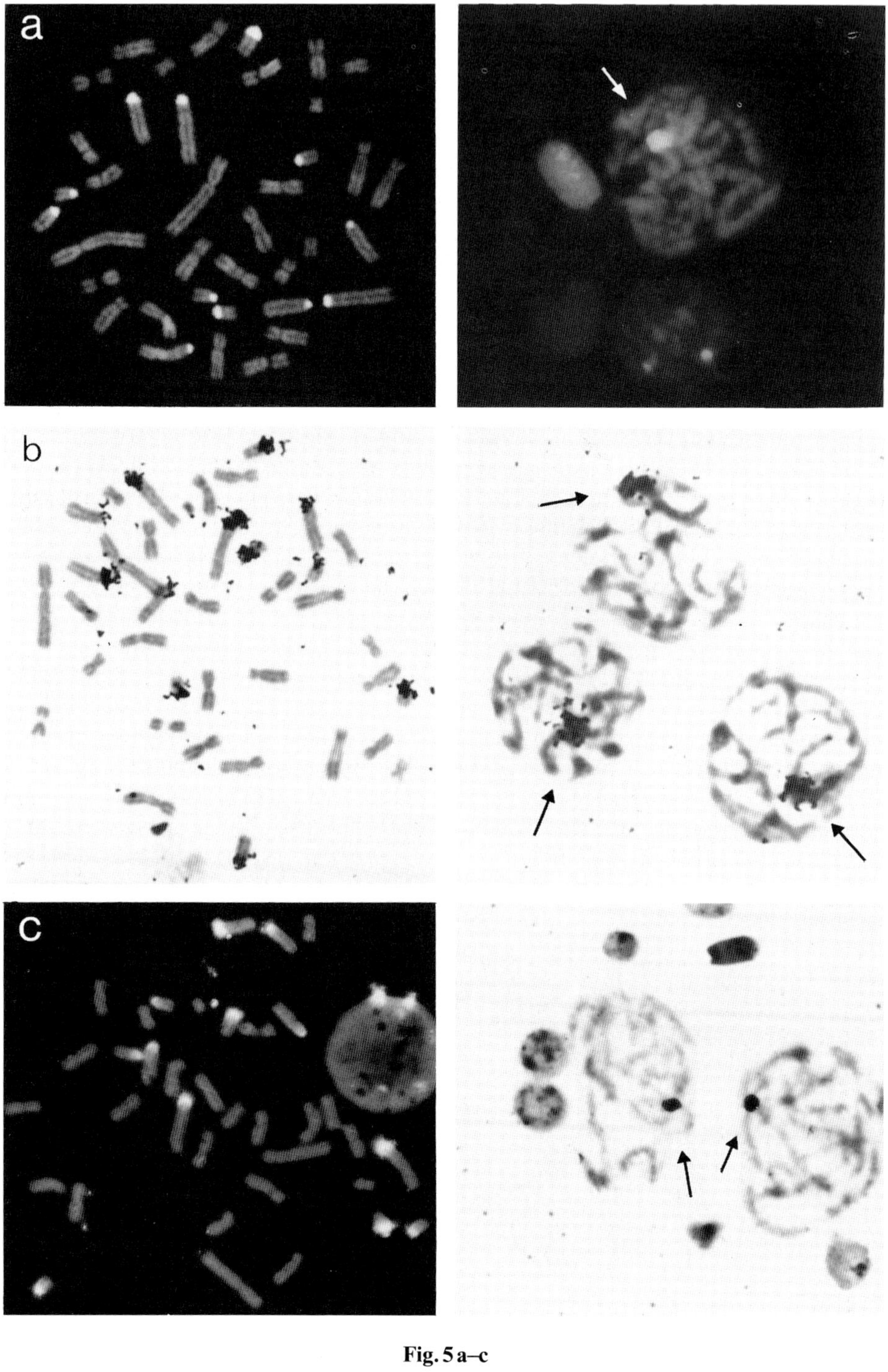

Fig. 5 a–c

thus reflects the bimodal karyotype differentiation with biarmed and telocentric chromosomes (Fig. 5a).

The experimental possibility to specifically differentiate the centromeres of the two chromosome groups in *Sus scrofa* provided a means for the cytological study of centromere arrangement at interphase and during the early stages of meiotic prophase. Fluorescence analysis of pachytene nuclei revealed that the DA/DAPI-positive C-bands of telocentrics form one or occasionally two large fusion chromocentres per cell, while the chromomycin-bright chromocentres, constituted by the centric regions of biarmed chromosomes, usually show little association, i.e., are well separated in the pachytene nucleus (Schwarzacher et al. 1984). From this observation it follows that the bivalents formed by the telocentric chromosome pairs are centromerically associated, while the biarmed bivalents are not. These results obtained by fluorescent staining techniques have been recently confirmed using cloned satellite DNA probes of *Sus scrofa* (B. Hamilton et al., in preparation; see also Figs. 5b, c).

Our model predicts that in *Sus scrofa domestica* all telocentrics should harbor similar centromeric satellite DNA being different from satellite DNA localized at centric regions of biarmed chromosomes. It may be further assumed that interchromosomal homology of heterochromatic DNA is greater in chromosomes Nos. 13–18 than in the biarmed chromosomes Nos. 1–12.

2.2 Functional Aspects of C-Heterochromatin

A large number of functions has been attributed to C-heterochromatin over the years (see Cooper 1959; Brown 1966; Britten and Kohne 1968; Walker et al. 1969; Yunis and Yasmineh 1971; Arrighi and Saunders 1973; Hsu 1975; Fry and Salser 1977; John and Miklos 1979; Brutlag 1980) but as many have been questioned or rejected. In the following, functional aspects of heterochromatin are discussed in the light of what we know about its chromosomal distribution.

2.2.1 C-Heterochromatin Function in Relation to Sequence Composition and Chromosomal Distribution

C-heterochromatic DNA is mainly composed of short, highly repeated, and tandemly arranged sequences. Although basic repeats are often remarkably conservative between related species, there can exist a large number of variants even

Fig. 5a–c. In situ hybridization of a cloned satellite III DNA probe (P 3.4) of *Sus scrofa domestica* to mitotic chromosomes (*left*) or meiotic pachytene nuclei (*right*) using either a ^{3}H-nick translated probe and autoradiography (**b**) or the biotin/streptavidin/peroxidase method (Manuelidis and Ward 1984) in conjunction with reflection contrast- (**c** *left*) or light microscopy (**c** *right*). All the 12 centric C-bands of acrocentric chromosome pairs Nos 13–18 which are highlighted by distamycin/DAPI fluorescent staining (see Schwarzacher et al. 1984) are labelled by probe P 3.4. Fluorescent staining (**a**) and in situ hybridization (**b, c**) of pachytene nuclei demonstrate association of these heterochromatic regions to a single chromocenter (cf. Fig. 4). Implications for heterochromatin composition and karyotype evolution are discussed in Section 2.1.5

within one species. As John and Miklos (1979) point out, this poses a serious problem for any theory of satellite function. If there exists a function for heterochromatin which is common to all higher organisms, it cannot be related to the sequence composition of heterochromatic DNA.

The amounts of C-band-positive material can vary considerably between species, ranging from over the half of the karyotype length to its virtual absence. This makes a vital function for heterochromatin unlikely. Its presence can also be variable between populations or individuals. Thus it seems that it is not indispensable for particular individuals, even in species where heterochromatin normally occurs. Therefore a function might rather be sought at a higher level than the individual organism, e.g., by creating differences between populations or individuals, extending the ecological spectrum of a species.

Not only the amount and sequence composition, but also the chromosomal distribution of C-positive material is subject to a great variation between species. As outlined in the previous section, the C-band distribution patterns can be explained in nucleotypic terms. The model by Schweizer and Loidl (1986) is based on a spread of heterochromatic sequences to numerous chromosomal locations which is governed by karyotype structure and not by a response to functional needs. According to this concept, the regions where C-bands are found are those where they are tolerated rather than those where they have been selected for functional reasons. The model, if correct, would mean that the chromosome and karyotype distribution of bands does not provide the clue to heterochromatin function.

Taken together, any hypothesis on heterochromatin function has to comply with the condition that it must be unrelated to both heterochromatin sequence and chromosomal location.

2.2.2 Functions Versus Effects

The idea that parts of the genome could be parasitic was expressed in view of the behavior of B-chromosomes by Östergren (1945). It was adopted by Doolittle and Sapienza (1980) and Orgel and Crick (1980) for repetitive DNA. They state that certain sequences could accumulate by virtue of an efficient mechanism for amplification which over-rides the selective forces against the maintenance of nonfunctional ("junk") DNA sequences. This concept of a "selfish" or "parasitic" heterochromatin agrees best with the proposed mode of infectious heterochromatin sequence spread throughout the karyotype, which by itself has parasitic characteristics.

However, a selfish mode of heterochromatin accumulation which is not governed by selection pressure does not rule out that heterochromatin has acquired functional significance. This may have happened via nucleotypic effects of heterochromatin which have turned out advantageous under some circumstances.

There are two main examples where the borderline between mere effects and functions (in the sense of a benefit to the host) is not clear: the effect on cell volume and the effect on recombination.

2.2.3 The Control of Cell Volume

Heterochromatin contains "non-transcribed nuclear skeleton DNA" (Cavalier-Smith 1978), whose presence increases the amount of nuclear DNA which is strongly positively correlated with cell size, cell cycle duration and time for growth of the organism (lit. in Cavalier-Smith 1978). The cause for slowing down of the cell cycle and growth due to larger DNA quantities may be the additional workload for the replication machinery.

On the other hand, Nagl (1974) found evidence that the nuclear DNA content can increase by the addition of heterochromatin without lengthening of the cell cycle. The cycle may even become shorter. Cavalier-Smith (1978) tried to explain this contradictory observation by an increase in the number of nuclear pores and an enlarged nuclear surface (which is favorable for nucleus/plasma interaction) in cells with higher DNA content.

Thus, even if it is expressed in different ways, there is an apparent nucleotypic effect of heterochromatin and it is likely that several organisms or populations take enough advantage of it to compensate for selection against heterochromatin.

2.2.4 Meiotic Effects of Heterochromatin

Probably the best-documented effect of C-heterochromatin is exerted on meiotic events. This is the influence on chiasma frequency and distribution. Crossing-over is absent from heterochromatin (see, e.g., John and Lewis 1965; Marks 1974; John 1976; Jones 1978; Loidl 1979). This reduces the mean number of chiasmata per karyotype length compared to an unbanded karyotype of the same length. However, since heterochromatin is devoid (or poor) in genes and is usually present as an additional material (i.e., it is not transformed euchromatin, see Loidl 1983), the ratio of crossing-over to unit length of euchromatin is not altered by the presence of heterochromatin. This could be different, of course, for heterochromatin which is transformed euchromatin. King (1980) claimed that this heterochromatin type occurs in some frogs. It is not known, however, whether or not this heterochromatin excludes chiasmata.

By contrast, it has also been claimed that the presence of heterochromatin increases the number of chiasmata in a karyotype, so that the number of chiasmata rises more than expected from the gain in length owing to the addition of heterochromatin (e.g., Riva et al. 1984).

In several grasshoppers a repulsion effect of C-heterochromatic bands on chiasmata was observed (e.g., Southern 1967; Miklos and Nankivell 1976; John and Miklos 1979). This means that if a terminal or proximal C-band is present in a chromosome arm, chiasmata form preferentially at its opposite end. De la Torre et al. (1986) found that this effect is strongest when the bands are heterozygous with respect to presence or size. This may provide the explanation for the phenomenon of heterochromatin-chiasma interference: Even if C-bands do not impair chromosome synapsis, the "effective pairing" (Chandley 1986) (i.e., the precise sequence matching which is necessary for crossing-over) could be out of phase in the neighborhood of heterozygous bands. An alternative explanation for the spatial separation of heterochromatin and chiasmata is that heterochromatin can originate only in regions with low chiasma frequencies (Charlesworth et al. 1986). However, if this were the case, chiasmata should occupy the same positions in the presence and absence of bands, which is not supported by observations.

In other organisms, by contrast, a close proximity between C-bands and chiasmata was found (e.g., Linnert 1955; Dyer 1963; Jones 1978; Friebe 1979; Loidl 1979, 1982; Murer-Orlando and Richer 1983). Of particular interest are the findings in the liliaceous plants *Allium flavum* and *A. carinatum* (Loidl 1979, 1982). In these species chiasmata always occur in euchromatin near the borders of the distally to terminally located bands, but this correlation is also retained when a band is exceptionally located in a more proximal position. Obviously, in *Allium* effective pairing near C-heterochromatin is achieved in spite of band heterozygosities, allowing crossing-over to occur in the vicinity of bands. It may be speculated that this is made possible by a high incidence of intercalary pairing initiation in the regions close to the bands. However, this does not explain the exclusive occurrence of chiasmata in the proximity of bands. The explanation is possibly found in a suggestion by Chandley (1986), according to which the eu-/heterochromatin junctions, which are junctions between early and late replicating regions, are especially susceptible to strand breaks which initiate crossing-over events.

The examples presented above show that the presence of heterochromatin can override a genetically determined chiasma distribution in some instances. Although achieved in different ways, by a redistribution of chiasmata either toward or away from the bands, the consequence of the presence of C-bands for the restriction of intrachromosomal recombination is apparent.

It is conceivable that the influence of heterochromatin on genetic recombination may be useful for the conservation of advantageous combinations of traits and hence being of benefit to a population. If this would turn out to be true, it would illustrate how heterochromatin as an originally neutral or parasitic element may have outwitted selection by adopting a function.

Acknowledgments. The authors wish to thank Dr. D. G. Bedo, Canberra, for reading the manuscript. Many thanks are expressed to Dr. G. Holmquist, Duarte, and Prof. B. John, Canberra, for discussion and comments. This study received financial support from the Österreichischer Fonds zur Förderung der wissenschaftlichen Forschung (Projekt P 5080).

References

Appels R (1981) The molecular cytology of wheat-rye hybrids. Int Rev Cytol 80:93–132

Appels R, Peacock WJ (1978) The arrangement and evolution of highly repeated (satellite) DNA sequences with special reference to *Drosophila*. Int Rev Cytol (Suppl) 8:69–126

Árnason Ú (1974) Comparative chromosome studies in Cetacea. Hereditas 77:1–36

Arnheim N, Krystal M, Schmickel R, Wilson G, Ryder O, Zimmer E (1980) Molecular evidence for genetic exchanges among ribosomal genes on nonhomologous chromosomes in man and apes. Proc Natl Acad Sci USA 77:7323–7327

Arrighi FE, Saunders GF (1973) The relationship between repetitious DNA and constitutive heterochromatin with special reference to man. In: Pfeiffer RA (ed) Modern aspects in cytogenetics: constitutive heterochromatin in man. Schattauer, Stuttgart New York, pp 113–133

Barnes SR, James AM, Jamieson G (1985) The organisation, nucleotide sequence, and chromosomal distribution of satellite DNA from *Allium cepa*. Chromosoma 92:185–192

Bernardi G, Olofson B, Filipski J, Zerial M, Salinas J, Cuny G, Meunier-Rotival M, Rodier F (1985) The mosaic genome of warm-blooded vertebrates. Science 228:953–958

Britten RJ, Kohne DE (1968) Repeated sequences in DNA. Science 161:529–540

Brown SW (1966) Heterochromatin. Science 151:417–425

Brutlag DL (1980) Molecular arrangement and evolution of heterochromatic DNA. Annu Rev Genet 14:121–144

Burkholder GD, Comings DE (1972) Do the Giemsa banding patterns of chromosomes change during embryonic development? Exp Cell Res 75:268–271

Caspersson T, Chapelle A de la, Schröder J, Zech L (1972) Quinacrine fluorescene of metaphase chromosomes. Identical patterns in different tissues. Exp Cell Res 72:56–59

Cavalier-Smith T (1978) Nuclear volume control by nucleosceletal DNA, selection for cell volume and cell growth rate, and the solution of the DNA C-value paradox. J Cell Sci 34:247–278

Chandley AC (1986) A model for effective pairing and recombination at meiosis based on early replicating sites (R-bands) along chromosomes. Hum Genet 72:50–57

Charlesworth B, Langley CH, Stephan W (1986) The evolution of restricted recombination and the accumulation of repeated DNA sequences. Genetics 112:947–962

Comings DE (1978) Mechanisms of chromosome banding and implications for chromosome structure. Annu Rev Genet 12:25–46

Cooper KW (1959) Cytogenetic analysis of major heterochromatic elements (especially Xh and Y) in *Drosophila melanogaster*, and the theory of "heterochromatin". Chromosoma 10:535–588

Cremer T, Cremer C, Baumann H, Luedtke E-K, Sperling K, Teuber U, Zorn C (1982) Rabl's model of the interphase chromosome arrangement tested in Chinese hamster cells by premature chromosome condensation and Laser-UV-microbeam experiments. Hum Genet 60:46–56

Doolittle WF, Sapienza C (1980) Selfish genes, the phenotype paradigm and genome evolution. Nature (London) 284:601–603

Dover GA, Flavell RB (1984) Molecular coevolution: DNA divergence and the maintenance of function. Cell 38:622–623

Dover GA, Brown S, Coen E, Dallas J, Strachan T, Trick M (1982) The dynamics of genome evolution and species differentiation. In: Dover GA, Flavell RB (eds) Genome evolution. Academic Press, London, New York, pp 343–372

Dyer AF (1963) Allocyclic segments of chromosomes and the structural heterozygosity that they reveal. Chromosoma 13:545–576

Friebe B (1979) Chiasmabildung in normalen und mutierten Karyotypen von *Vicia faba*. Biol Zentralbl 98:37–53

Fry K, Salser W (1977) Nucleotide sequences of HS-α satellite DNA from kangaroo rat *Dipodomys ordii* and characterization of similar sequences in other rodents. Cell 12:1069–1084

Ganner E, Evans HJ (1971) The relationships between patterns of DNA replication and of quinacrine fluorescence in the human chromosome complement. Chromosoma 35:326–341

Gartler SM, Riggs AD (1983) Mammalian X chromosome inactivation. Annu Rev Genet 17:153–190

Goldman MA, Holmquist GP, Gray MC, Caston LA, Nag A (1984) Replication timing of genes and middle repetitive sequences. Science 224:686–692

Greilhuber J (1984) Chromosomal evidence in taxonomy. In: Heywood VH, Moore DM (eds) Current concepts in plant taxonomy. Academic Press, London New York, pp 157–179

Greilhuber J, Loidl J (1983) On regularities of C-banding patterns, and their possible cause. In: Brandham PE, Bennett MD (eds) Kew Chromos Conf II. Allen & Unwin, London, p 344

Heitz E (1933) Die Herkunft der Chromocentren. Planta 18:571–636

Holmquist G (1979) The mechanism of C-banding: depurination and β-elimination. Chromosoma 72:203–224

Holmquist G (1986a) DNA sequences in G-bands and R-bands. In: Adolph KW (ed) Chromosome and chromatin structure. CRC Press (in press)

Holmquist G (1986b) Role of replication time in the control of tissue specific gene expression. (in press)
Holmquist G (1986c) Mobile genetic elements in G-band and R-band DNA. In: Daniel A (ed) The cytogenetics of mammalian autosomal rearrangements. Liss, New York (in press)
Holmquist G, Caston LA (1986) Replication time of interspersed repetitive DNA sequences in hamsters. Biochim Biophys Acta (in press)
Holmquist G, Gray M, Porter T, Jordan J (1982) Characterization of Giemsa dark and light-band DNA. Cell 31:121–129
Hsu TC (1975) A possible function of constitutive heterochromatin: the bodyguard hypothesis. Genetics (Suppl) 79:137–150
Jablonka E, Goitein R, Marcus M, Cedar H (1985) DNA hypomethylation causes an increase in DNase-I sensitivity and an advance in the time of replication of the entire inactive X chromosome. Chromosoma 93:152–156
John B (1976) Myths and mechanisms of meiosis. Chromosoma 54:295–325
John B, Lewis KR (1965) The meiotic system. Protoplasmatologia (Wien) VI/F/1
John B, Miklos GLG (1979) Functional aspects of satellite DNA and heterochromatin. Int Rev Cytol 58:1–114
John B, King M, Schweizer D, Mendelak M (1985) Equilocality of heterochromatin distribution and heterochromatin heterogeneity in acridid grasshoppers. Chromosoma 91:185–200
Jones GH (1978) Giemsa C-banding of rye meiotic chromosomes and the nature of "terminal" chiasmata. Chromosoma 66:45–57
Kerem B-S, Goitein R, Diamond G, Cedar H, Marcus M (1984) Mapping of DNAase I sensitive regions on mitotic chromosomes. Cell 38:493–499
Kiel K von, Hameister H, Somssich IE, Adolph S (1985) Early replication banding reveals a strongly conserved functional pattern in mammalian chromosomes. Chromosoma 93:69–76
King M (1980) C-banding studies on Australian hylid frogs: secondary constriction structure and the concept of euchromatin transformation. Chromosoma 80:191–217
King M, John B (1980) Regularities and restrictions governing C-band variation in Acridoid grasshoppers. Chromosoma 76:123–150
Kroisel PM, Rosenkranz W, Schweizer D (1985) Simultaneous production of R-bands and either replication patterns or sister chromatid differentiation. Hum Gen 71:333–341
Kurnit DM, Brown FL, Maio JJ (1978) Mammalian repetitive DNA sequences in a stable Robertsonian system: II. Characterization, in situ hybridization and cross species hybridization of repetitive DNAs in calf, sheep and goat chromosomes. Cytogenet Cell Genet 21:145–167
Latt SA (1973) Microfluorometric detection of deoxyribonucleic acid replication in human metaphase chromosomes. Proc Natl Acad Sci USA 70:3395–3399
Linnert G (1955) Die Struktur der Pachytänchromosomen in Euchromatin und Heterochromatin und ihre Auswirkung auf die Chiasmabildung bei Salvia-Arten. Chromosoma 7:90–128
Loidl J (1979) C-band proximity of chiasmata and absence of terminalisation in *Allium flavum* (Liliaceae). Chromosoma 73:45–51
Loidl J (1982) Further evidence for a heterochromatin-chiasma correlation in some *Allium* species. Genetica 60:31–35
Loidl J (1983) Some features of heterochromatin in wild *Allium* species. Plant Syst Evol 143:117–131
Manuelidis L, Ward DC (1984) Chromosomal and nuclear distribution of the HindIII 1.9-kb human DNA repeat segment. Chromosoma 91:28–38
Marks GE (1975) The Giemsa-staining centromeres of *Nigella damascena*. J Cell Sci 18:19–25
Mayr B, Schweizer D, Geber G (1984) NOR activity, heterochromatin differentiation, and the Robertsonian polymorphism in *Sus scrofa* L. J Hered 75:79–80
Mayr B, Schweizer D, Mendelak M, Krutzler J, Schleger W, Kalat M, Auer H (1985) Levels of conservation and variation of heterochromatin and nucleolus organizers in the Bovidae. Can J Genet Cytol 27:665–682
Miklos GLG, John B (1979) Heterochromatin and satellite DNA in man: properties and prospects. Am J Hum Genet 31:264–280

Miklos GLG, Nankivell RN (1976) Telomeric satellite DNA functions in regulating recombination. Chromosoma 56:143–167
Murer-Orlando M, Richer C-L (1983) Heterochromatin heterogeneity in Chinese hamster sex bivalents. Cytogenet Cell Genet 35:195–199
Nagl W (1974) Role of heterochromatin in the control of cell cycle duration. Nature (London) 249:53–54
Orgel LE, Crick FHC (1980) Selfish DNA: the ultimate parasite. Nature (London) 284:604–607
Östergren G (1945) Parasitic nature of extra fragment chromosomes. Bot Not 1945/2:157–163
Paris Conference (1971) 1972. Standardization in human cytogenetics. Cytogenetics 11:317–362
Redi CA, Garagna S, Mazzini G, Winking H (1986) Pericentromeric heterochromatin and A-T contents during Robertsonian fusion in the house mouse. Chromosoma 94:31–35
Riva E, Fox DP, Giraldez R, Santos JL (1984) Chiasma frequency and distribution in the presence and absence of supernumerary chromosome segments in the grasshopper *Euchorthippus pulvinatus* gallicus. Heredity 53:101–106
Sahar E, Latt SA (1978) Enhancement of banding patterns in human metaphase chromosomes by energy transfer. Proc Natl Acad Sci USA 75:5650–5654
Sahar E, Latt SA (1980) Energy transfer and binding competition between dyes used to enhance staining differentiation in metaphase chromosomes. Chromosoma 79:1–28
Sande JH van de, Lin CC, Jorgenson KF (1977) Reverse banding on chromosomes produced by a guanosine-cytosine specific DNA binding antibiotic: olivomycin. Science 195:400–402
Schempp W (1980) Asynchrony in late replication between homologous autosomes in primary cultures of Chinese hamster fibroblasts. Chromosoma 79:199–206
Schnedl W, Abraham R, Förster M, Schweizer D (1981) Differential fluorescent staining of porcine heterochromatin by chromomycin A_3/distamycin A/DAPI and D287/170. Cytogenet Cell Genet 31:249–253
Schwarzacher T, Schweizer D (1982) Karyotype analysis and heterochromatin differentiation with Giemsa C-banding and fluorescent counterstaining in Cephalanthera (Orchidaceae). Plant Syst Evol 141:91–113
Schwarzacher T, Mayr B, Schweizer D (1984) Heterochromatin and nucleolus-organizer-region behaviour at male pachytene of *Sus scrofa domestica*. Chromosoma 91:12–19
Schweizer D (1976) Reverse fluorescent chromosome banding with chromomycin and DAPI. Chromosoma 58:307–324
Schweizer D (1977) R-banding produced by DNase I digestion of chromomycin-stained chromosomes. Chromosoma 64:117–124
Schweizer D (1980) Simultaneous fluorescent staining of R bands and specific heterochromatic regions (DA-DAPI bands) in human chromosomes. Cytogenet Cell Genet 27:190–193
Schweizer D (1981) Counterstain-enhanced chromosome banding. Hum Genet 57:1–14
Schweizer D, Ehrendorfer F (1983) Evolution of C-band patterns in Asteraceae – Anthemideae. Biol Zentralbl 102:637–655
Schweizer D, Loidl J (1986) A model for heterochromatin dispersion and the evolution of C-band patterns. Chromosomes Today 9 (in press)
Sims SH, Macgregor HC, Pellatt PS, Horner HA (1984) Chromosome 1 in crested and marbled newts (*Triturus*). An extraordinary case of heteromorphism and independent chromosome evolution. Chromosoma 89:169–185
Singer MF (1982) Highly repeated sequences in mammalian genomes. Int Rev Cytol 76:67–112
Southern DI (1967) Pseudo-multiple formation as a consequence of prolonged non-homologous chromosome association in *Metrioptera brachyptera*. Chromosoma 21:272–284
Sumner AT, Buckland RA (1976) Relative DNA contents of somatic nuclei of ox, sheep and goat. Chromosoma 57:171–175
Torre J de la, López-Fernández C, Nichols R, Gosálvez J (1986) Heterochromatin readjusting chiasma distribution in two species of the genus *Arcyptera*: the effect among individuals and populations. Heredity 56:177–184
Viegas-Péquignot E, Derbin C, Malfoy B, Taillandier E, Leng M, Dutrillaux B (1983) Z-DNA immunoreactivity in fixed metaphase chromosomes of primates. Proc Natl Acad Sci USA 80:5890–5894

Walker PMB, Flamm WG, McLaren A (1969) Highly repetitive DNA in rodents. In: Lima-de-Faria A (ed) Handbook of molecular cytology. Elsevier/North-Holland, Amsterdam New York, pp 52–66

White MJD (1973) Animal cytology and evolution, 3rd edn. Cambridge Univ Press, Cambridge

Yunis JJ (1976) High resolution of human chromosomes. Science 191:1268–1270

Yunis JJ, Yasmineh WG (1971) Heterochromatin, satellite DNA, and cell function. Science 174:1200–1209

Multiple Nonhistone Protein-DNA Complexes in Chromatin Regulate the Cell- and Stage-Specific Activity of an Eukaryotic Gene

A. E. SIPPEL, U. BORGMEYER, A. W. PÜSCHEL, R. A. W. RUPP, A. STIEF, U. STRECH-JURK, and M. THEISEN [1]

1 Introduction

The general mechanisms underlying cell differentiation are still widely unknown. At present, we are forced to reduce the complexity of problems to more basic questions. In a first approximation, cell differentiation must be considered to be the result of differential expression of the same genomic information present in all cell types of an organism. In regards to this, we have asked the question: "What mechanism during cell differentiation prepares an eukaryotic gene so that it is activated only in a specific cell type and only at the right period of time?"

The first cell-specific acting DNA elements in the vicinity of eukaryotic genes have been recently discovered (for review, see Serfling et al. 1985). Originally identified in viral genomes, so-called transcriptional enhancers were found to potentiate transcription from RNA polymerase II-transcribed promoters in an orientation- and distance-independent manner (Khoury and Gruss 1983). Later, cis-acting DNA elements of this type were also found in cellular DNA and it could be shown that they exert a high degree of cell-specific function (Gillies et al. 1983; Banerji et al. 1983; Queen and Baltimore 1983). Today regulatory sequences responsible for the tissue-specific and developmentally regulated transcriptional activation have been found for many genes in the immediate 5′-flanking region of the promoter as well as further upstream or downstream (for review, see Serfling et al. 1985). In order to understand the function of cis-acting DNA elements in the transcriptional regulation of eukaryotic genes, it is necessary to identify factors which interact with these sequences and to elucidate the detailed molecular structure of these genomic switches in the chromatin.

For the study of the molecular mechanisms involved in cell and stage-specific transcriptional activation, we have concentrated on the gene for chicken lysozyme. The gene itself, its transcriptional products and the general regulatory physiology of its activity are well characterized (Sippel et al. 1978; Nguyen-Huu et al. 1979; Jung et al. 1980; Grez et al. 1981; Schütz et al. 1981; Sippel and Nowock 1982). In contrast to other genes which are studied for their mechanism of cell-specific expression, like the globin genes, the immunoglobulin genes, the ovalbumin genes or the vitellogenin genes, the gene for chicken lysozyme offers a certain advantage. The same gene, which is present only once per haploid genome, is expressed in two widely different cell types and its activity is regulated differently in these cells.

[1] Zentrum für Molekulare Biologie der Universität Heidelberg (ZMBH), Im Neuenheimer Feld 282, 6900 Heidelberg 1, FRG.

Results and Problems in Cell Differentiation 14
Structure and Function of Eukaryotic Chromosomes
Edited by W. Hennig

1. Lysozyme is one of the four major egg white proteins produced in the tubular gland cells of the laying hen oviduct. The transcriptional activity of the gene in these cells is strictly dependent on the presence of at least one type of the naturally occurring steroid hormones (Schütz et al. 1978).

2. As part of the bactericidal strategy of the organism, lysozyme is also expressed in mature macrophages. In these cells, the gene is constitutively active and completely independent of steroid hormones (Sippel et al. 1986).

When total cellular RNA was used to map the site of transcriptional initiation in oviduct cells and peripheral blood-derived macrophages, it could be shown that the same promoter site is used in both cell types (Grez et al. 1981; Sippel et al. 1986; Theisen et al. 1986).

In liver cells, as in most other cells of the organism, the lysozyme gene is inactive. It is not steroid inducible, even though chicken liver cells are excellent estrogen target cells in which the yolk protein genes are regulated via a functional estrogen receptor (Ryffel and Wahli 1983; Burch and Weintraub 1983).

In order to approach the mechanism of differential gene regulation, it is necessary to find the cis- and trans-acting components which determine the various modes of activity of the gene. It is our hope that by doing so, we will obtain clues as to which general mechanisms control cell differentiation.

2 DNase Hypersensitive Chromatin Sites are Clustered Around Genes

It is widely assumed that the structural organization of chromatin contributes to the state of activity of eukaryotic genes (Weintraub 1985). Rather than initially concentrating on the structure and function of the promoter region we have started out looking at the chromatin structure of a wider region flanking the lysozyme gene. DNase I was used to probe the nuclease accessibility of DNA in the chromatin of isolated nuclei. When we mapped DNase I hypersensitive sites by the indirect end-labelling method, originally developed by Wu (1980) and Nedospasov and Georgiev (1980), we found that nuclease hypersensitivity is not restricted to promoter proximal regions of the chromatin.

Figure 1 summarizes the location of all hypersensitive sites mapped in laying hen oviduct nuclei in the region of 50 kb encompassing the lysozyme gene (Fritton et al. 1983; 1986) and in which no other gene could be detected. In oviduct chromatin, in which the gene is fully induced by steroid hormones, seven hypersensitive sites are clustered in the flanking DNA region from -8 kb to $+8$ kb with respect to the promoter of the gene. The cluster is framed on both sides by chromatin regions free of hypersensitive sites for at least 15 kb of DNA. No hypersensitive site could be detected within the transcriptional unit. All seven sites are located in a domain of general DNase sensitivity spanning approximately 24 kb from around -13 kb to $+11$ kb (Jantzen et al. 1986; Strätling et al. 1986). Since the extension of the cluster of hypersensitive sites was found to be the same also in other chicken cell types (Fritton et al. 1986), we conclude that it comprises

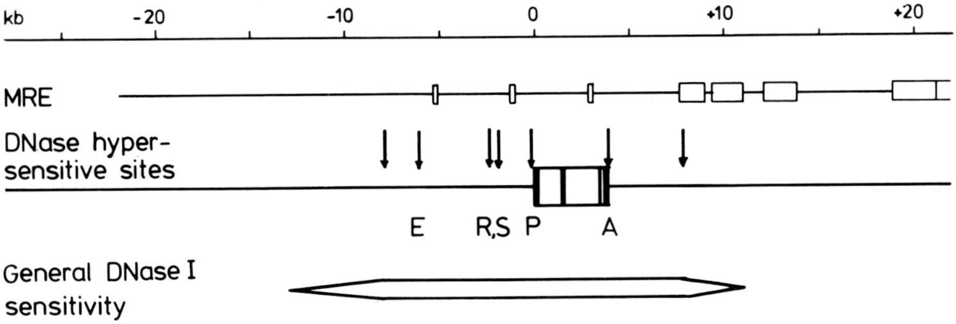

Fig. 1. The chromatin domain of the lysozyme gene in laying hen oviduct. The diagram shows the chicken lysozyme gene with its four exons (*filled bars*) and three introns (*open bars*). *Vertical arrows* indicate the positions of DNase I hypersensitive sites. *Open boxes* show locations of middle repetitive DNA elements (MRE). The extension of the 24 kb region of general DNase I sensitivity (Jantzen et al. 1986) is indicated by a *horizontal open bar*. E = enhancer; R = location of a putative repressing function; S = steroid-dependent chromatin site; P = promoter; A = poly(A) point

the complete chromatin domain of the chicken lysozyme gene, possibly the structural correlate of a chromatin loop as suggested in the radial loop model of chromosome organization (Paulson and Laemmli 1977; Igo-Kemenes et al. 1982).

Similar results emerged in a comparable extensive mapping of hypersensitive sites done for the human α 1(I) collagen gene (Barsh et al. 1984). In this case, the cluster of hypersensitive sites spans approximately 65 kb, indicating a more extended chromatin domain. It is conceivable that most eukaryotic genes or groups of related genes are organized in such a domain-like structure.

3 Alternative Chromatin Structures Characterize the Various Transcriptional States of a Gene

The domain organization of eukaryotic chromatin implies that neighbouring domains are regulated relatively independent of each other and that all cis-acting components involved in the control of a gene are located within the limits of its domain. We therefore tested the pattern of DNase I hypersensitive sites of the entire domain in different cell types in which the lysozyme gene is either hormone responsive as in oviduct cells, nonresponsive as in macrophages or inactive as in liver cells.

Table 1 summarizes mapping data of DNase I hypersensitive sites for the chromatin of five different states of activity of the gene (Fritton et al. 1984, 1986). Differences in chromatin structure can be clearly seen, indicating that these sites may contribute to the way the gene is transcriptionally expressed. The following features are noteworthy:

1. The hypersensitive site at the promoter, HS-0.1, is only present in cells which either express or have the potential to express the lysozyme gene.

Table 1. DNase I hypersensitive sites in the chromatin of the chicken lysozyme gene region at various states of activity

Cell type	Functional state of the lysozyme gene	Transcriptional activity	Presence of DNase I hypersensitive site (position in kb)[a]								
			−7.9	−6.1	−2.7	−2.4	−1.9	−0.7	−0.1	+3.9	+7.9
Oviduct	Steroid-induced	+++++	+	+	−	+	+	−	+	+	+
Oviduct	Steroid-deinduced	−	+	+	−	+	−	−	+	+	+
Macrophage	Constitutive	+	+	+	+	−	−	+	+	+	−
Liver, kidney	Inactive	−	+	−	−	+	−	−	−	+	+
Erythrocyte	Dormant	−	−	−	−	−	−	−	−	+	−

[a] + Denotes hypersensitive site present; − denotes closed chromatin structure at this position; the transcription unit of the lysozyme gene starts at 0 and ends at +3.9.

2. The site at −1.9 kb, HS-1.9, is present only in steroid-induced oviduct cells.

3. In liver cells, as in other nonexpressing cells, only a subset can be detected of the hypersensitive sites present in oviduct cells. The hypersensitive sites HS-0.1, HS-1.9 and HS-6.1 are closed. This difference in chromatin structure is most likely responsible for the inability of the gene to be activated by steroids. The chromatin structure looks underdeveloped.

4. The most informative result is that marked differences exist in the structural organization of chromatin between nuclei from the two lysozyme-producing cell types. In macrophages HS-1.9 and HS-2.4 are absent and two other sites, HS-0.7 and HS-2.7, can be mapped.

5. The presence of HS-2.4 in all nonproducing cells, except in terminally differentiated erythrocytes, indicates that it might contain a suppressive function. Such a function would be required also in oviduct cells in which transcription is reversibly changed, depending on the hormonal state, but would be unnecessary in constitutively expressing mature macrophages.

6. At −6.1 kb, far away from the promoter, HS-6.1 is the only hypersensitive site which is strictly correlated to the state of the promoter. The open state of this site is as characteristic for the active chromatin as is the open state of the promoter.

These findings show that different modes of transcriptional regulation of an eukaryotic gene correlate with a different chromatin organization and support the notion that the hypersensitive structures themselves may determine the functional state of a gene. Such a thorough cataloguing of the hypersensitive sites and their correlation with the transcriptional activity of the gene makes it possible to venture an educated guess as to their function (Reudelhuber 1984).

4 DNase Hypersensitive Sites Mark the Position of Cis-Acting DNA Elements

DNase hypersensitive sites in chromatin are believed to be short nucleosome-free regions which mark positions where nonhistone proteins have access to specific DNA sequences (McGhee et al. 1981; Emerson and Felsenfeld 1984). We assume that the recognition of specific signal sequences by DNA-binding proteins must be at the basis of regulatory protein-DNA complexes in chromatin. In a first approach to strengthen this hypothesis, we attempted to show in a functional assay system the cis-regulatory potential of DNA sequences contained in the regions of DNase hypersensitive chromatin. Transient gene transfer has proven to be an important means to test the regulatory function of DNA sequences in vivo. With this method, the first cellular enhancer sequences were detected in the immunoglobulin gene regions (Gillies et al. 1983; Banerji et al. 1983; Queen and Baltimore 1983).

Figure 2 shows an outline of plasmids which were constructed to test the function of the DNA sequence in the −6.1 kb region of the lysozyme gene. In a vector

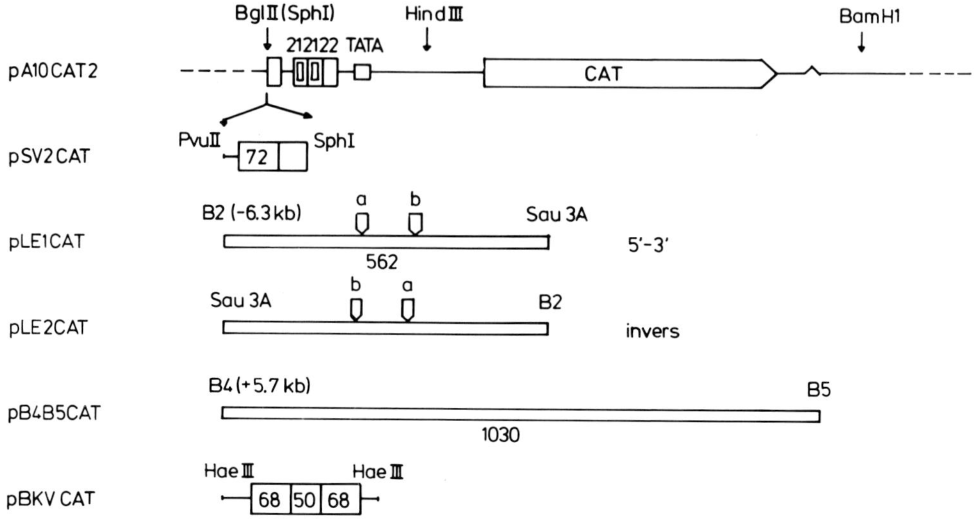

Fig. 2. The constructed plasmids to show transcriptional enhancer activity of the lysozyme −6.1 kb region. The basic vector (pA10CAT2) contained the early promoter of SV40 fused to the bacterial chloramphenicol transacetylase (CAT) gene (Laimins et al. 1983). Into its BglII site a 562 bp DNA fragment from base pair −6331 (B2 = BamH1 No. 2; cf. Fig. 4) to base pair −5772 (Sau 3A) of lysozyme 5′-flanking DNA was introduced in both orientations (pLE1CAT, pLE2CAT). For a positive control, an HaeIII fragment containing the BK virus enhancer (pBKVCAT) and a PvuII-SphI fragment reconstituting the full SV40 enhancer-early promoter region (pSV2CAT) were introduced into the BglII site. For a negative control, an unrelated 1030 bp fragment, starting at BamH1 site No. 4 (B4 at +5.7 kb) of the 3′-flanking region of the lysozyme gene (pB4B5CAT) was introduced into the BglII site of the vector. *Open arrows* indicate the binding sites for the TGGCA protein on the lysozyme enhancer (Borgmeyer et al. 1984). *72* = 72 bp repeat; *21,22* = 21,22 bp repeat; *TATA* = TATA box of the SV40 enhancer-early promoter region; *68,50* = 68,50 bp repeats of the BK virus enhancer

containing the bacterial chloramphenicol-acetyl-transferase (CAT) gene fused to the early promoter of SV40 (pA10CAT2; Laimins et al. 1983), a 562 bp BamHI-Sau3A fragment from approximately −6.35 to −5.80 kb of lysozyme 5′-flanking DNA was put in place of the SV40 72 bp repeat enhancer. DNA of these constructs and of control plasmids (Fig. 2) containing viral enhancers, unrelated cellular DNA or no foreign DNA were introduced into the macrophage-like cells of the chicken myelomonocytic cell line HBC-1 (Beug et al. 1979). Transient expression of the CAT gene constructs was tested 48 h after DNA transfer.

The results presented in Fig. 3 and SI mapping data not presented show that the −6.1 kb DNA region of the lysozyme gene is able to activate CAT expression from the SV40 early promoter in an orientation-independent fashion similar to but not as potent as the viral enhancer elements of SV40 and BKV. The lysozyme −6.1 kb DNA element also activates transcription from the homologous lysozyme promoter (Theisen et al. 1986). Again this cis activation is orientation-independent and could be shown to be relatively distance-independent. The lysozyme promoter alone is transcriptionally inactive even if 1200 bp of 5′-flanking promoter proximal DNA is used. The cis activation in HBC-1 cells, which are lyso-

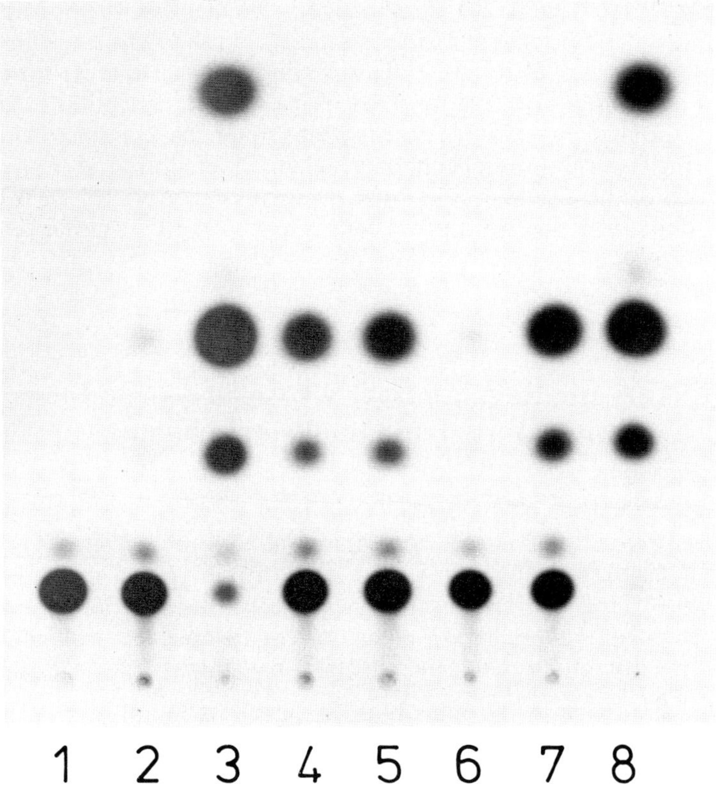

Fig. 3. Transient chloramphenicol transacetylase (CAT) activity of HBC1 cells transfected with plasmids shown in Fig. 2. MC29-transformed, macrophage-like monocytes (HBC1; Beug et al. 1979) were transfected as described by Theisen et al. (1986) with (*lane 2*) pA10CAT2, (*3*) pSV2CAT, (*4*) pLE1CAT, (*5*) pLE2CAT, (*6*) pB4B5CAT and (*7*) pBKVCAT. CAT activity was measured 48 h after transfection (Theisen et al. 1986). The autoradiogram shows the radioactive reaction components after thin layer chromatography. From *bottom to top: lane 1,* chloramphenicol; *lane 8,* two different monoacetylated and one diacetylated chloramphenicol

zyme-producing transformed macrophage precursor cells, is not seen, however, in chicken embryo fibroblasts (Theisen et al. 1986). Since the cell-specific action of the transfected −6.1 kb element extends also to peripheral blood-derived primary macrophages, on the one hand, and primary chicken fibroblast cells, on the other hand (Theisen, unpublished results), it nicely reflects the physiological cell-specific activity of the endogenous lysozyme gene. Recently, we were able to show that this cis activation also works when the −6.1 kb element close to the lysozyme promoter is stably integrated into the genome of the macrophage-like cells (Stief 1985; Stief, unpublished results).

With these features, the lysozyme −6.1 kb DNA element contains all the characteristics of a cellular transcriptional enhancer element. We therefore call this element the "lysozyme enhancer." The modular action of the lysozyme enhancer in gene transfer experiments (it acts in the absence of all the other possible elements in the lysozyme flanking region) plus the fact that the open chromatin state in the −6.1 kb lysozyme region is strictly correlated to the active state of the promoter (Sippel et al. 1986), makes us confident that this far upstream element is the physiological activator of lysozyme gene transcription in vivo. With this finding, the chicken lysozyme gene is added to the few examples of eukaryotic genes for which transcriptional enhancer elements were found. It is possible that one or several of these cell-specific transcriptional regulators will be found in all chromatin domains if the search for them is extended to promoter distal DNA sequences.

5 Eukaryotic Regulatory DNA Elements can Interact with More than One Protein Factor

A considerable mechanistic problem has been posed by the finding that some of the transcriptional regulatory elements in eukaryotes act relatively independent of their distance to the start site for transcription (for a recent review, see Nasmyth 1986). How do long-distance-acting elements function? Since the discovery of transcriptional enhancer elements 4 years ago, the problem remained unsolved. Two general mechanisms must be considered.

1. A direct interaction of nonhistone protein-DNA complexes via strong protein-protein contacts could bring distant DNA elements in close spatial position. Such a "DNA loop-out model" would very well explain the striking distance independency of action.
2. Alternatively, promoter distal elements could be sites of a (enzymatic) function which progressively alters chromatin structure in such a way that the promoter is made available for transcription initiation.

In order to approach this question, we have started to analyze the molecular structure of the lysozyme enhancer as an example of a long-distance-acting regulatory switch in cellular chromatin. Originally looking for individual, sequence-specific, DNA-binding proteins in nuclear protein extracts (Nowock and Sippel 1982; Borgmeyer et al. 1984), we recently attempted to reconstitute in vitro the lysozyme enhancer protein complex from crude nuclear extracts by incubation with isolated DNA fragments from the −6.1 region of the gene. DNA-binding proteins were detected by one of the most sensitive assay systems for sequence-specific protein-DNA interaction, the exonuclease III assay (Shalloway et al. 1980). When asymmetrically 5′-labelled linear double-stranded, enhancer-containing DNA fragments were digested with *E. coli* exonuclease III after incubation with crude protein extracts from isolated nuclei of oviduct cells, macrophage-like HBC-1 cells or liver cells, identical complex patterns of electrophoretically visualized "stop-bands" indicated that more than one sequence-specific, DNA-binding protein reassociates with this DNA in each case (Püschel 1986; Püschel, Borgmeyer and Sippel, in preparation). Imprints of individual DNA-binding proteins or groups of interacting DNA-binding proteins could be identified by addition of various competitor DNA sequences to the exonuclease III assay system. Figure 4 shows schematically the result of our structural analysis of the in vitro reconstituted enhancer DNA-protein complex (Püschel 1986; Püschel, Borgmeyer and Sippel, in preparation). The following features should be pointed out:

1. The borders of the reconstituted protein-DNA complex are roughly identical with the functional minimal enhancer DNA fragment active in transient transfection experiments with the macrophage cell line (Theisen, unpublished result).
2. The enhancer protein complex shows a tripartite structure. Each domain consists of more than one protein component. Exonuclease III-DNA competition assays show protein-protein interaction to occur between components of the two flanking domains.
3. The lysozyme enhancer-binding protein with the highest affinity to double-stranded DNA has two DNA recognition sequences within the −6.1 kb region, their centers being 92 bp apart (Borgmeyer et al. 1984; Sippel et al. 1986). This protein functions as an anchor protein for other factors with less affinity for their adjacent DNA recognition sequences.

The anchor protein for the lysozyme enhancer protein complex is identical with our previously discovered "TGGCA protein" (Nowock and Sippel 1982; Borgmeyer et al. 1984), the chicken homologue to HeLa cell "nuclear factor I" (NFI) (Nowock et al. 1985; Leegwater et al. 1986). NFI is a member of a family of ubiquitous nuclear DNA-binding proteins involved in the initiation reaction of adenovirus replication (Nagata et al. 1983).

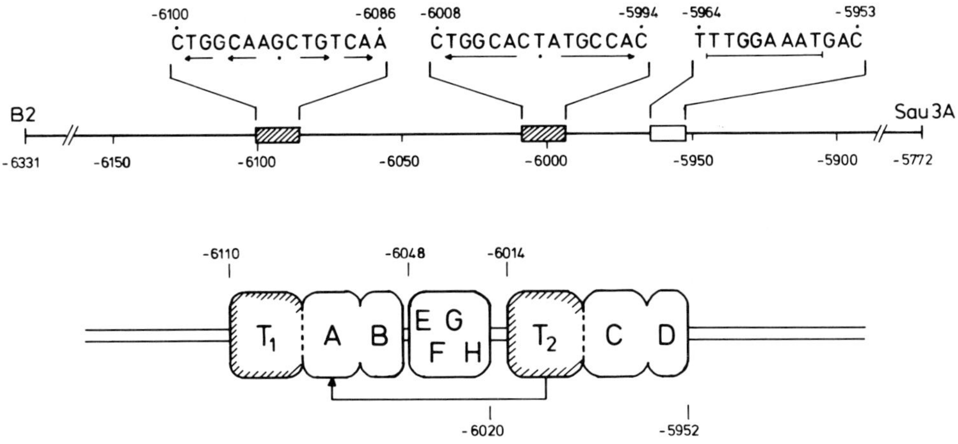

Fig. 4. The in vitro reconstituted lysozyme enhancer: nonhistone protein complex. At the *top* the positions are shown of the palindromic TGGCA protein recognition sites (BS1a, BS1b; Borgmeyer et al. 1984) plus the sequence resembling the "deca/octamere motif" (Falkner and Zachau 1984; Singh et al. 1986). *Numbers* indicate base pairs with respect to the cap site of the lysozyme gene; *B2* = BamH1 site No. 2 at −6331 bp. The *bottom part* shows the approximate positions of oviduct, macrophage and liver nuclear protein components *A* to *H*, T_1 and T_2 mapped by exonuclease III DNA competition assays (Püschel 1986). T_1, T_2 = TGGCA protein; *arrow* indicates protein-protein interaction

4. The most promoter proximal protein component in the complex covers a DNA sequence motif which was also found in other DNA elements known to cis-regulate RNA polymerase II promoters (Mattaj et al. 1985; Singh et al. 1986). The respective sequence motif for this second DNA-binding protein is very similar to the highly conserved decanucleotide sequence 5'-ATGCAAATNA-3' which was first discovered to be present in the enhancer plus the promoter region of mouse and human immunoglobulin heavy chain and kappa light chain genes (Falkner and Zachau 1984).

In conclusion, considering the in vitro reconstitution data, we found that the lysozyme enhancer complex is most likely a multifactorial nonhistone protein-DNA structure. This finding is in accordance with the general picture arising from the structural and functional analysis of other viral and cellular regulatory elements (Serfling et al. 1985). Currently, however, we are unable to pinpoint those components of the enhancer structure which determine its cell specificity of action in vivo.

Whether all nonhistone protein-DNA complexes, indicated by DNase hypersensitive sites in chromatin, are multifactorial structures has to be shown. However, the multiple DNA-binding sites for steroid receptors at the promoter region of the lysozyme gene (Renkawitz et al. 1984; von der Ahe et al. 1985) already indicate that also the complex at 0.1 kb is a multifactorial element.

6 The Active Chromatin Structure of a Gene Develops in Successive Steps During Cell Differentiation

For the full understanding of the role of regulatory nonhistone protein-DNA complexes in eukaryotic chromatin it will be ultimately necessary to understand when they originate during development. To study this question, it is necessary to select a system in which enough homogeneous cells of different developmental stages can be obtained to analyze chromatin structure in specific gene regions. Only very few systems are suitable for this purpose. One of them is the hematopoietic system of vertebrates.

In hematopoietic cells, lysozyme is expressed only in late myeloid cell stages. The lysozyme gene can be considered a marker gene for the myeloid branch, like the globin genes for the erythroid branch or the immunoglobulin genes for the lymphoid branch. Graf and Beug (Beug et al. 1979) could show that certain acute retroviruses transform hematopoietic bone marrow cells into leukemic cells with phenotypes close to natural precursors of the mature blood cells. Instead of the natural precursors, which are more difficult to obtain in the necessary quantities, we have used several well-characterized transformed cell types of this kind to look for the structure of the 5'-flanking chromatin in the lysozyme gene region. Again, the pattern of DNase I hypersensitive sites was taken as an indicator for functional aspects.

Table 2 summarizes our mapping data in the chromatin of transformed cells, which resemble early and late myeloid and erythroid stages of cell differentiation

Table 2. State of chromatin at the enhancer, the promoter and the "intermediate region" of the lysozyme gene in various hematopoietic cells[a]

Chicken cell	Lysozyme mRNA[b]	DNase I hypersensitive site at (position in kb)[c]				
		En-hancer	Intermediate region		Promoter	
		−6.1	−2.7	−2.4	−0.7	−0.1
E26-transformed myeloblast	+/−	+	−	+	+	+
MC29-transformed macrophage-like monocyte	+	+	+	+	+	+
Nature blood-derived macrophage	++	+	+	−	+	+
AEV-transformed erythroblast	−	−	−	+	−	−
Mature erythrocyte	−	−	−	−	−	−

[a] Strech-Jurk et al. (1986), Strech-Jurk U, Beug H, Theisen M, Graf T, Sippel AE, in preparation.
[b] Relative cellular steady-state level of lysozyme-specific transcripts.
[c] + Denotes "open" chromatin structure; − denotes closed chromatin structure, DNase I hypersensitive site not present.

(Strech-Jurk et al. 1986; Strech-Jurk, Beug, Theisen, Graf and Sippel, in preparation). The results confirm that the final structure of macrophage-specific constitutive expressing chromatin is the result of two successive steps which alter chromatin structure during cell differentiation.

1. Already in early myeloid stages, the enhancer (−6.1 kb) and the promoter (−0.7 kb, −0.1 kb) are in the same open configuration as in mature macrophages. Unfortunately, it is currently impossible to map the chromatin structure in bone marrow stem cells as they comprise roughly only 1 in 10,000 bone marrow cells. However, since early stages of erythroid cells (AEV-transformed erythroblasts) do not show open enhancer and promoter chromatin, we conclude that the formation of these DNase I sensitive nonhistone protein-DNA complexes must be an early event in myeloid determination.
2. The final, full transcriptional activation of the lysozyme gene in macrophages is accompanied by a rearrangement of the chromatin structure at an intermediate position between enhancer and promoter. As discussed earlier, the presence of a hypersensitive site at −2.4 kb in all inactive cell types may mean that it contains a suppressive regulatory element. The observed chromatin structural change in late macrophage differentiation would then be consistent with the idea that a suppressive chromatin conformation is replaced by the macrophage-specific active conformation, indicated by a hypersensitive site at −2.7 kb, which then allows constitutive expression of the gene. MC29-transformed macrophage-like monocytes in culture show at the "intermediate region" a resemblance to a transitional state between the early and late myeloid states of chromatin.

What is seen here can be considered a structural correlative for cell differentiation at the level of nuclear chromatin. We must await further research to see whether the two-step process of gene activation, as it could be shown for the lysozyme gene during macrophage development, is of more general importance for the activation of other genes. A cell-lineage-specific, but cell-stage-independent function, as it is expected for the lysozyme enhancer in myeloid cells, was found for the immunoglobulin heavy chain gene (IgM) enhancer in lymphoid cells (Grosschedl and Baltimore 1985; Gerster et al. 1986). Our results are consistent with the assumption that a developmental program of formation and decomposition of individual regulatory elements in a chromatin domain prepares each gene or groups of genes for their final state of expression. The developmental history of each gene can thus be obtained by looking at the molecular structure of its regulatory switches.

7 Conclusions and Future Directions

In our work over the past years, we have deepened our understanding of the gene for chicken lysozyme with the aim of drawing general conclusions for the functioning of other eukaryotic genes. Thus, the lysozyme gene can be considered a small window to the structure and function of a large part of the genome, the euchromatin.

From the results obtained, we conclude that the tissue- and stage-specific activation of eukaryotic genes involves more than the highly specific function of transcriptional promoter regions. Our chromatin structural studies imply that an entire chromatin domain with extensive flanking sequences on both sides of the transcription units contain a number of loosely spaced cis acting DNA elements which are involved in the various aspects of regulation of the respective gene. Although we have concentrated on the function and molecular structure of one far distant enhancer element upstream from the promoter, it became increasingly evident that the full cell- and stage-specific regulation of the lysozyme gene can only be explained by the combined action of all the protein-DNA complexes indicated as DNase hypersensitive sites in the chromatin domain.

It remains to be elucidated how alternative subsets of elements interact to lock the gene either in the steroid-regulated active mode, the constitutive active, macrophage-specific mode or the repressed mode, as seen for example in liver cells.

A system in which the respective elements could act together independently of their distance, their orientation and their strict linear sequence to each other would be of considerable evolutionary advantage over a system in which the spatial relation to each other would be rigid. A "DNA loop-out" structure, in which functional interaction of distant elements are due to direct protein-protein interactions between multiple nonhistone protein-DNA complexes would give a new perspective to the spatial organization of eukaryotic genes. Nonhistone protein-DNA complexes could create a transcriptional complex involving parts of or all of the known functional elements, thereby preparing the correct target structure for RNA polymerase and/or some additional factors. In such a globular DNA loop-out model of gene structure the organization of regulatory elements could be considered somewhat like an extension of the exon-intron organization of the transcriptional unit.

We are intrigued by the complexity underlying the regulation of a single gene. The regulatory complexity of a chromatin domain comes close to the regulatory complexity encountered in viral genomes. Multiple controlling elements, each one alone possibly a composite structure consisting of multiple (protein) factors, are found at the molecular basis of gene regulation on the level of the genome. Comparable work on other genes and their controlling elements will show whether the spatiotemporal regulation of all genes during cell differentiation can be explained by a low number of components which work together in a combinatorial fashion. In this respect, future work on the characterization of the components of the molecular switches in eukaryotic chromatin will be most fruitful. Cloning of the genes for these proteins will bring the next step of experimental sophistication. Ultimately, a molecular understanding of a set of regulatory modules will enable us to construct vector systems for reinsertion of individual genes in cells and organisms in such a way that they can be predictably controlled in a cell- and stage-specific manner.

Acknowledgements. Many former and present colleagues and collaborators of our research team, not shown as authors of this chapter, have contributed substantially to our current picture of cell- and stage-specific regulation of lysozyme gene activity. We thank them for practical help, discussions and suggestions and hope that we have cited them correctly. We also thank Gaik Pin Wee and Sigrid Marga-Henzel for secretarial support. Research from our group, covered in this contribution, was supported by grants from the Deutsche Forschungsgemeinschaft (SFB 74, Si 165/4-1, SFB 229), the Bundesministerium für Forschung und Technologie (BCT 0364/1, BCT 0381/5) and the Fonds der Chemischen Industrie.

References

Banerji J, Olson L, Schaffner W (1983) A lymphocyte-specific cellular enhancer is located downstream of the joining region in immunoglobulin heavy chain genes. Cell 33:729–740

Barsh GS, Roush CL, Gelinas RE (1984) DNA and chromatin structure of the human α1(I) collagen gene. J Biol Chem 259:14906–14913

Beug H, von Kirchbach A, Doederlein G, Conscience JF, Graf T (1979) Chicken hematopoietic cells transformed by seven strains of defective avian leukemia viruses display three distinct phenotypes of differentiation. Cell 18:375–390

Borgmeyer U, Nowock J, Sippel AE (1984) The TGGCA-binding protein: a eucaryotic nuclear protein recognizing a symmetrical sequence on double-stranded linear DNA. Nucleic Acids Res 12:4295–4311

Burch JB, Weintraub H (1983) Temporal order of chromatin structural changes associated with activation of the major chicken vitellogenin gene. Cell 13:65–76

Emerson BM, Felsenfeld G (1984) Specific factor conferring nuclease hypersensitivity at the 5′ end of the chicken adult β-globin gene. Proc Natl Acad Sci USA 81:95–99

Falkner FG, Zachau HG (1984) Correct transcription of an immunoglobulin κ gene requires an upstream fragment containing conserved sequence elements. Nature 310:71–74

Fritton HP, Sippel AE, Igo-Kemenes T (1983) Nucleasehypersensitive sites in the chromatin domain of the chicken lysozyme gene. Nucleic Acids Res 11:3467–3485

Fritton HP, Igo-Kemenes T, Nowock J, Strech-Jurk U, Theisen M, Sippel AE (1984) Alternative sets of DNAase I-hypersensitive sites characterize the various functional states of the chicken lysozyme gene. Nature 311:163–165

Fritton HP, Igo-Kemenes T, Nowock J, Strech-Jurk U, Theisen M, Sippel AE (1986) Chromatin structure of the lysozyme gene in steroid hormone target and non-target cells. Biol Chem Hoppe-Seyler 368:111–119

Gerster T, Picard D, Schaffner W (1986) During B-cell differentiation enhancer activity and transcription rate of immunoglobulin heavy chain genes are high before m-RNA accumulation. Cell 45:45–52
Gillies SD, Morrison SL, Oi VT, Tonegawa S (1983) A tissuespecific transcription enhancer element is located in the major intron of a rearranged immunoglobulin heavy chain gene. Cell 33:717–728
Grez M, Land H, Giesecke K, Schütz G, Jung A, Sippel AE (1981) Multiple m-RNAs are generated from chicken lysozyme gene. Cell 25:743–752
Grosschedl R, Baltimore D (1985) Cell-type specificity of immunoglobulin gene expression is regulated by at least three DNA sequence elements. Cell 41:885–897
Igo-Kemenes T, Hörz W, Zachau HG (1982) Chromatin. Annu Rev Biochem 51:89–121
Jantzen K, Fritton HP, Igo-Kemenes T (1986) The DNAase I sensitive domain of the chicken lysozyme gene spans 24 kb. Nucleic Acids Res 14:6085–6099
Jung A, Sippel AE, Grez M, Schütz G (1980) Exons encode functional and structural units of chicken lysozyme. Proc Natl Acad Sci USA 77:5759–5763
Khoury G, Gruss P (1983) Enhancer elements. Cell 33:313–314
Laimins LA, Khoury G, Gorman CM, Howard B, Gruss P (1983) Hostspecific activation of transcription by tandem repeats from simian virus 40 and Molony murine sarcoma virus. Proc Natl Acad Sci USA 79:6453–6457
Leegwater PAJ, van der Vliet PC, Rupp RAW, Nowock J, Sippel AE (1986) Functional homology between the sequence-specific DNA binding protein nuclear factor I from Hela cells and the TGGCA protein from chicken liver. EMBO J 5:381–386
Mattaj IW, Lienhard S, Jiricry J, DeRobertis EM (1985) An enhancer-like sequence within the Xenopus U2 gene promoter facilitates the formation of stable transcription complexes. Nature 316:163–167
McGhee JK, Wood WI, Dolan M, Engel JD, Felsenfeld G (1981) A 200 base pair region of the 5′ end of the chicken adult β-globin gene is accessible to nuclease digestion. Cell 27:45–55
Nagata K, Guggenheimer RA, Hurwitz J (1983) Specific binding of a cellular DNA replication protein to the origin of replication of adenovirus DNA. Proc Natl Acad Sci USA 80:6177–6181
Nasmyth K (1986) A U-turn in the regulation of transcription. Trends Genet 2:115–116
Nedopasov SA, Georgiev GP (1980) Non-random cleavage of SV40 DNA in the compact minichromosome and free in solution by micrococcal nuclease. Biochem Biophys Res Commun 92:532–539
Nguyen-Huu MC, Stratmann M, Groner B, Wurtz T, Land H, Giesecke K, Sippel AE, Schütz G (1979) The chicken lysozyme gene contains several intervening sequences. Proc Natl Acad Sci USA 76:76–80
Nowock J, Sippel AE (1982) Specific protein-DNA interaction at four sites flanking the chicken lysozyme gene. Cell 30:607–615
Nowock J, Borgmeyer U, Püschel AW, Rupp RAW, Sippel AE (1985) The TGGCA protein binds to the MMTV-LTR, the adenovirus origin of replication and the BK virus enhancer. Nucleic Acids Res 13:2045–2061
Paulson JR, Laemmli UK (1977) The structure of histone-depleted metaphase chromosomes. Cell 12:817–828
Püschel AW (1986) In vitro Analyse der Protein-DNA Wechselwirkungen am Lysozymgen-"enhancer" des Huhns. Diploma Thesis, Univ Heidelberg
Queen C, Baltimore D (1983) Immunoglobulin gene transcription is activated by downstream sequence elements. Cell 33:741–748
Renkawitz R, Schütz G, von der Ahe D, Beato M (1984) Sequences in the promoter region of the chicken lysozyme gene required for steroid regulation and receptor binding. Cell 37:503–510
Reudelhuber T (1984) Gene regulation – A step closer to the principles of eucaryotic transcriptional control. Nature 311:301
Ryffel GU, Wahli W (1983) Regulation and structure of the vitellogenin genes. In: Maclean N, Gregory SP, Flawell RA (eds) Eukaryotic genes: Their structure, activity and regulation. Butterworths, London, pp 329–341

Schütz G, Nguyen-Huu MC, Giesecke K, Hynes NE, Groner B, Wurtz T, Sippel AE (1978) Hormonal control of egg white protein messenger RNA synthesis in the chicken oviduct. Cold Spring Harbor Symp Quant Biol 42:617–624

Schütz G, Lindenmaier W, Grez M, Giesecke K, Hauser H-J, Blin N, Land H, Sippel AE (1981) Structure and expression of the chicken egg white lysozyme gene. In: Sauer HW (ed) Progress in developmental biology. Fischer, Stuttgart New York (Fortschritte der Zoologie 26:257–270)

Serfling E, Jasin M, Schaffner W (1985) Enhancers and eukaryotic gene transcription. Trends Genet 1:224–230

Shalloway D, Kleinberger T, Livingston DM (1980) Mapping of the SV40 DNA replication origin binding sites for the SV40 T antigen by protection against exonuclease III digestion. Cell 20:411–422

Singh H, Sen R, Baltimore D, Sharp PA (1986) A nuclear factor that binds to a conserved sequence motif in transcriptional control elements of immunoglobulin genes. Nature 319:154–158

Sippel AE, Nowock J (1982) The gene for chicken lysozyme: Structure and expression of a steroid regulated gene. In: Jaenicke L (ed) Biochemistry of differentiation and morphogenesis. Springer, Berlin Heidelberg New York, pp 177–183

Sippel AE, Land H, Lindenmaier W, Nguyen-Huu MC, Wurtz T, Timmis KN, Giesecke K, Schütz G (1978) Cloning of chicken lysozyme structural gene sequences synthesized in vitro. Nucleic Acids Res 5:3275–3294

Sippel AE, Fritton HP, Theisen M, Borgmeyer U, Strech-Jurk U, Igo-Kemenes T (1986) The TGGCA protein binds in vitro to DNA contained in a nuclease-hypersensitive region that is present only in active chromatin of the lysozyme gene. In: Grodzicker T, Sharp PA, Botchan MR (eds) DNA tumor viruses: Gene expression and replication. Cold Spring Harbor Press, Cancer Cells 4:155–162

Stief A (1985) Stabile Integration von rekombinanter Plasmid-DNA in transformierte Makrophagenzellen des Huhns. Diploma Thesis, Univ Heidelberg

Strätling WH, Dölle A, Sippel AE (1986) Chromatin structure of the chicken lysozyme gene domain as determined by chromatin fractionation and micrococcal nuclease digestion. Biochemistry 25:495–502

Strech-Jurk U, Beug H, Theisen M, Graf T, Sippel AE (1986) The chromatin structure of the lysozyme gene region in myeloid cells of different developmental stages. In: The twelfth EMBO symposium: Macromolecular structure in cellular organization

Theisen M, Stief A, Sippel AE (1986) The lysozyme enhancer: Cell specific activation of the chicken lysozyme gene by a far upstream DNA element. EMBO J 5:719–724

von der Ahe D, Janich S, Scheidereit C, Renkawitz R, Schütz G, Beato M (1985) Glucocorticoid and progesteron receptors bind to the same sites in two hormonally regulated promoters. Nature 313:706–709

Weintraub H (1985) Assembly and propagation of repressed and derepressed chromosomal states. Cell 42:705–711

Wu C (1980) The 5′ ends of Drosophila heat shock genes in chromatin are hypersensitive to DNase I. Nature 286:854–860

Genetics of Sex Determination in Eukaryotes

ROLF NÖTHIGER and MONICA STEINMANN-ZWICKY [1]

1 What is Sex?

"... the most important inventions (of evolution) are sex and death" (F. Jacob 1982)

The biologically significant consequence of sex is that a genetic variant is no longer the unique property of an individual and its *clonal descendents,* but becomes part of a gene pool that is shared by the members of a *population.* Variants can spread through the population, leading, without additional mutations, to the rapid production of new genetic combinations. The essence of sexual processes thus is the recombination of existing genetic information.

In principle, the exchange of genetic material could occur between any two individuals of a population, as it happens between viruses and many hermaphrodites. But in eukaryotes, a species is usually split into subgroups of complementary mating partners. Genetic exchange is possible between the subgroups, but not within a subgroup. Even when the mating partners look identical, as in isogamous algae, physiological tests reveal two mating types, designated as + and −. In the course of evolution, the subgroups or sexes became progressively and visibly different until conspicuous morphological, physiological, and behavioral traits characterized and distinguished males and females.

In our article, we will review the genetic devices that lead to differential gene activity in males and females of a species. We will try to reveal the parallels behind the apparent differences in sex-determining mechanisms. Our goal is an attempt to reduce the multitude of phenomena and to look for a common principle behind the ostensible variety.

2 What Determines Sex? – From Chromosomes to Molecules

This question has puzzled laymen and scientists alike throughout human history. Many theories have been put forward, some weird, some amusing, but rarely useful. The Greek philosopher Anaxagoras (around 450 B.C.) contended that the products of the right testis gave male offspring, those of the left testis females. But

[1] Zoological Institute, University of Zurich, Winterthurerstraße 190, 8057 Zurich, Switzerland.

Results and Problems in Cell Differentiation 14
Structure and Function of Eukaryotic Chromosomes
Edited by W. Hennig

the determinative effect was also ascribed to the female, stating that ova from the left ovary were determined for one sex, those of the right ovary for the other sex. The great Aristotle taught that very young and very old couples would preferentially have girls because the reproductive fluids were immature or degenerating. In 1889, Geddes and Thomson (quoted in Crew 1965) listed some 260 "theories" that tried to explain how sex is determined.

With the advent of genetics and the discovery of sex-linked inheritance, a scientific approach to the mechanisms of sex determination became possible. The combined efforts of geneticists and cytologists identified heteromorphic chromosomes that were associated with sex and sex-linked inheritance. It looked as if the age-old problem had found its solution. But soon, the picture became blurred and confusing, since no consistent pattern emerged. Mammals showed male heterogamety; in birds and butterflies the female was the heterogametic sex; the Y-chromosome was sometimes missing, and XO karyotypes were males in some species, females in others; some species showed identical chromosomes in males and females, whose sex was nevertheless genetically determined as judged by sex-linked inheritance of fortuitous mutations; and some species had their sex determined by environmental factors; in many vertebrates, sex could be reversed by hormones; and eventually, mutations were discovered that could override the genetic instructions provided by the sex chromosome constitution.

Where was the principle? The wealth of genetic variations and the analytical power of Bridges's experiments soon made the sex-determining mechanism of *Drosophila* the best-known case, and the balance concept became *the* paradigm that was applied to all other systems, if necessary by brute force. Its power (and its weakness) resided in the quantitative formulation of a model that assumed an (undefined) number of (ill-defined) male- and female-determining genes, (intangibly) scattered over the X-chromosome and the autosomes. The model developed into an iron concept that prevented even great minds from seeing the meaning of discrete genes and their mutations, such as "transformer" in *Drosophila melanogaster*.

A new mystery was met by Beermann (1955) when he discovered that in *Chironomus tentans* a dominant male-determiner could be associated with chromosome 2 in Holstein and with chromosome 1 in Stockholm. Thus, what was a sex chromosome in one population was an autosome in another. Later, Thompson and Bowen (1972) found strains of this species with a dominant male determiner on the left arm of chromosome 1, others with a dominant female-determiner on the right arm of chromosome 1. Martin and Lee (1984) recently extended these studies to Australasian *Chironomus* species and found the male sex-determiner located on various chromosomal arms, as if it had moved around in the genome during evolution of the genus *Chironomus*. Direct evidence for mobile male-determining elements, as noted by Green (1980), was provided by Mainx (1964) for *Megaselia scalaris* and by Denholm et al. (1986) for *Musca domestica,* both Diptera, as is *Drosophila*.

Heteromorphic sex chromosomes thus are not of primary importance in the process of sex determination. They have evolved because the chromosome carrying a dominant sex-determining factor could never become homozygous, so that no selection could operate on defective genes that were allowed to accumulate near the sex factor(s) (Lucchesi 1978). Different mechanisms arose to compensate for the loss of gene function in one sex (X-inactivation in mammals, regulation of transcription in *Drosophila* and *Caenorhabditis*). For sex determination, the sex factor alone is important, and the sex chromosome is only its carrier.

Whereas Beermann had identified the sex chromosomes of Chironomids by the presence of inversions with which sex segregated, Ribbert (1967) and Hägele (1985) found that in *Calliphora* and *Chironomus thummi thummi* sex is correlated with the presence or absence of a single chromomere. This band is only present in males and only in one member of a chromosome pair. If this band harbors the male-determining gene(s), the modern techniques of microdissection and cloning (Scalenghe et al. 1981) open up the molecular approach to sex determination. We are now able to isolate and characterize sex-determining genes, to analyze their transcripts and proteins, and to study the interaction between regulating and regulated genes.

3 Genetics of Sex Determination

In this chapter we want to present the state of research on sex determination for four paradigmatic cases that are especially well analyzed. These are *Saccharomyces cerevisiae, Caenorhabditis elegans, Drosophila melanogaster* (and some other insects), and mammals. Our goal is to emphasize the main points and to draw the picture in bold lines, neglecting the details. The reader who wants to penetrate deeper is referred to recent reviews mentioned in the text.

In dealing with the genetic control of developmental pathways, it is important to distinguish between *regulatory* genes which dictate a specific developmental pathway, e.g., male vs. female or mesothorax vs. metathorax, and *differentiation* genes which code for specific products characteristic of one developmental pathway, e.g., yolk polypeptides in the female sex or sperm-specific proteins in the male. When studying the mechanisms of sex determination in various organisms, the multitude and variety of phenomena is overwhelming and perplexing. We will try to show that a largely invariant principle is used in all systems of sex determination, namely a primary signal that can be of genetic or environmental nature and that is transmitted through a small number of regulatory genes down to the differentiation genes.

3.1 *Saccharomyces cerevisiae*

Among the lower eukaryotes, the genetic control of sex determination has been best analyzed for the yeast *Saccharomyces cerevisiae* (for review see Nasmyth 1982; Hicks et al. 1985; Herskowitz 1985). A yeast population can grow by cell division, small daughter cells budding from mother cells. One cell cycle later, the daughter cells are themselves mothers and divisions can go on without sexual processes. Nevertheless, haploid cells display either of two mating types, *a* or α. These are characterized by the expression of different sets of differentiation genes.

Cells of opposite types recognize each other and mate to form diploid *a*/α cells that can undergo meiosis and thus rearrange their genetic information. We want to emphasize that determination of mating type is formally analogous to sex determination of individual cells in an animal.

One gene, the mating type locus (*MAT*), governs the choice between the two different pathways. There are two alleles, *MATa* which confers the *a* cell type, and *MAT*α which confers the α cell type. Each of the alleles codes for two products, *a*1 and *a*2 or α1 and α2, respectively. These products regulate other genes, leading to the expression of a set of genes specific to a given mating type.

The *a* cell-specific genes are constitutively expressed unless they are repressed by α2; the α cell-specific genes are inactive in *a* cells and activated by α1 in α cells (Fig. 1). In diploid *a*/α cells, *a*1 and α2 act together to repress α1 and the haploid-specific genes, i.e., genes not needed in diploids (Strathern et al. 1981). The product α2 is a site-specific DNA-binding protein. It recognizes and binds to a 31 bp site located some 135 bp upstream of the *a*-specific gene *STE6*, and a strongly homologous sequence is found upstream of four other *a*-specific genes that are repressed by α2 (Wilson and Herskowitz 1986). These 31 bp are sufficient to place *CYC1*, a gene that is normally not controlled by mating type, under negative control by α2 (Johnson and Herskowitz 1985). Similarly, a 20 bp "operator" sequence was identified in the 5′ flanking sequences of *MAT*α*1*, and of haploid-specific genes. This site is sufficient for a gene to become repressed in *a*/α diploid cells

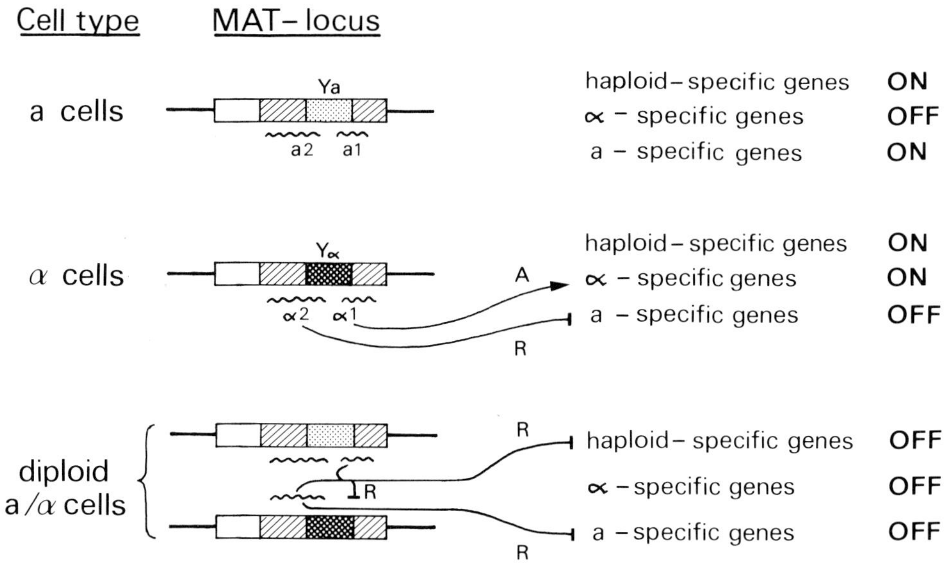

Fig. 1. The *MAT*-locus in its different allelic states. The figure shows the genetic structure of the locus in haploid (*a* and α) and diploid cells (*a*/α), the gene products (*wavy lines*) *a*1 *a*2 and α1 α2, and the consequences on sets of subordinate genes ("differentiation genes"). The function of the products is to activate (*A*) or repress (*R*) subordinate genes, specifying the cell type. The function of *a*2 is unknown; the haploid-specific and the *a*-specific genes work constitutively unless repressed. The difference between *a* and α lies within a DNA segment of about 700 base pairs designated Y*a* or Yα

(Miller et al. 1985). A specific sequence of the α2 product, weakly homologous to a conserved DNA-binding protein of prokaryotes and to the eukaryotic homeo-domain, was shown to be essential for the two repressor activities of α2, repression of *a*-specific genes and of haploid-specific genes (Porter and Smith 1986).

Thus, sex determination is a matter of gene regulation, *MAT* being the master regulatory locus dictating one of two alternative pathways, by repressing or activating specific sets of genes. Interestingly, this locus is itself under the influence of *trans*-regulatory genes whose action is responsible for the switching of mating type, a phenomenon that takes place with great regularity in growing populations of yeast cells. Although sex is determined by an allelic difference *a* vs. α, at a single locus *MAT*, both mating types are encountered among the members of a haploid clone. A haploid cell, after it has budded off its first daughter cell, changes its mating type before it buds again so that it and its daughter cell are now both of the opposite mating type. This phenomenon rapidly and reliably provides the opportunity for sexual processes within a clone. The change from α to *a* or vice versa is possible because copies of α and *a* are stored in silent "cassettes" some 150 to 180 kb to the left (*HML*α) and to the right (*HMRa*) of the *MAT* locus (Fig. 2). The switching is initiated by the activity of the haploid-specific gene *HO* which codes for an endonuclease. This enzyme cuts within the *MAT*-locus from which the expressed *a* or α sequences are removed and replaced by sequences copied from *HML*α or *HMRa*. The outcome of the copying process is reminiscent of gene conversion or transposition, inasmuch as an unaltered copy remains in the silent "cassette" at *HML*α or *HMRa*. A group of four genes, called *SIR* (*s*ilent

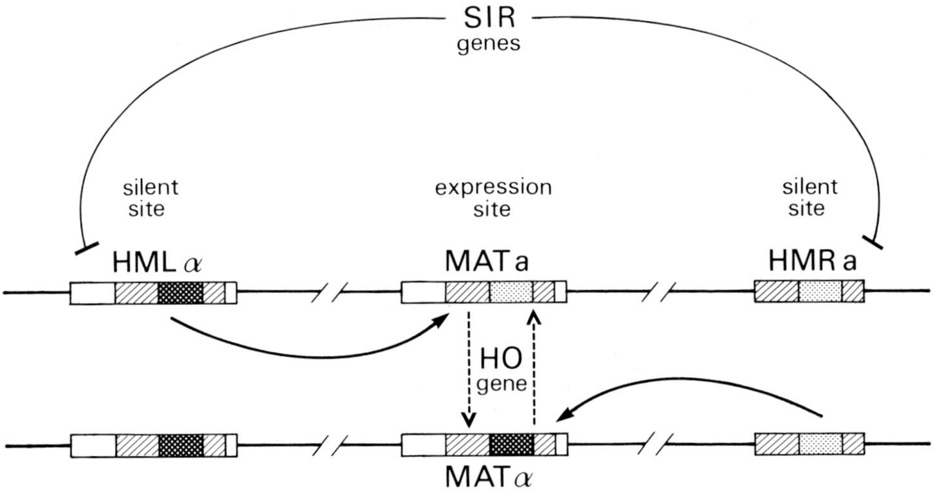

Fig. 2. The "cassette" model for the switching of mating type. The figure shows the MAT locus where one of two allelic forms, *a* or α, is present and expressed, and the silent sites where unexpressed copies of α and *a* are stored. The cassettes at *HML*α and *HMRa* are repressed by the *SIR* genes. *Solid arrows* indicate the copying of α or *a* from the silent site into the expression site so that switching of mating type occurs (*broken arrows*). The process is initiated by the gene *HO*

*i*nformation *r*egulator) encode products that prevent transcription of *HMLα* and *HMRa* and thus guarantee that only the "cassette" inserted at the *MAT* locus is expressed.

The picture emerging from this short and simplified description is that a *primary signal,* through the onset of *HO* transcription, triggers a process that leads to the expression of a different allele of a *key gene, MAT,* and the different products of the two alleles control the sex-specific *differentiation genes*. We can distinguish levels of control forming a cascade through which information is passed down, thus achieving the specification of the sexual phenotype.

The unique and peculiar aspect of sex in yeast is the switching of mating type. A mutation in a single gene, *HO,* however, can abolish this ability and thus create a situation as it is commonly found among haplontic organisms, e.g., in the unicellular alga *Chlamydomonas*. Here, sex is also determined by an allelic difference, mt^+ vs. mt^-, but contrary to yeast all the members of a clone have the same sex, dictated by mt^+ or mt^-, and propagate it faithfully through indefinite cell divisions (Galloway and Goodenough 1985). This shows that seemingly different mechanisms of sex determination may result from a simple change in one of the control elements that forms part of an invariant basic structure governing a developmental pathway.

3.2 *Caenorhabditis elegans*

This worm, a free-living nematode of about 1 mm in length, is a self-fertilizing hermaphrodite that produces some 300 sperm and 2000 oocytes in its gonad. Males are rare among the self-progeny of a hermaphrodite and are the result of nondisjunction of the X-chromosomes. But they compose half of the cross-progeny of a mating between a hermaphrodite and a male. A male produces some 3000 sperm in his testis. Hermaphrodites are chromosomally XX and males are XO. Consequently, they have the same genes whose information, however, is differentially used in the two sexes, as seen from a pronounced sexual dimorphism manifesting itself at the morphological, behavioral and biochemical level (Fig. 3 and Table 1).

Madl and Herman (1979) showed that the ratio of X chromosomes to sets of autosomes (X:A) is the primary signal that initiates the sexual pathway. Using a series of deletions and duplications, they scanned a large part of the X-chromosome for the presence of a major sex-determining gene. None was found. Instead, different X-chromosomal regions turned out to be about equally strong in promoting hermaphrodite development in aneuploid genotypes, suggesting dispersed genes with quantitative and additive effects.

Discrete genes with sex-determining functions, however, were identified, but they are all located on autosomes. Mutations were found that transform XX animals into males or into females, or XO animals into hermaphrodites or into females. Mainly through the work of Hodgkin and collaborators, who ingeniously exploited these mutations, we have gained important insights into the genetic control of sex determination in *C. elegans* (for reviews see Hodgkin 1984, 1985; Hodgkin et al. 1985).

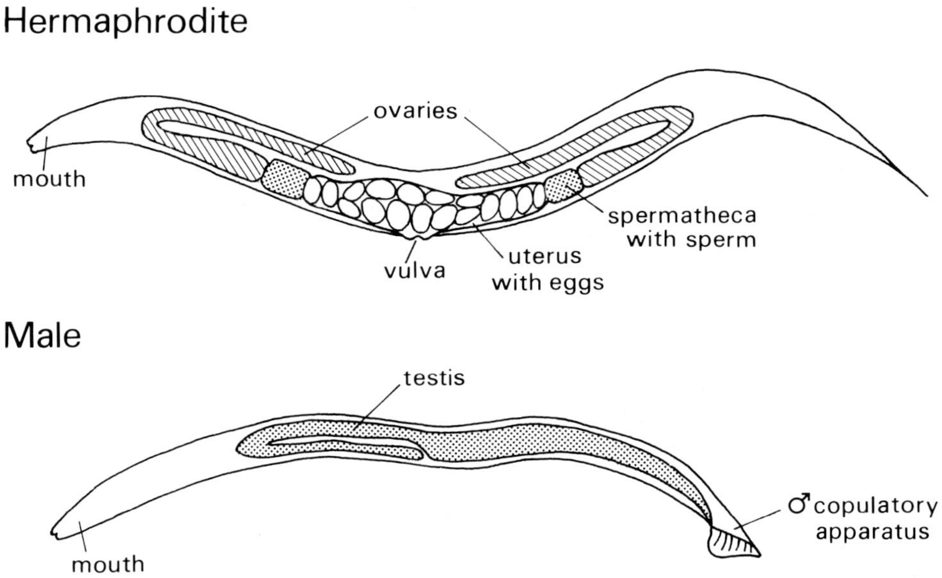

Fig. 3. Schematic drawing of a hermaphrodite and male of *Caenorhabditis elegans*. (After Hodgkin 1985)

Table 1. Sexually dimorphic features in *Caenorhabditis*

	Hermaphrodite	Male
Karyotype	2n=12, XX	2n=11, XO
Germ line	~ 300 sperm ~2000 eggs	~3000 sperm
Gonad	Two arms (143 nuclei)	One arm (55 nuclei)
Somatic nuclei (excl. gonad)	816	976
Sex-specific epidermal structures	Vulva	Copulatory apparatus with sense organs
Sex-specific muscles	Uterus, vulva (16 nuclei)	Copulatory app. (41 nuclei)
Sex-specific neurons	8	87
Intestine	Yolk polypeptides	–
Behavior	Oviposition	Copulation

According to a proposed model, the still mysterious quantitative signal of the X:A ratio is read by a key gene, *her-1,* whose state of activity, symbolized as ON or OFF in Fig. 4, controls the next two genes, *tra-2* and *tra-3*. These in turn control the three *fem* genes which then regulate *tra-1* on whose activity the sexual pathway finally depends. The hierarchy of gene control is essentially negative, forming a cascade in which each gene, when active, represses the next one. The cascade may be largely complete as it stands now since extensive mutagenesis screens, while yielding new alleles of the known loci, failed to bring forth mutations in new genes.

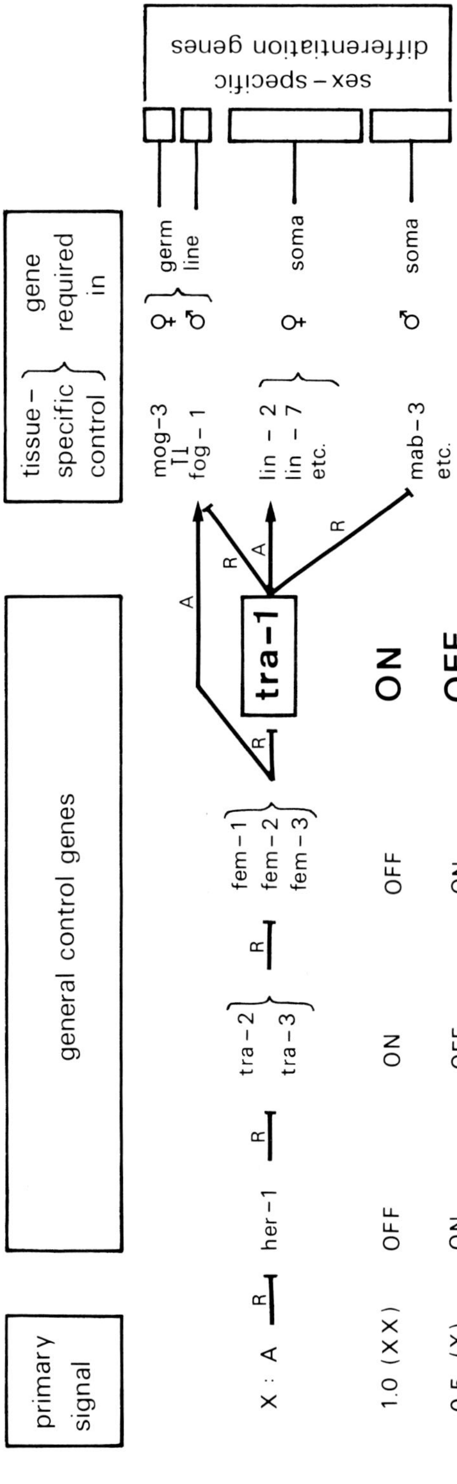

Fig. 4. Genetic control of sex determination in *Caenorhabditis elegans*. The figure shows the four levels of the genetic hierarchy in the sexual pathway. Drawn are the primary signal, the seven genes with general control functions, a few of the tissue-specific control genes, and the blocks of sex-specific differentiation genes. Gene regulation is essentially achieved by repression (*R*), rarely by activation (*A*). OFF and ON are simplifying terms to indicate that activity of the gene is required (*ON*) or must be absent (*OFF*), as deduced from the phenotype of recessive and dominant mutations. Tissue-specific control genes: *mog* and *fog* stand for *m*asculinization *o*f *g*erm line and *f*eminization *o*f *g*erm line (denoting the respective mutant phenotypes); the *lin* genes control fixed cell *lin*eages that differ in the two sexes; *mab* stands for *m*ale *ab*normal. (After Hodgkin et al. 1985)

The establishment of the hierarchical sequence became possible because recessive (lack of function) and dominant (constitutive or overproducing) alleles were generated for most of the seven genes and because the mutations could be arranged in an epistatic series. In combinations of two or more mutant genes, it was generally found that a mutation in one gene was epistatic over a mutation in another gene. In a hierarchy of genetic control, the mutation in the gene that comes last in the cascade determines the phenotype. A crucial criterion for the definition of a control gene is that recessive and dominant mutations give opposite transformations. This is in fact the case: *tra-1*$^-$ is male, and *tra-1*^{D} is female, irrespective of the X:A ratio.

Mutations in the seven genes affect all aspects of sexual development, suggesting that these genes are general regulators of sex in all cells. Thus, an *XX; tra-1*$^-$*/tra-1*$^-$ animal does not have a vulva, nor does it make eggs nor yolk polypeptides, but instead produces sperm, a male copulatory apparatus, and male muscles and neurons. Such an animal is almost indistinguishable from a normal XO male. Conversely, the genotype *XO; her-1*$^-$*/her-1*$^-$ is a functional hermaphrodite with all its characters as listed in Table 1. These examples show that the activities of these control genes regulate entire batteries of subordinate genes. The last gene in the cascade with this global property is *tra-1*. It seems that this gene does not directly reach the differentiation genes, but uses tissue-specific control genes to transmit the signal. A few of these have been identified and are shown in Fig. 4. Recessive mutations in *fog-1*, for example, prevent spermatogenesis and shift the germ cells into the oogenic pathway; males now make apparent oocytes in their testis. Similarly, mutations in *mab-3* fail to repress the yolk polypeptide genes so that males synthesize these proteins in their intestine.

We can thus distinguish four levels in the genetic hierarchy that governs sex (Fig. 4): the primary signal *X:A*, the general control genes, the tissue-specific control genes, and the differentiation genes that transform the abstract genetic signals into the concrete phenotype of the male or female sex.

The reader will have noticed that a simple ON-OFF situation at the control locus *tra-1* should produce males when OFF and females when ON. How then are hermaphrodites made? These are in essence females whose germ line produces a few sperm before it switches to oogenesis. This is clearly demonstrated by dominant mutations at *tra-1* that apparently render the gene insensitive to repression by the *fem*-genes. The constitutive *tra-1*^{D} mutations transform XX and XO into fertile females that do not produce any sperm. The phenotype suggests that the activity of *tra-1* must be modulated so that the gene is transiently shut down in hermaphrodites to allow a short phase of spermatogenesis. The scheme in Fig. 4 does not incorporate this temporal regulation in the germ line. It is achieved through *tra-2*, that is itself modulated by a yet unknown mechanism operating specifically in the germ line (Doniach 1986).

Mutations in the *fem* genes reveal another interesting facet of sex determination. All recessive mutations cause a complete transformation of XX and XO into fertile females; they are epistatic over *tra-2* and *tra-3*. Relative to *tra-1* they are epistatic in the germ line, but hypostatic in the soma, since genotype *XX;fem-1*$^-$*; tra-1*$^-$ is a male with oocytes in its testis. The function of the *fem* genes thus must be to repress *tra-1* activity and to promote spermatogenesis by activating the germ line-specific control gene *fog-1*.

A similar dual function is also postulated and indicated in Fig. 4 for *tra-1*. Its product serves to *activate* the female-specific control genes (*lin-2*, etc.) and to *repress* the male-specific control genes (*fog-1*, *mab-3*, etc.). If a single genetic unit has to exert activating and repressing functions, we should be prepared that the molecular analysis may reveal a more complex picture than the geneticist can draw.

The hermaphrodite represents a more complicated situation than the female. Most other nematodes, including close relatives such as *C. remanei*, have males and females. Hermaphroditism in *C. elegans* probably evolved from gonochorism through modulation and fine tuning in the cascade of genes controlling sex. We therefore expect to find the same genes and the same basic scheme in *C. remanei*, but without the complications of transient modulation as it is seen in the hermaphroditic *C. elegans*.

3.3 *Drosophila* and Some Other Insects

We will start this chapter by a summary of the principles as they are known for *Drosophila*. We will then apply this scheme to other species of insects about which some information on sex determination is available. We hope to demonstrate that a unifying principle emerges which uses allelic variations in three genes to create the seemingly different mechanisms observed in insects.

3.3.1 *Drosophila*

Sex determination in *Drosophila* has recently been reviewed by Baker and Belote (1983), Nöthiger and Steinmann-Zwicky (1985a), and Cline (1985). Among the various insects studied, *Drosophila melanogaster* is the species whose sex-determining mechanism is best known. Since the pioneering work of Bridges (1921, 1925), the ratio of X chromosomes to sets of autosomes (X:A) is recognized as the primary signal that governs sex determination in the soma and in the germ line and also regulates dosage compensation. Genotype X; 2A is male, 2X; 2A is female. The Y chromosome has no sex-determining function. The difference in X-chromosomal gene dosage between the sexes is compensated by a regulatory process so that the single X chromosome of the male is transcribed at twice the rate of an X of the female.

Over the years, mutations became known that change the sexual phenotype of XX or XY flies. They identify a small number of genes that are instrumental in the genetic control of sex determination. Mutations in these genes have profound effects on every aspect of somatic sex, changing morphological, physiological, and behavioral traits in all sexually dimorphic tissues. Only the germ line appears to escape the action of these mutations (Marsh and Wieschaus 1978; Schüpbach 1982). Some mutant genotypes, e.g., *tra/tra*, *tra-2/tra-2*, or $dsx^{D}/-$ lead to perfect transformation of XX zygotes into sterile males (pseudomales); others (*ix*, *dsx*) produce intersexes whose cells manifest a sexually intermediate phenotype. The global effects of the mutations indicate that these genes control entire batteries of subordinate sex-specific genes. So far, five genes have been described which, when mutant, change the sexual phenotype of the zygote. These genes, *Sxl, tra, tra-2, ix, dsx* act downstream of the X:A ratio since mutations in them are epistatic over the chromosomal primary signal. An important characteristic of the mutations is that they affect either only XX or only XY zygotes. From this and from the mutant phenotypes we can infer that two complementary patterns of gene activity must exist in the wild type, one specifying the female, the other the male pathway (Fig. 5). Combinations of the various mutations revealed epistatic relations that suggest a hierarchical arrangement of the genes, with *Sxl* at the beginning and *dsx* at the end of the cascade. The X:A ratio, together with the maternally acting gene *da,* sets the state of activity of *Sxl,* and this in turn, via *tra, tra-2,* and *ix* regulates *dsx,* the gene that finally dictates female or male sexual differentiation. A constitutive mutation at this locus, dsx^{D}, will dictate the male pathway despite two X chromosomes and despite active *Sxl, tra, tra-2* and *ix* genes. No direct influence of the X:A ratio on the sex differentiation genes is observed (Steinmann-Zwicky and Nöthiger 1985a). This present view of the genetic control of sex determination is summarized in Fig. 5. The reader will notice the surprising similarity with the scheme of *Caenorhabditis* (Fig. 4).

The gene *Sxl* occupies a special position at the top of the hierarchy. It not only controls sex determination, but also appears to be involved in dosage compensation: lack-of-function mutations (Sxl^{f}) are lethal in XX, but viable in XY animals; constitutive gain-of-function mutations (Sxl^{M}), have no effect in XX, but kill XY animals. This implies that the activity of *Sxl* is required in females to keep the rate of X-chromosomal transcription low, and must be absent in males so that the single X is transcribed at a high rate. The sex-transforming effect of the mutations is only seen in cell clones of genetic mosaics where clones of Sxl^{M}/O form female structures, whereas clones of Sxl^{f}/Sxl^{f} form male structures. The state of activity of *Sxl* is set very early, around the blastoderm stage, and irreversibly. Later, the primary signal becomes dispensable. On the other hand, an active *Sxl* gene is continuously required to maintain the cells in the female pathway. This conclusion derives from the observation that XX clones made homozygous for Sxl^{f} during larval development differentiate male structures.

The most mysterious component in the regulatory pathway is the X:A ratio. How does this signal achieve that *Sxl* is ON in females and OFF in males? A maternal factor, provided by the gene da^{+} of the mother, is a necessary prerequisite for the activation of *Sxl,* but it cannot play a discriminative role since it is present in all eggs. The discriminating factors then must be zygotic. Using defined deletions and duplications of X-chromosomal regions we could identify a small distal segment that when present in two doses functions as a major activator of *Sxl* (Steinmann-Zwicky and Nöthiger 1985b). Another element, *sis-a,* was recently localized near the middle of the X chromosome (Cline 1986). Thus, the quantitative impact of the X chromosome on sex determination appears to result from major and minor factors involved in the activation of *Sxl.* Chandra and his coworkers (Gadagkar et al. 1982; Chandra 1985) proposed a model to explain how the quantitative signal of the X:A ratio could be transformed into differential activity at the *Sxl*-locus. The model, which assumes that a molecular signal is titrated by the X chromosomes, has three components: (1) a factor *R* that is produced by the autosomes in limited, but equal amounts in males and females and

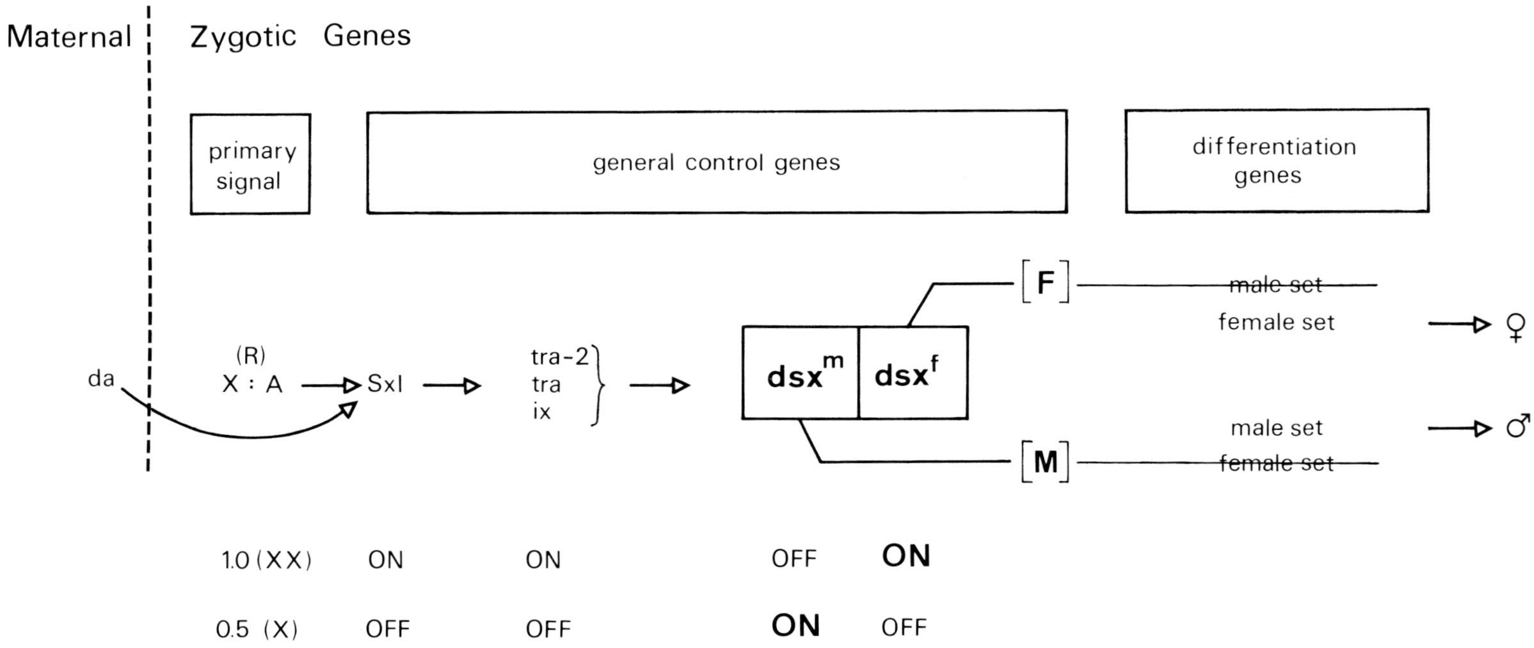

Fig. 5. Genetic control of sex determination in *Drosophila*. The X : A ratio acts as the primary signal to control the key gene *Sxl* which can only be activated when a maternal gene product (*da*) is present in the zygote. The state of *Sxl* is communicated to the double switch *dsx*^{m} *dsx*^{f} via the genes *tra-2*, *tra* and *ix* as intermediaries. The double switch produces either [F] or [M] that act to prevent either the male set or the female set of differentiation genes from being expressed. (*R*) symbolizes the postulated *R* genes whose products may be titrated by the number of X chromosomes (see text)

that can repress *Sxl;* (2) several X-chromosomal sites *S* that bind *R* with high affinity; (3) the gene *Sxl* that binds *R* with low affinity. In females with two X chromosomes, all *R* molecules are bound by the *S* sites so that *Sxl* can be transcribed which dictates female development. In males with only half the number of *S* sites, some *R* molecules remain free to bind to *Sxl* and thus to prevent its activity. We will return to this model when we discuss other insects.

Lack of *dsx* function results in an intersexual phenotype suggestive of both sets of sex differentiation genes being expressed in the same cell. This implies that the differentiation genes are basically constitutive and have to be differentially repressed by *dsx*. Genetic data reveal that *dsx* harbors two functions, dsx^{m} and dsx^{f}, that can independently mutate: recessive lack-of-function mutations in dsx^{m} transform XY, those in dsx^{f} transform XX animals into intersexes. In the wild type, the two functions dsx^{m} and dsx^{f} mutually exclude each other, so that either dsx^{m} is ON and dsx^{f} is OFF, or vice versa. The locus thus operates as a double-switch specifying either of two products, [M] or [F], that act to prevent either the male set or the female set of sex differentiation genes from being expressed (Fig. 5). As depicted in this figure, *dsx* is under the control of *Sxl, tra, tra-2,* and *ix*. It is through *dsx* that the simple binary code of these genes achieves a choice between two mutually exclusive developmental pathways.

Determination and Reversibility – a Paradox? A powerful tool to study the roles of genes in *Drosophila* is the production of genetic mosaics, animals that are composed of genetically different cells. The first mosaics described were *gynandromorphs*. These flies consist of XX and XO cells, resulting from loss of an X chromosome in a nucleus during the first zygotic divisions. Such mosaics differentiate female and male traits side by side, thereby showing that each cell assesses the X:A ratio in an autonomous manner and follows its sexual pathway as dictated by its own genotype and independently of neighboring cells of the other sex. Sexual development in *Drosophila* is a cellular process with no long-range effects, such as hormones, being involved.

Another type of mosaic can be generated by *mitotic recombination*. This process, that can be experimentally induced any time during proliferative development, in essence results in the elimination of the wild-type allele from a heterozygous cell which will then give rise to a clone of homozygous mutant cells. Animals of the genotype *XX;* tra^{-}/tra^{+} or *XX;* $tra\text{-}2^{-}/tra\text{-}2^{+}$ develop as females. When the wild-type allele, tra^{+} or $tra\text{-}2^{+}$, is removed from a cell in an embryo or larva, the resulting homozygous tra^{-} or $tra\text{-}2^{-}$ clone differentiates male structures, again showing cell-autonomous sexual differentiation.

The experiment just described reveals a remarkable feature of the genetic system controlling sex determination in *Drosophila*. Although *Sxl* becomes irreversibly set at the blastoderm stage, and thus sex is “determined” in terms of gene activities, the pathway remains flexible and reprogrammable throughout development. A change from a female (e.g., tra^{+}/tra^{-}) to a male genotype (tra^{-}/tra^{-}) in a clone results in a corresponding change in sexual differentiation even late in larval development when the gonadal and genital primordia are already typically female. The previous female history is forgotten after a few cell divisions and the members of the clone follow their new genetic instructions (Epper and Nöthiger

1982). Even more impressive are results obtained with a temperature-sensitive allele of *tra-2,* called *tra-2*^{ts} (Belote et al. 1985a). At the permissive temperature (16 °C), *XX; tra-2*^{ts}/*tra-2*^{ts} develop into females that start to synthesize yolk polypeptides (YP) some 2 days after the flies had emerged. At 29 °C, the same genotype produces a male that does not synthesize YP. When this pseudomale, however, is shifted to 16 °C, its genes coding for YP become active and the transcription is followed by translation of the mRNA into YP. Conversely, when the females are brought to the restrictive temperature, their YP genes are gradually shut down. Since *tra* and *tra-2* control sexual differentiation via *dsx,* we conclude that this locus can change from expressing the female function to expressing the male function, and vice versa, whenever the signal of *tra* or *tra-2* is changed, even in adult flies. Imagine the psychological situation of an animal that has lived as a male at 29 °C and now, at 16 °C, gradually starts to feel like a female!

In summary, the experiments have shown that sex becomes determined at a very early developmental stage when the X:A ratio and *da* irreversibly fix the state of activity of *Sxl* which then becomes independent of the X:A ratio and is faithfully propagated during all cell divisions. The other elements of the pathway, however, down to the differentiation genes, continuously depend on *Sxl* and can switch from ON to OFF and from OFF to ON whenever the experimenter changes a signal higher up in the hierarchy.

Molecular Analysis. The isolation and characterization of the sex-determining genes is presently under way in several laboratories. Until now, the successful cloning of *Sxl* (Maine et al. 1985), of *dsx* (Belote et al. 1985b) and of *tra* (Belote et al. 1985b; Butler et al. 1986) has been reported. The size of the genes varies from 2 kb for *tra* to some 30 kb for *dsx*. Using a transposable vector, we were able to reintroduce the *tra*$^{+}$ gene into the germ line and to show that 2 kb can restore femaleness to *XX; tra*/*tra* zygotes (Butler et al. 1986). Such experiments now make it possible to study the anatomy of a control gene and to identify functional domains. Northern analyses have revealed that the transcriptional pattern is more complex than the genetic data had led to anticipate. The molecular probes will be very important tools to unravel the actual threads in the genetic network and to determine which gene is regulating and which is regulated.

Analyses of products of sex-specific differentiation genes have shown that their presence or absence depends on the state of activity at the sex-determining genes rather than on the primary signal of the X:A ratio (Postlethwait et al. 1980; Bownes and Nöthiger 1981; Belote et al. 1985a; Schäfer 1986a, 1986b; DiBenedetto et al. 1987). Determining the cis-acting sequences of sex-specifically transcribed genes, combined with molecular analysis of the control genes, will hopefully reveal how downstream genes are regulated (Garabedian et al. 1986).

3.3.2 Other Insects: Variety and a Unifying Concept

A survey of sex determination in insects reveals a perplexing multitude of phenomena listed in Table 2. Even among the order Diptera the variety is confusing: to determine sex, several species use an allelic difference at a single locus, *Drosophila* and *Sciara* the enigmatic ratio of X chromosomes to sets of autosomes (X:A); in *Anopholes, Calliphora* and *Musca,* the Y chromosome determines maleness; but some strains of *Musca* and *Chironomus* have no heteromorphic sex chromosomes and instead carry dominant male determiners (*M*) anywhere in their genome, sometimes in multiple copies, other strains dominant female determiners (*F*), whereby *M* may be epistatic over *F*, or vice versa; in *Chrysomyia* a maternal gene and in *Heteropeza* a nutritional factor in the hemolymph of the mother determine the sex of the offspring. Among the order *Hymenoptera*, the haplo-diplo mechanism is prevalent.

We want to propose that these seemingly different mechanisms represent simple variations (mutations) in only three genetic elements that form the top of a hierarchical control system: a primary signal (*R*), present or absent, and a maternal gene (*da*) are used to regulate a key gene (*Sxl*) (Fig. 6). These three genetic elements form the variables with which evolution played to create variations in the sex-determining system by introducing either an allelic difference or a conditional mutation at any of these genes. The state of activity of the key gene *Sxl* then controls the sex differentiation genes through a genetic double switch (*dsx*). The double switch can operate in two mutually exclusive modes so that either only the male set or only the female set of sex differentiation genes is expressed. This simple system forms the basis of sex determination in insects.

The primitive state is probably represented by those species in which the males are heterozygous at a single sex-determining locus, e.g., *Culex* where *M*/*m* is a male, and *m*/*m* is a female (Lucchesi 1978). We now assume that *M* corresponds to the primary signal *R* whose product acts to repress the key gene *Sxl*, thus implementing the male pathway. In genotype *m*/*m*, no product R is formed and *Sxl* can become active, thus implementing the female pathway. In our terminology, the genotypic formula for *Culex* would then be $R/r;\ Sxl^+/Sxl^+$ for males, and $r/r;\ Sxl^+/Sxl^+$ for females. As shown for *Drosophila*, a prerequisite for the activation of *Sxl* may be the presence of a maternal component specified by a gene *da*. The Calliphorid fly *Chrysomyia rufifacies* in fact uses an allelic difference at a maternally acting locus to determine the sex of the offspring (see Table 2).

In Table 2, we demonstrate how simple mutations in the three elements *R*, *Sxl*, *da* can generate mechanisms of sex determination that phenomenologically look very different from one another. All of them, however, achieve the same result, namely that an active product of *Sxl* is formed in what is to become a female, and that an inactive or no product is formed in what is to become a male. We have listed concrete cases together with our interpretation of how a particular mechanism can be represented as a simple variation of the primitive type.

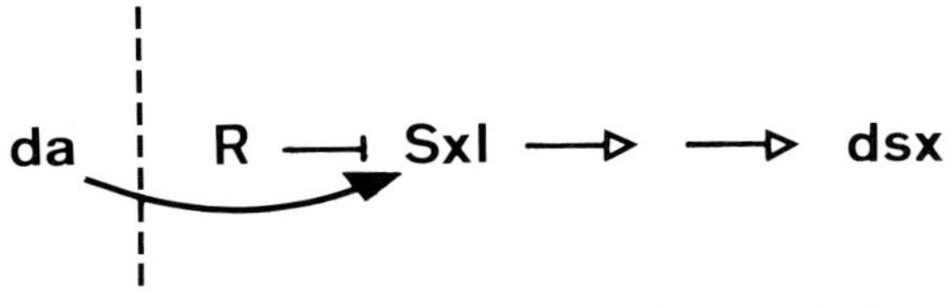

Fig. 6. Basic scheme for sex determination in insects. The three genetic elements *R*, *da*, and *Sxl* are used to regulate the double switch *dsx* (see Fig. 5 and text). *R*, *da*, and *Sxl* form the variables for evolution (see Table 2)

Table 2. Sex-determining mechanisms in insects: examples and interpretations.
The table lists concrete cases taken mainly from among the Diptera. It describes the phenomena and gives our interpretation in terms of the signal gene(s) *R*, the key gene *Sxl*, and the maternally acting gene *da*. *R* produces a dominant male-determining signal, *r* means absence of this signal. *Mutations:* Sxl^a, Sxl^b, alleles that yield nonfunctional, but complementing products; Sxl^c, dominant (constitutive) allele that is not suppressible by *R*; Sxl^-, recessive loss-of-function allele; Sxl^s, R^s, alleles whose products are sensitive to environmental factors (e.g., temperature, nutrition)

Description		Interpretation			Genus	Reference
Genotype, sex	Phenomenon		Genotype	Product of Sxl		
M/m = ♂	M, dominant male determiner	M = R	R/r Sxl^+/Sxl^+	Absent	*Aëdes*	McClelland (1962)
m/m = ♀			r/r Sxl^+/Sxl^+	Active	*Calliphora*	Ribbert (1967)
	Locus of M varies from strain to strain	R on a transposable element?			*Culex*	Gilchrist and Haldane (1947)
					Chironomus	Martin and Lee (1984); Hägele (1985)
					Megaselia	Mainx (1964)
					Musca	Franco et al. (1982)
Y/X = ♂	Y, dominant male determiner	Y carrying R, heteromorphic	R/r Sxl^+/Sxl^+	Absent	*Anopheles*	Baker and Sakai (1979)
					Calliphora	Ullerich (1983)
X/X = ♀			r/r Sxl^+/Sxl^+	Active	*Musca*	Franco et al. (1982)
X;AA = ♂	X:A ratio	R is produced by autosomes, absorbed by X	R/R Sxl^+/Y	Absent	*Drosophila*	Baker and Belote (1983) Nöthiger and Steinmann-Zwicky (1985a)
XX;AA = ♀			R/R Sxl^+/Sxl^+	Active	*Sciara*	Metz (1938); Crouse (1960)
haploid = ♂	Multiple complementing alleles	Mutations at Sxl lead to inactive product	r Sxl^a or Sxl^b	Inactive	*Apis*	Rothenbuhler (1957)
					Habrobracon	Whiting P.W. (1943)
diploid = ♀			r/r Sxl^a/Sxl^b	Active (complementation)	*Mormoniella*	Whiting A.R. (1967)

MMM f/f=♂ MMM F/f=♀	Several dominant male determiners M, one dominant female determiner F, F epistatic over M	M=R on several chromosomes; F=Sxl^c, not repressible (constitutive)	R/R Sxl^+/Sxl^+ R/R Sxl^c/Sxl^+	Absent Active	*Musca*	Franco et al. (1982)
f/f=♂ F/f=♀	F, dominant female determiner, M of other strains epistatic over F	f=Sxl^- F=Sxl^+	r/r Sxl^-/Sxl^- r/r Sxl^+/Sxl^-	Absent Active	*Chironomus*	Thompson and Bowen (1972)
			Maternal genotype			
f/f=♀ arrhenogenic	F′, maternal factor, dominant female	f=da^-	da^-/da^- r/r Sxl^+/Sxl^+	Absent	*Chrysomyia*	Ullerich (1973, 1984)
F′/f=♀ thelygenic	determiner	F′=da^+	da^+/da^- r/r Sxl^+/Sxl^+	Active		
♂ and ♀ have same genotype	Environment determines sex	R or Sxl conditional	R^s/R^s Sxl^+/Sxl^+ r/r Sxl^s/Sxl^s	Depends on conditions		
	– temperature				*Aëdes*	Horsfall and Anderson (1963)
	– nutrition				*Heteropeza*	Went and Camenzind (1980)

The male determiner *R* has characteristics of a mobile element. In *Megaselia, Chironomus,* and most spectacularly in *Musca, R* can change its chromosomal location. It can also increase in number as shown by certain strains of *Musca domestica* with multiple male determiners. The mobility of *R* may also be the basis for the evolution of the balance type of *Drosophila* and *Sciara*. In these species, multiple *R* factors may have spread on the autosomes, while sites on the X chromosomes simultaneously acquired the capacity to bind and neutralize the products of *R* (see model by Chandra and coworkers cited earlier). More detailled arguments for a single principle of sex determination in insects can be found in Nöthiger and Steinmann-Zwicky (1985b).

3.4 Mammals

3.4.1 Evidence for a Hierarchical Control System

In mammals the Y chromosome determines the male sex. Genotypes lacking a Y (XX or XO) develop as females. This developmental program is considered to represent the ground state. The Y chromosome acts as a genetic signal that leads to differentiation of testes and secondary male characteristics, irrespective of the number of X chromosomes (XY, XXY, XXXY, XXXXY). Therefore, the Y carries a gene or genes directing the formation of testes.

Analyses of human males and females with abnormal karyotypes allowed a cytogenetic localization of this postulated gene, called *TDF*, for "testis-determining factor." Genotypes carrying only Yq, the long arm of the Y chromosome (XYq, XXYq) develop as females; presence of the short arm (Yp), on the other hand, leads to male development (see Gordon and Ruddle 1981). This points to Yp as the carrier of *TDF*.

The relatively frequent occurrence of XX human males (one in 20000 to 30000 newborn boys, see Guellaen et al. 1984) provided further information. One such male failed to express the paternal allele for *Xg*, an X-linked gene, but did express his father's allele for 12E7, a Y-linked marker (de la Chapelle et al. 1984). Another study showed that among 12 analyzed XX males, one was lacking one dose of the X-linked gene STS which had been present on the father's X chromosome (Wieacker et al. 1983). These findings and others suggest that crossing-over between the Y and the X took place and that portions of varying size were exchanged (Ferguson-Smith 1966; Burgoyne 1982). In fact, during meiosis, large parts of Yp and Xp pair to form a synaptonemal complex (Solari 1980). The pairing segments are at least in part homologous at the DNA level. Several DNA probes were shown to hybridize both to human X and Y chromosomes, but not to autosomes (Buckle et al. 1985; Cooke et al. 1985; Simmler et al. 1985; Rouyer et al. 1986). Closer analysis localized some of these sequences near the tip of Yp and Xp. During male meiosis, crossing-over can occur between the X and the Y within the homologous pairing segment. This recombination event leads to exchange of DNA sequences from the tip of X and Y chromosomes, but as a rule the exchange does not include the postulated gene *TDF*. If, however, the crossing-over occurs at a site proximal to *TDF*, the testis-determining gene will be transferred to the X chromosome, and XX males as well as XY females will result. Molecular analysis provided evidence for such a hypothesis. Several XX males were shown to carry DNA segments specific to the Y chromosome (Page et al. 1985), and more precisely to the short arm of the Y chromosome (Vergnaud et al. 1986). In three cases, these sequences were mapped by in situ hybridization to the short arm of one of the two X chromosomes (Andersson et al. 1986). Conversely, XY women were found to lack corresponding Y-specific sequences (Disteche et al. 1986a). Thus, XX males and XY women can be explained as complementary products of a crossing-over having occurred proximal to *TDF*. This, however, does not exclude that some cases of sex reversal could be due to some other mutational event. A Y-autosome translocation was recently discovered in a 45, X male (Disteche et al. 1986b). Using finer probes and analyzing more of the aberrant genotypes should eventually identify the Y-chromosomal region that harbors the primary male-determining genetic signal (for review see Page 1986).

Evidence for a testis-determining gene on the Y chromosome also exists for mice. A dominant mutation was identified that causes sex reversal of XX animals. The progeny of XY males carrying the mutation Sex reversal (*Sxr*) fall into four about equally frequent classes: normal XY males, carrier XY males, normal XX females, and XX males. Carrier males were shown to have an aberrant Y chromosome. DNA sequences normally located near the centromere are present as a duplicate at the distal tip. These sequences are also found on one X of the transformed XX males (Singh and Jones 1982). In carrier males, an obligatory crossing-over event between the aberrant Y and a normal X leads to transfer of the extra sequences onto the X in 50% of the gametes (Fig. 7) (Evans et al. 1982). Since the duplicated sequences are masculinizing, they must include *TDF*, the Y-chromosomal gene that induces testis formation. In mice, this gene is called *Tdy*, for "testis-determining gene on the Y chromosome."

Although the Y chromosome gives the primary signal for sex determination, other genes are needed to respond. Evidence for such genes involved in testis differentiation comes from studies with mice and wood lemmings. Males from various wild mouse strains were crossed to females from a laboratory strain (C57BL/

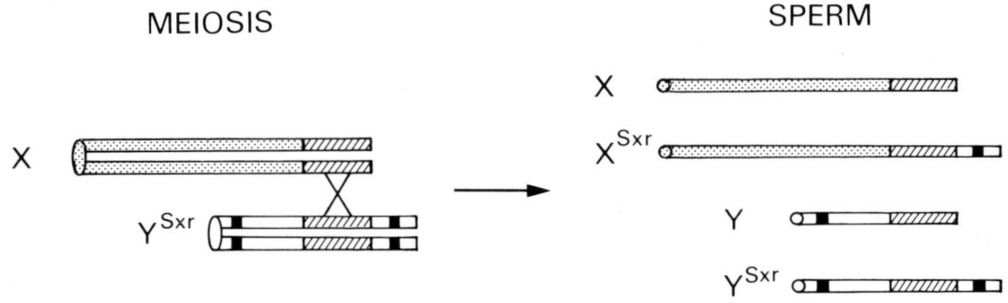

Fig. 7. During meiosis of male mice, an obligatory crossing-over takes place between X and Y, within a segment homologous to both chromosomes (*hatched*). Nonhomologous sequences are dotted for the X and white for the Y that also carries the postulated male determining gene *Tdy* (*black square*). The abnormal Y^{Sxr} chromosome carries distally a duplication of Y sequences including *Tdy*. These are transferred to the X chromosome in 50% of the gametes as a result of crossing-over. (In the mouse, X and Y chromosomes are acrocentric)

6J). Resulting male progeny were again crossed to C57BL/6J females. Most XY progeny from this and from following back-crosses were either female or hermaphrodite, some of them fertile (Eicher et al. 1982). These animals had their Y chromosome from one population (e.g., Y^{POS} from *Mus poschiavinus* or Y^{ORB} from *Mus orobis*) and most of their other chromosomes from the laboratory strain C57BL/6J. Thus, although the primary signal in the form of a Y chromosome is present, the interaction with autosomal or X-linked testis-determining genes is disturbed. We conclude that the Y chromosome is not sufficient to determine the male sex, because the primary signal has to be transmitted through other genes.

In the wood lemming, *Myopus schisticolor,* an altered X chromosome was identified that overrides the Y chromosomal signal. In a population, four different genotypes can be found; XX, X*X and X*Y females, and XY males (Herbst et al. 1978). X* can be cytologically distinguished from a normal X, since the length and the G-banding pattern of the short arms differ. Since the Y chromosome of the X*Y females derives from a normal father, the current hypothesis is that an X chromosomal gene is needed to read and transmit the Y chromosomal signal. This gene, which we may call *Tdx* (for testis-determining gene on the X chromosome), is mutated in X*. As expected, X*O and X*YY animals are females, whereas X*XY can be males, females, or hermaphrodites depending on which of the two X chromosomes is inactivated and which is expressed (Fredga 1983; Schempp et al. 1985). In view of the high degree of homology between the X chromosomes of different mammalian species (Sawyer and Hozier 1986), we may extrapolate that a gene homologous to *Tdx* exists in other mammals as well.

Autosomal mutations interfering with normal testis differentiation have also been described, but the roles of the genes they identify are unclear. A dominant mutation *Tas* on chromosome 17 of the mouse produces XY hermaphrodites with ovaries or ovotestes (Washburn and Eicher 1983). *Tas* might be a loss-of-function mutation of a gene required for normal testicular differentiation; alternatively, *Tas* could be a constitutive mutation, which would mean that the wild-type product is needed in females for ovarian differentiation. In goats, homozygosity for an autosomal mutation, Polled, causes development of testicular tissue in XX animals (Hamerton et al. 1969). As with *Tas,* it is unclear whether this points to a gene needed for ovary formation or testis formation.

The primary testis-determining signal dictates the male pathway to the indifferent gonads, so that these develop as testes that differentiate two specific somatic cell types, first the Sertoli cells and then the Leydig cells. The Sertoli cells produce the anti-Müllerian factor to prevent the differentiation of the female gonoducts. The Leydig cells synthesize the hormone testosterone whose release leads to differentiation of the male secondary sex characteristics. Human XY patients have been described who have testes without Leydig cells; they develop female external genitalia, but since Sertoli cells are present and functional, no female gonoducts are formed (see Simpson 1982).

A mutation is known that affects a hormonal step of sex determination. X-linked *Tfm* mutations (Testicular feminization) were described for mice, rats, and humans. Chromosomal males carrying the mutation (*Tfm*/*Y*) develop as females

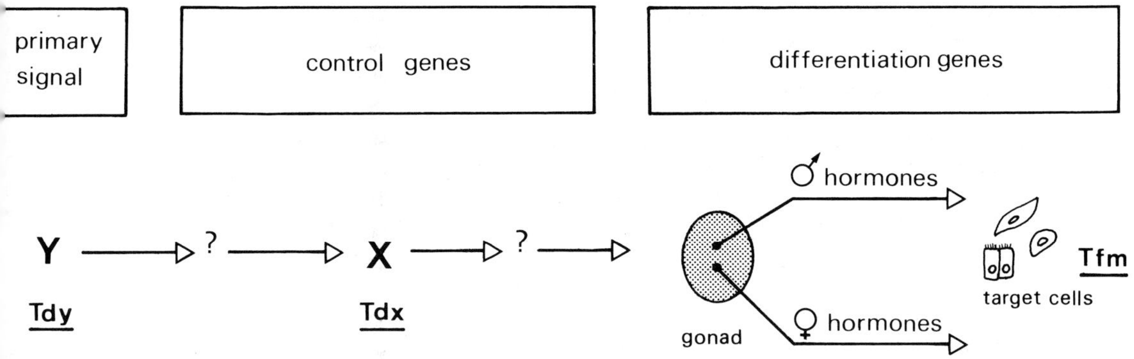

Fig. 8. A simplified model of sex determination in mammals. The Y chromosome carries DNA (*Tdy*$^+$) that acts as a genetic signal by which *Tdx*$^+$ on the X chromosome is controlled. *Tdx* directs the differentiation of the gonad into a testis. In the absence of this genetic signal, the indifferent gonad develops into an ovary. Male sex hormones (androgens), produced by the testis, achieve the expression of male-specific differentiation genes in the target cells which leads to male development. *Tfm*$^+$ codes for an androgen receptor, the lack of which results in female development. *Question marks* indicate that yet unknown genes may be involved in the transmission of the signal emanating from *Tdy*

with testes producing normal amounts of testosterone. The female phenotype results from the lack of the specific hormone receptor protein that normally is present on all male and female cells. Thus, *Tfm* affects the ability of all tissues to respond to testosterone (Meyer et al. 1975; Ohno 1979).

In conclusion, this chapter shows that a hierarchy of genetic and hormonal signals controls the sexual pathway of mammals (Fig. 8). The deduction of this cascade is mainly based on four different kinds of females with XY karyotypes: *XYq* are females because they lack the primary signal *Tdy* located on *Yp;* the X*Y females of the wood lemming are defective in *Tdx* that reads and transmits the primary signal; at a third level, a genetic defect can eliminate the Leydig cells, thus removing the source of testosterone; and finally, *Tfm* mutations abolish the ability of the cells to respond to testosterone.

3.4.2 The Role of the Y Chromosome: Source or Sink?

There is no doubt that the Y chromosome of mammals provides the primary signal that initiates the male pathway; but although we can localize the gene or genes (*Tdy* or *TDF*) responsible for this signal to a specific region on the Y chromosome, its nature is entirely unknown. In principle, we can imagine two alternative modes of action that are very different in molecular terms and that we may paraphrase as the "source" or the "sink" model: (i) the Y chromosome could *produce* a positive signal in the form of a transcript or protein that *activates Tdx,* the testis-determining gene on the X chromosome; or (ii) Y-chromosomal sequences could bind or inactivate molecules that are produced (by autosomal genes) in limited amounts and whose function is to *repress Tdx;* in XX zygotes, these molecules remain free to bind to *Tdx,* thus repressing this gene. We will now present published reports, some of which favor the role of the Y chromosome as a source, others as a sink. As we will see, however, none of the evidence convincingly supports either hypothesis.

i. *The Y as a Source: H-Y Antigen?* The H-Y antigen was first defined as a male-specific histocompatibility antigen discovered in skin grafting experiments that involved males and females of a highly inbred strain of mice. Later, it was found that the antigen was extremely conserved among all vertebrates tested, and that it was always associated with the heterogametic sex. For many years, it was postulated to be encoded by genes on the Y chromosome and to be the key substance that induces the gonadal primordium to differentiate as a testis in mammals, or as an ovary in birds (see Wachtel et al. 1975; Ohno 1979). The evidence was weak and controversial, and the issue was further confounded because most studies monitored a serologically detectable male antigen, now called SDM (Silvers et al. 1982), instead of the H-Y transplantation antigen.

Recent evidence now indicates that the H-Y antigen is neither sufficient nor necessary for male development. The hybrid XY^{POS} or XY^{ORB} mice mentioned earlier and T(X; 16)/X^{Sxr} mice in which the X^{Sxr} is inactivated are H-Y positive, but nevertheless female (see Simpson 1986). Conversely, McLaren and coworkers (1984) described $XX^{Sxr'}$ mice that are males, but negative for H-Y antigen. These mice carry an X chromosome with a modified *Sxr,* termed *Sxr',* that apparently still carries *Tdy,* but has lost the information for H-Y antigen (see also McLaren 1985).

ii. *The Y as a Sink: Bkm Sequences? Bkm* sequences were isolated as a sex-specific minor satellite of the Indian banded krait, a snake with heteromorphic sex chromosomes and female heterogamety (ZW). A high accumulation of these sequences was found on the W chromosome of snakes as well as on the Y chromosome and on the *Sxr* segment of mice which prompted speculations about a possible role of *Bkm* sequences in sex determination (Singh et al. 1981; Jones 1983; Epplen et al. 1983a). No accumulation of *Bkm* was found on Y chromosomes of other mammals. Nevertheless, Chandra (1985) suggested that the *Bkm* sequences could function as binding sites for repressor molecules so that *Tdx* could be transcribed, thus promoting testicular differentiation.

The *Bkm* sequences are very conserved and found among many vertebrates and even flies and slime molds. Molecular data show that cross-hybridization is due to two tandem repeats (GATA)n and (GACA)n (Epplen et al. 1983b; Singh et al. 1984). Sequences flanking these tandem repeats, however, are not related among different species. Therefore, the widespread presence of *Bkm* sequences may be fortuitous (Levinson et al. 1985). Tandem reiterations of short sequences are frequent in eukaryotes, and their accumulation on the W or Y chromosomes may be a phenomenon accompanying, or resulting from, heterochromatinization, i.e., a consequence of heterogametic sex determination rather than its cause (Lucchesi 1978). The *Bkm* sequences are apparently transcribed (Singh et al. 1984), but their function is presently unknown.

Although we do not know the actual molecules involved in mammalian sex determination, the logic of the genetic system, namely a hierarchical flow of information, remains unaffected by the controversy. There is good hope that the molecular analyses will soon identify the Y-chromosomal DNA sequences that are responsible for the primary male-determining signal, and from this information, we may eventually proceed to an understanding of the function.

4 Concluding Remarks

4.1 A General Principle?

The sexual phenotype with all the differences between males and females is the product of the action of differentiation genes. Molecular biology has shown that there are male-specific and female-specific gene products, which indicates the existence of two sets of genes that are expressed in a complementary pattern. To achieve the mutually exclusive expression of the differentiation genes, a minimum of two properly coordinated genetic functions is required, one to control the male set and one to control the female set. A clear case for this type of control is the bifunctional *dsx*-locus of *Drosophila,* where the two functions are wired in such a way that they exclude each other: when dsx^{m} is ON, dsx^{f} is OFF; and when dsx^{m} is OFF, dsx^{f} is ON. Two functions, one activating and one repressing a set of genes, are also exerted by *MATα* of *Saccharomyces* (Fig. 1) and *tra-1* of *Caenorhabditis* (Fig. 4). Whether the products act by repressing a basically constitutive set of genes, as seems to be the case for [M] and [F] of *Drosophila* (Fig. 5), or by activating a silent set of genes, as does α1 of yeast, is irrelevant for the logic of the system.

The synoptic presentation of the cases discussed in this article (Fig. 9) reveals a difference between unicellular and multicellular organisms. The latter seem to require additional general regulatory genes that mediate between the primary sig-

	Saccharomyces	Caenorhabditis	Drosophila	Mammals
Signal ↓	HO transcription	X : A	X : A	Tdy (Y)
Key gene ↓	MAT (α / a)	her (ON / OFF)	Sxl (ON / OFF)	Tdx (ON / OFF)
Control genes ↓	—	tra-2 tra-3 fem tra-1	ix, tra-2, tra dsx	? (hormones)
Differentiation genes (♂, ♀)	ON / OFF	ON / OFF	ON / OFF	ON / OFF

Fig. 9. Synopsis of the presented cases. The analogies are formal, pointing out the parallels in the logic of the systems; they do not imply homologies at the DNA level

nal and the differentiation genes (control genes in Fig. 9). Since mutations in these control genes affect every aspect of sex, they cannot function as tissue-specific regulators. The existence of these genes, therefore, is not easily understood, at least not with our current knowledge. They may allow for more evolutionary flexibility, but of course they also present more opportunities for things to go wrong. For *Caenorhabditis elegans,* he have seen that the more complex task of building a hermaphrodite requires temporal and spatial modulation of the terminal control gene *tra-1,* and that *tra-2* may be used to achieve this. In *Drosophila, tra-2, tra,* and *ix* mediate between *Sxl* and *dsx*. Their products together achieve the female-specific expression of *dsx* by transcriptional or posttranscriptional regulation.

Descriptions of the phenomena of sex determination emphasize the variability and the differences that exist sometimes even between closely related species. What superficially looks very different, however, may be the result of only minor changes in an invariant set of a few control elements. Consider, for example, *Caenorhabditis elegans,* whose sex is normally determined by a chromosomal signal, the X : A ratio (Fig. 4). Introducing a constitutive (*tra-1*^{D}) and a null (*tra-1*$^{-}$) mutation at the *tra-1* locus, Hodgkin (1983) constructed a fertile strain where the females are *X/X; tra-1*^{D}*/tra-1*$^{-}$ and the males are *X/X; tra-1*$^{-}$*/tra-1*$^{-}$. This simple "manipulation" transformed a system with male heterogamety and chromosomal sex determination into one with female heterogamety and homomorphic "sex" chromosomes where now an allelic difference at a single locus determines the sex.

In the above example, the mechanism is still genetic. How a genetic system can be converted into one with environmental sex determination is demonstrated by an arctic mosquito, *Aëdes stimulans* (Horsfall and Anderson 1963). In this species, males are genotypically *M/m,* and females are *m/m.* The genetic system works at low temperature, but at high temperature all animals develop into females. A mutation may have rendered the product of *M* thermosensitive, thus transforming the genetic system into one where an environmental factor can determine the sex, and thus assumes the role of the discriminating signal. In *Drosophila,* the temperature-sensitive mutation *tra-2*^{ts} in principle achieves the same result: *X/X; tra-2*^{ts}*/tra-2*^{ts} become females at 16 °C and males at 29 °C.

Genetic and environmental sex determination occur naturally among the reptiles (Bull 1980). The more highly evolved snakes have heteromorphic sex chromosomes and genetic sex determination. In many turtles and some alligators and lizards, however, ambient temperature is used as the discriminator to determine the sex of the developing embryos. For the turtle *Emys orbicularis,* all animals are males when raised at 28 °C and females when raised at 30 °C. The case can be explained by assuming that this species is homozygous for a male-determining gene whose product is thermosensitive with a very narrow zone of transition of only 2 °C.

We wrote this review with the intention of revealing a common principle hidden under the multitude of phenomena. Figure 10 summarizes the results of our efforts. When we generously neglect the details, we can recognize a strategy that nature uses to make males and females. A primary signal that can be genetic or environmental acts as a discriminator to assure that a key gene is either differ-

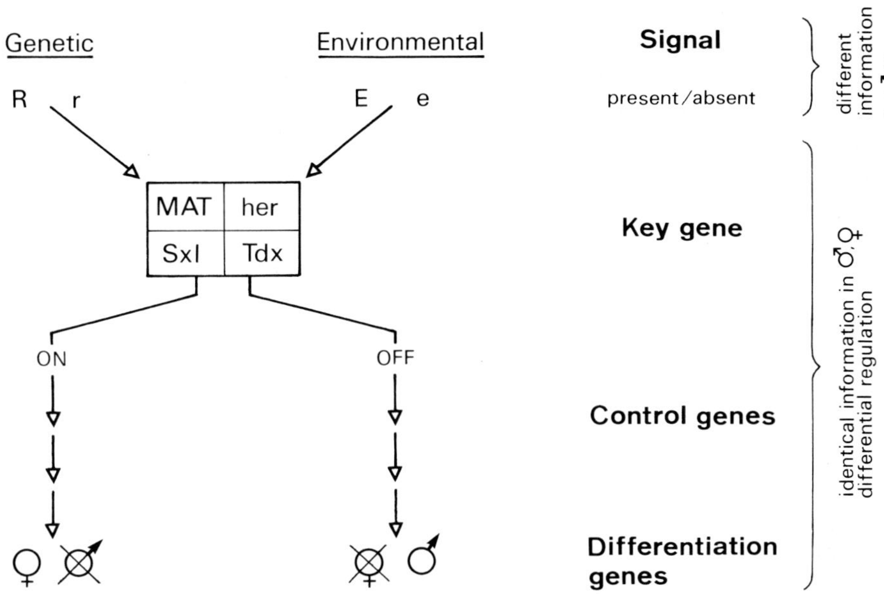

Fig. 10. Generalized scheme showing the levels of control in the sexual pathway. Differences in genetic (R vs. r) or environmental (E vs. e) information are used to achieve expression of either the male set or the female set of differentiation genes. *Symbols* in the box (*MAT, her, Sxl, Tdx*) correspond to the key genes of *Saccharomyces, Caenorhabditis, Drosophila,* and *Mammals*

entially expressed, or its products are differentially or conditionally functional. The differential information of the key gene reaches a double control which secures that only the male or only the female set of differentiation genes can be expressed. By mutational changes, the discriminating event may be delegated from the primary signal down to a control gene, as exemplified by the experimentally created situation just described for the *tra-1* locus of *C. elegans*. In this case, only one gene is left to determine sex. All other genes acting upstream in the hierarchy are dispensable and their existence, even if they mutated, would go unnoticed. Almost certainly, evolution has used such mutational changes to create the great diversity of mechanisms that are observed today.

4.2 What Does "Sex" Mean for a Cell?

The products of *dsx* in *Drosophila* or of *tra-1* in *Caenorhabditis* are abstract molecular signals for "femaleness" or "maleness," and they are probably synthesized and present in every cell of the animal. The same result is achieved in mammals, although in a different way. Here, the molecular information for sex, the sex hormones, is produced by a group of specialized cells from where it is sent through the blood stream to reach every cell of the organism.

The molecular signal, although the same for all cells, elicits different responses that depend on the developmental history of the cells. For cells of the fat body of *Drosophila,* for example, or of the liver of mammals, femaleness means synthesis of yolk polypeptides; for the cells of the brain it means establishing female behavior, but the yolk polypeptide genes remain silent in the nervous system of females. In multicellular organisms, the task is not just to select the female set or the male set of differentiation genes. An additional regulatory system must choose a subset that is to be, or can be, expressed in a particular tissue. The developmental fate of a cell thus defines the concrete sexual phenotype to be differentiated. In other words, the molecular information "maleness" or "femaleness" is interpreted by the cell within the context of its ontogenetic history. In *Drosophila,* the fate of a cell is encoded in a unique combination of active and inactive "selector" genes (Garcia-Bellido 1975; Struhl 1982). The sexual phenotype then may be visualized as the result of two regulatory systems – one defining the developmental fate, the other defining the sex – superimposed on each other and interacting with each other.

Acknowledgments. We want to thank Drs. K. Nasmyth, J. Hodgkin, I. Greenwald, A. McLaren and P. Goodfellow for critically reading different sections of this article. We also thank Dr. D. Page for sending preprints, A. Kohl for drawing the figures, and S. Schlegel for typing the manuscript. This work was supported by the Swiss National Science Foundation.

References

Andersson M, Page DC, de la Chapelle A (1986) Chromosome Y-specific DNA is transferred to the short arm of X chromosome in human XX males. Science 33:786–788

Baker BS, Belote JM (1983) Sex determination and dosage compensation in *Drosophila melanogaster*. Annu Rev Genet 17:345–393

Baker RH, Sakai RK (1979) Triploids and male determination in the mosquito, *Anopheles culicifacies*. J Hered 70:345–346

Beermann W (1955) Geschlechtsbestimmung und Evolution der genetischen Y-Chromosomen bei Chironomus. Biol Zentralbl 74:525–544

Belote JM, Handler AM, Wolfner MF, Livak KJ, Baker BS (1985a) Sex-specific regulation of yolk protein gene expression in *Drosophila*. Cell 40:339–348

Belote JM, McKeown MB, Andrew DJ, Scott TN, Wolfner MF, Baker BS (1985b) Control of sexual differentiation in *Drosophila melanogaster*. Cold Spring Harbor Symp Quant Biol 50:605–614

Bownes M, Nöthiger R (1981) Sex determining genes and vitellogenin synthesis in *Drosophila melanogaster*. Mol Gen Genet 182:222–228

Bridges CB (1921) Triploid intersexes in *Drosophila melanogaster*. Science 54:252–254

Bridges CB (1925) Sex in relation to chromosomes and genes. Amer Nat 59:127–137

Buckle V, Mondello C, Darling S, Craig IW, Goodfellow PN (1985) Homologous expressed genes in the human sex chromosome pairing region. Nature 317:739–741

Bull JJ (1980) Sex determination in reptiles. Quart Rev Biol 55:3–21

Burgoyne PS (1982) Genetic homology and crossing-over in the X and Y of mammals. Hum Genet 61:85–90

Butler B, Pirrotta V, Irminger-Finger I, Nöthiger R (1986) The sex determining gene *tra* of *Drosophila:* Molecular cloning and transformation studies. EMBO J 5:3607–3613

Chandra HS (1985) Sex determination: A hypothesis based on noncoding DNA. Proc Natl Acad Sci USA 82:1165–1169

Cline TW (1985) Primary events in the determination of sex in *Drosophila melanogaster*. In: Halvorson H, Munroy A (eds) The origin and evolution of sex. Liss, New York, pp 301–327

Cline TW (1986) A female-specific lethal lesion in an X-linked positive regulator of the *Drosophila* sex determination gene, Sex-lethal. Genetics 113:641–663

Cooke HJ, Brown WRA, Rappold GA (1985) Hypervariable telomeric sequences from the human sex chromosomes are pseudoautosomal. Nature 317:687–692
Crew FAE (1965) Sex determination. Methuen, London
Crouse HV (1960) The controlling element in sex chromosome behaviour in *Sciara*. Genetics 45:1429–1443
de la Chapelle A, Tippett PA, Wetterstrand G, Page D (1984) Genetic evidence of X-Y interchange in a human XX male. Nature 307:170–171
Denholm I, Franco MG, Rubini PG, Vecchi M (1986) Geographical variation in house-fly (*Musca domestica* L.) sex determinants within the British Isles. Genet Res 47:19–27
DiBenedetto AJ, Lakich DM, Kruger WD, Belote JM, Baker BS, Wolfner MF (1987) Sequences expressed sex-specifically in *D. melanogaster* adult. Dev Biol 119:242–251
Disteche CM, Casanova M, Saal H, Friedman C, Sybert V, Graham J, Thuline H, Page DC, Fellous M (1986 a) Small deletions of the short arm of the Y chromosome in 46, XY females. Proc Natl Acad Sci USA 83:7841–7844
Disteche CM, Brown L, Saal H, Friedman C, Thuline HC, Hoar DI, Pagon RA, Page DC (1986 b) Molecular detection of translocation (Y:15) in a 45, X male. Hum Genet 74:372–377
Doniach T (1986) Activity of the sex-determining gene *tra-2* is modulated to allow spermatogenesis in the *C. elegans* hermaphrodite. Genetics 114:53–76
Eicher EM, Washburn LL, Whitney JB, Morrow KE (1982) *Mus poschiavinus* Y chromosome in the C57BL/6J murine genome causes sex reversal. Science 217:535–537
Epper F, Nöthiger R (1982) Genetic and developmental evidence for a repressed genital primordium in *Drosophila melanogaster*. Dev Biol 94:163–175
Epplen JT, Cellini A, Shorte M, Ohno S (1983 a) On evolutionarily conserved simple repetitive DNA sequences: Do "sex-specific" satellite components serve any sequence dependent function? Differentiation 23(Suppl):60–63
Epplen JT, Cellini A, Romero S, Ohno S (1983 b) An attempt to approach the molecular mechanisms of primary sex determination: W- and Y-chromosomal conserved simple repetitive DNA sequences and their differential expression in mRNA. J Exp Zool 228:305–312
Evans EP, Burtenshaw MD, Cattanach BM (1982) Meiotic crossing-over between the X and Y chromosomes of male mice carrying the sex-reversing (Sxr) factor. Nature 300:443–445
Ferguson-Smith MA (1966) X-Y chromosomal interchange in the aetiology of true hermaphroditism and of XX Klinefelter's syndrome. Lancet 2:475
Franco MG, Rubini PG, Vecchi M (1982) Sex-determinants and their distribution in various populations of *Musca domestica* L of Western Europe. Genet Res 40:279–293
Fredga K (1983) Aberrant sex chromosome mechanisms in mammals. Evolutionary aspects. Differentiation 23(Suppl):23–30
Gadagkar R, Najundiah V, Joshi NV, Chandra HS (1982) Dosage compensation and sex determination in *Drosophila:* mechanism of measurement of the X:A ratio. J Biosci 4:377–390
Galloway RE, Goodenough UW (1985) Genetic analysis of mating locus linked mutations in *Chlamydomonas reinhardii*. Genetics 111:447–461
Garabedian MJ, Shepherd BM, Wensink PC (1986) A tissue-specific transcription enhancer from the Drosophila yolk protein 1 gene. Cell 45:859–867
Garcia-Bellido A (1975) Genetic control of wing disc development in *Drosophila*. Ciba Found Symp 29:161–182
Gilchrist BM, Haldane JBS (1947) Sex linkage and sex determination in a mosquito, *Culex molestus*. Hereditas 33:175–190
Gordon JW, Ruddle FH (1981) Mammalian gonadal determination and gametogenesis. Science 211:1265–1271
Green MM (1980) Transposable elements in *Drosophila* and other Diptera. Annu Rev Genet 14:109–120
Guellaen G, Casanova M, Bishop C, Geldwerth D, Andre G, Fellous M, Weissenbach J (1984) Human XX males with Y single-copy DNA fragments. Nature 307:172–173
Hägele K (1985) Identification of a polytene chromosome band containing a male sex determiner of *Chironomus thummi thummi*. Chromosoma 91:167–171
Hamerton JL, Dickson JM, Pollard CE, Grieves SA, Short RV (1969) Genetic intersexuality in goats. J Reprod Fertil (Suppl 7):25–51

Herbst EW, Fredga K, Frank F, Winking H, Gropp A (1978) Cytological identification of two X-chromosome types in the wood lemming (*Myopus schisticolor*). Chromosoma 69:185–191

Herskowitz I (1985) Master regulatory loci in yeast and lambda. Cold Spring Harbor Symp Quant Biol 50:565–574

Hicks JB, Ivy JM, Strathern JN, Klar AJS (1985) Sex in budding yeast: How and why. The origin and evolution of sex. Liss, New York, pp 97–111

Hodgkin J (1983) Two types of sex determination in a nematode. Nature 304:267–268

Hodgkin J (1984) Switch genes and sex determination in the nematode *C. elegans*. J Embryol Exp Morphol 83s:103–117

Hodgkin J (1985) Males, hermaphrodites and females: Sex determination in *Caenorhabditis elegans*. Trends Genet 1:85–88

Hodgkin J, Doniach T, Shen M (1985) The sex determination pathway in the nematode *Caenorhabditis elegans:* Variations on a theme. Cold Spring Harbor Symp Quant Biol 50:585–593

Horsfall WR, Anderson JF (1963) Thermally induced genital appendages on mosquitoes. Science 141:1183–1184

Jacob F (1982) The Logic of Life. Pantheon Books, New York

Johnson AD, Herskowitz I (1985) A repressor (MATα 2 product) and its operator control expression of a set of cell type specific genes in yeast. Cell 42:237–247

Jones KW (1983) Evolutionary conservation of sex specific DNA sequences. Differentiation 23(Suppl):56–59

Levinson G, Marsh JL, Epplen JT, Gutman GA (1985) Cross-hybridizing snake satellite, *Drosophila,* and mouse DNA sequences may have arisen independently. Mol Biol Evol 2(6):494–504

Lucchesi JC (1978) Gene dosage compensation and the evolution of sex chromosomes. Science 202:711–716

Madl JE, Herman RK (1979) Polyploids and sex determination in *Caenorhabditis elegans*. Genetics 93:393–402

Maine EM, Salz HK, Cline TW, Schedl P (1985) The Sex-lethal gene of *Drosophila:* DNA alterations associated with sex-specific lethal mutations. Cell 43:521–529

Mainx F (1964) The genetics of *Megaselia scalaris* Loew (*Phoridae*): A new type of sex determination in Diptera. Am Nat 98:415–430

Marsh JL, Wieschaus E (1978) Is sex determination in germ line and soma controlled by separate genetic mechanisms? Nature 272:249–251

Martin J, Lee BTO (1984) A phylogenetic study of sex determiner locations in a group of Australasian *Chironomus* species (Diptera, Chironomidae). Chromosoma 90:190–197

McClelland GAH (1962) Sex-linkage in *Aëdes aegypti*. Trans R Soc Trop Med Hyg 56:4

McLaren A (1985) Sex determination in mice. Cold Spring Harbor Symp Quant Biol 50:623–625

McLaren A, Simpson E, Tomonari K, Chandler P, Hogg H (1984) Male sexual differentiation in mice lacking H-Y antigen. Nature 312:552–555

Metz CW (1938) Chromosome behavior, inheritance and sex determination in *Sciara*. Am Nat 72:485–520

Meyer WJ, Migeon BR, Migeon CJ (1975) A locus on the human X-chromosome for dihydrotestosterone receptor and androgen insensitivity. Proc Natl Acad Sci USA 72:1469–1472

Miller AM, MacKay VL, Nasmyth KA (1985) Identification and comparison of two sequence elements that confer cell-type specific transcription in yeast. Nature 314:598–603

Nasmyth KA (1982) Molecular genetics of yeast mating type. Annu Rev Genet 16:439–500

Nöthiger R, Steinmann-Zwicky M (1985a) Sex determination in *Drosophila*. Trends Genet 1:209–215

Nöthiger R, Steinmann-Zwicky M (1985b) A single principle for sex determination in insects. Cold Spring Harbor Symp Quant Biol 50:615–621

Ohno S (1979) Major sex-determining genes. Monographs on endocrinology, vol II. Springer, Berlin Heidelberg New York

Page DC (1986) Sex reversal: deletion mapping the male-determining function of the human Y chromosome. Cold Spring Harbor Symp Quant Biol 51 (in press)

Page DC, de la Chapelle A, Weissenbach J (1985) Chromosome Y-specific DNA in related human XX males. Nature 315:224–226
Porter SD, Smith M (1986) Homoeo-domain homology in yeast MATα 2 is essential for repressor activity. Nature 320:766–768
Postlethwait JH, Bownes M, Jowett T (1980) Sexual phenotype and vitellogenins in *Drosophila*. Dev Biol 79:1205–1216
Ribbert D (1967) Die Polytänchromosomen der Borstenbildungszellen von *Calliphora erythrocephala*. Chromosoma 21:296–344
Rothenbuhler WC (1957) Diploid male tissue as new evidence on sex determination in honey bees *Apis mellifera* L. J Hered 48:160–168
Rouyer F, Simmler MC, Johnsson C, Vergnaud G, Cooke HJ, Weissenbach J (1986) A gradient of sex linkage in the pseudoautosomal region of the human sex chromosomes. Nature 319:291–295
Sawyer JR, Hozier JC (1986) High resolution of mouse chromosomes: banding conservation between man and mouse. Science 232:1632–1635
Scalenghe F, Turco E, Edström J-E, Pirrotta V, Melli ML (1981) Microdissection and cloning of DNA from a specific region of the *Drosophila melanogaster* polytene chromosomes. Chromosoma 82:205–216
Schäfer U (1986a) Genes for male-specific transcripts in *Drosophila melanogaster*. Mol Gen Genet 202:219–225
Schäfer U (1986b) The regulation of male-specific transcripts by sex determining genes in *Drosophila melanogaster*. EMBO J 5:3579–3582
Schempp W, Wiberg U, Fredga K (1985) Correlation between sexual phenotype and X-chromosome inactivation pattern in the X*XY wood lemming. Cytogenet Cell Genet 39:30–34
Schüpbach T (1982) Autosomal mutations that interfere with sex determination in somatic cells of *Drosophila* have no direct effect on the germ line. Dev Biol 89:117–127
Silvers WK, Gasser DL, Eicher EM (1982) H-Y antigen, serologically detectable male antigen and sex determination. Cell 28:439–440
Simmler MC, Rouyer F, Vergnaud G, Nyström-Lahti M, Ngo KY, de la Chapelle A, Weissenbach J (1985) Pseudoautosomal DNA sequences in the pairing region of the human sex chromosomes. Nature 317:692–697
Simpson E (1986) The H-Y antigen and sex reversal. Cell 44:813–814
Simpson JL (1982) Abnormal sexual differentiation in humans. Annu Rev Genet 16:193–224
Singh L, Jones KW (1982) Sex reversal in the mouse (*Mus musculus*) is caused by a recurrent nonreciprocal crossover involving the X and an aberrant Y chromosome. Cell 28:205–216
Singh L, Purdom IF, Jones KW (1981) Conserved sex chromosome associated nucleotide sequences in eukaryotes. Cold Spring Harbor Symp Quant Biol 45:805–813
Singh L, Phillips C, Jones KW (1984) The conserved nucleotide sequences of *Bkm*, which define *Sxr* in the mouse, are transcribed. Cell 36:111–120
Solari AJ (1980) Synaptonemal complexes and associated structures in microspread human spermatocytes. Chromosoma 81:315–337
Steinmann-Zwicky M, Nöthiger R (1985a) The hierarchical relation between X-chromosomes and autosomal sex determining genes in *Drosophila*. EMBO J 4:163–166
Steinmann-Zwicky M, Nöthiger R (1985b) A small region on the X chromosome of *Drosophila* regulates a key gene that controls sex determination and dosage compensation. Cell 42:877–887
Strathern J, Hicks J, Herskowitz I (1981) Control of cell type in yeast by the mating type locus. The α1–α2 hypothesis. J Mol Biol 147:357–372
Struhl G (1982) Genes controlling segmental specification in the *Drosophila* thorax. Proc Natl Acad Sci USA 79:7380–7389
Thompson PE, Bowen JS (1972) Interactions of differentiated primary sex factors in *Chironomus tentans*. Genetics 70:491
Ullerich FH (1963) Geschlechtschromosomen und Geschlechtsbestimmung bei einigen Calliphorinen (*Calliphoridae, Diptera*). Chromosoma 14:45–110
Ullerich FH (1973) Die genetische Grundlage der Monogenie bei der Schmeissfliege *Chrysomya rufifacies (Calliphoridae, Diptera)*. Mol Gen Genet 125:157–172

Ullerich FH (1984) Analysis of sex determination in the monogenic blowfly *Chrysomya rufifacies* by pole cell transplantation. Mol Gen Genet 193:479–487

Vergnaud G, Page DC, Simmler M-C, Brown L, Rouyer F, Noel B, Botstein D, de la Chapelle A, Weissenbach J (1986) A deletion map of the human Y chromosome based on DNA hybridization. Am J Hum Genet 38:109–124

Wachtel SS, Ohno S, Koo GC, Boyse EA (1975) Possible role for H-Y antigen in the primary determination of sex. Nature 257:235–236

Washburn LL, Eicher EM (1983) Sex reversal in XY mice caused by dominant mutation on chromosome 17. Nature 303:338–340

Went DF, Camenzind R (1980) Sex determination in the dipteran insect *Heteropeza pygmaea*. Genetica 52/53:373–377

Whiting AR (1967) The biology of the parasitic wasp *Mormoniella vitripennis* (Walker). Quart Rev Biol 42:333–406

Whiting PW (1943) Multiple alleles in complementary sex determination of *Habrobracon*. Genetics 28:365–382

Wieacker P, Voiculescu J, Müller CR, Ropers HH (1983) An XX male with a single STS gene dose. Cytogenet Cell Genet 35:72–74

Wilson KL, Herskowitz I (1986) Sequences upstream of the STE6 gene required for its expression and regulation by the mating type locus in *Saccharomyces cerevisiae*. Proc Natl Acad Sci USA 83:2536–2540

Applications of Basic Chromosome Research in Biotechnology and Medicine

P. M. M. RAE [1]

1 Introduction

The aim of this chapter is to summarize some recent applications of basic chromosome research to medical genetics, particularly in the area of diagnosis of genetic disorders. The aspects covered include (1) the utility and limitations of restriction fragment length polymorphism (RFLP) analysis; (2) the application of synthetic oligonucleotide probes to the diagnosis of genetic disorders; and (3) karyotype analysis using DNA probes. Examples are drawn from hemoglobinopathies and some other heritable disorders, and from cancer biology.

2 Molecular Approaches to Human Genetic Disease Detection

Recombinant DNA technologies have allowed the isolation and characterization of a considerable number of human genes (Schmidtke and Cooper 1983), and there is also detailed genetic information on the molecular bases of a small, but increasing number of genetic disorders. This information is sometimes directly at the level of DNA, where DNA sequence comparisons of genes isolated from normal and mutant individuals have allowed the pinpointing of a genetic defect. In other cases, inference of the DNA change involved in a mutation has been made from amino acid sequence comparisons of protein products of genes.

When such details of a mutation are available, highly specific and sensitive nucleic acid hybridization procedures can be applied to DNA from amniocytes or chorionic villus material for prenatal diagnosis. DNA from small amounts of blood or other tissue can be examined for other genetic screening purposes, such as the determination of disease carrier status in prospective parents whose families have a history of expressing a particular disorder. Such testing of white blood cell DNA from parents can determine whether or not the risk of collection of fetal material for analysis is necessary.

Cloned segments of normal genes have been used in the isolation of some mutant genes from the DNAs of patients having genetic disorders. From studies of such DNAs it is clear that the genetic basis of a particular disorder can range from one or a number of single base-pair substitutions that alter amino acid codons or

[1] Molecular Diagnostics, Inc., 400 Morgan Lane, West Haven, Connecticut 06516, USA.

Results and Problems in Cell Differentiation 14
Structure and Function of Eukaryotic Chromosomes
Edited by W. Hennig

Table 1. Molecular bases of hemoglobinopathies

1. *Sickle cell anemia*	
Due to a single base-pair change in the gene for the β-chain of hemoglobin.	
2. *Thalassemias*	
Due to defective synthesis of either the α- or the β-type subunits of the hemoglobin tetramer.	
α^+: Low level of α-chain synthesis	Some due to gene deletions that leave only one of the two α-genes intact; some due to small intragenic deletions or point mutations that affect mRNA.
α^0: No α-chain synthesis	All due to complete or partial deletion of both chains.
β^+: Low level of β-chain synthesis	A point mutation in an intron splice junction leads to mostly incorrect mRNA production.
β^0: No β-chain synthesis	Some due to nonsense (chain terminating) mutations; one point mutation severely alters a splice junction; one due to a frameshift; one may be due to a point mutation in the promoter.
HPFP: Hereditary persistence of fetal hemoglobin $\delta\beta$: No δ- nor β-chain synthesis	Both involve deletions of various amounts of the ε-Gγ-Aγ-δ-β gene cluster, and are characterized by large amounts of γ-chain in adult red cells.
$\gamma\delta\beta$: No γ-, δ-, or β-chain synthesis	One form involves a large inversion and two deletions; two other forms lack the δ- and β-genes, and part or all of the Aγ-gene.

RNA splice sites, through single- or oligo-base pair insertions or deletions that cause translation reading frame shifts, to deletions or rearrangements of whole genes. An especially instructive genetic disease family is that of β-thalassemia (Table 1). A number of β-thalassemia alleles have been characterized at the DNA level, and analysis of cloned alleles of the β-globin gene has shown that virtually all of the kinds of DNA sequence alterations that could affect gene expression appear as the cause of one or another of the thalassemias (Table 1). For example, a form of β-thalassemia that is prevalent in Sardinia and is common in the Mediterranean region, β^{39}-thalassemia, is due to an alteration of codon 39 from CAG to the translation terminating TAG (Pirastu et al. 1983). Another, less severe, β-thalassemia results from a single base change within the first intervening sequence of β-globin that generates an alternative splice site causing abnormal processing of globin messenger RNA (Orkin et al. 1983). This form of β^+-thalassemia is the one most frequently encountered in Greek and Italian populations. A common Asian Indian β-thalassemia is due to deletion of 600 bp of the 3′ end of the β-globin gene (Orkin 1984).

Less direct information on defective genes has come from studies of the molecular hybridization of cloned, but unidentified ("anonymous") segments of DNA from a human recombinant DNA library to restriction enzyme digests of DNA prepared from blood or other tissue samples from individuals carrying mutant alleles of genes. These experiments take advantage of spontaneous base pair changes that have occurred along DNA, some of which involve restriction enzyme cleavage sites and introduce polymorphisms in the lengths of DNA restriction fragments in the human population (Botstein et al. 1980).

Screening a number of genomic DNAs with cloned segments from human DNA libraries has allowed the identification of cloned probes that detect polymorphisms along the lengths of each of the chromosomes (White et al. 1983). Application of such RFLP probes in hybridization analysis of restriction enzyme digests of DNA from patients and members of their families has in a number of instances allowed informative correlations to be drawn within a family between the presence in an individual's DNA of particular lengths of a restriction enzyme cleavage product and the presence of one or two copies of a disease allele. The power of this diagnostic technology is that one need only establish genetic linkage between a marker and a disorder, and one need know nothing about the affected gene itself. This is discussed further in Sect. 4.

3 Southern Blot Analysis with Cloned Gene Segments

The first human gene to be isolated and thoroughly characterized with respect to nucleotide sequence in normal and mutant alleles was that for β-globin (Lawn et al. 1980). Focus on the β-globin gene was first for technical reasons, because the abundant messenger RNA in reticulocytes had been well studied and could be used as a specific gene probe for the screening of recombinant DNA libraries. However, the β-hemoglobinopathies are manifold and of considerable medical significance so that attention was given early to the molecular characterization of these disorders. In particular, sickle cell anemia and some thalassemias are major health problems in the Third World, and the bases of these diseases are understood in great detail.

Through simple hybridization of cloned segments of the normal β-globin gene to gel electrophoresis displays of restriction enzyme-treated DNA from hemoglobinopathy patients, it was determined that discrimination of normal, heterozygous, and homozygous mutant genotypes can be made in the cases of sickle cell disease and some β-thalassemias (Orkin 1984).

In some thalassemias, deletions of gene sequence alter the length of a DNA segment defined as being flanked by particular restriction enzyme cleavage sites, or even completely remove sequences complementary to a probe (Orkin et al. 1982; Boehm et al. 1983). In the diagnosis of other forms of thalassemia, advantage has been taken of a very tight genetic linkage between globin gene mutations and alterations in restriction enzyme cleavage sites also within or close to the gene, such as a β°-thalassemia that is associated with an altered BamHI site (Kan et al. 1980).

In the case of sickle cell disease, a point mutation in the protein-coding portion of the gene fortuitously alters one of several sites in the normal gene recognized by the restriction enzymes Dde I and Mst II, so that a DNA fragment of the sickle cell allele that is generated by one or the other enzyme has a length that is the sum of the lengths of two contiguous fragments that are generated from the normal allele of the β-globin gene (Chang and Kan 1982; Wilson et al. 1982).

The single base pair change in the β-globin gene that results in the production of hemoglobin S is an A to T transversion in the sixth codon (Fig. 1). The nucleo-

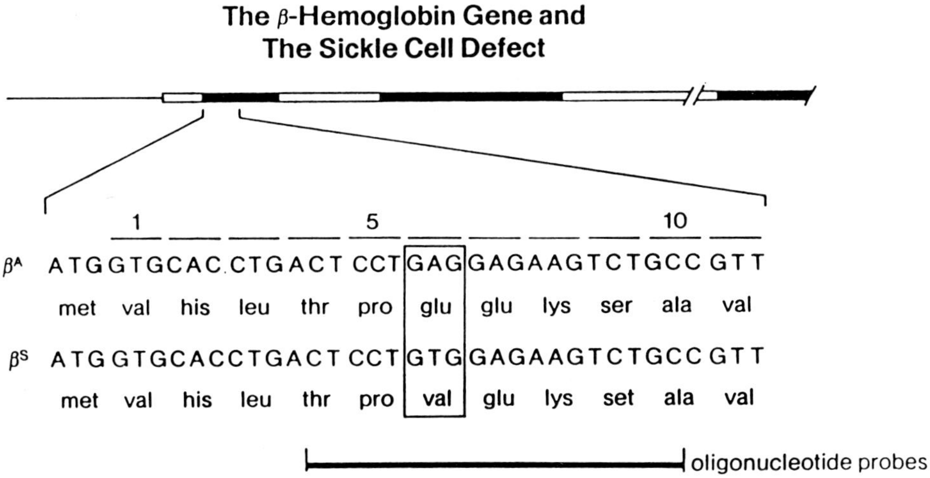

Fig. 1. The 5′ end of the beta-globin gene, showing the nucleotide and amino acid sequence difference between normal (β^A) and sickle cell anemia (β^S) alleles. The numbers are of the amino acids in the mature polypeptide. The restriction enzymes Dde I and Mst II recognize and cleave at the sequences CTNAG and CCTNAGG, respectively. These sequences include the codon for amino acid 6 (box), and the A to T transversion in this codon, responsible for sickle cell anemia, causes the site to be insensitive to the restriction enzymes. This is the basis of the Dde I and Mst II Southern blot hybridization tests for the sickle cell allele described in the text. Indicated by the position of the bar at the bottom of the figure are the sequences of 19-mer oligonucleotides used as DNA hybridization probes for genotype determinations such as those shown in Fig. 2. The alternating unfilled and solid portions of the gene diagram at the top of the figure are untranslated and translated regions, respectively

tide in question in the normal globin allele lies within the sequence CTGAG, which is the recognition and cleavage site for the restriction enzyme Dde I, and within the sequence CCTGAGG, which is attacked by Mst II. The sickle mutation alters these sequences to CTGTG and CCTGTGG, which are not sensitive to the restriction enzyme activities. The consequence of this is that Dde I or Mst II can be used to determine if a DNA sample is from an individual that is homozygous or heterozygous for the sickle cell allele, or is free of the allele. There are diagnostic β-globin gene restriction fragment lengths for normal and sickle alleles, and genotype determinations can be made by hybridization of labeled cloned β-globin DNA to Southern blots of Dde I- or Mst II-digested sample DNA. For example, an Mst II digest of DNA from a carrier of sickle cell anemia will produce fragments of 1340, 1140, and 201 bp that hybridize with a β-globin gene probe, while DNA from an individual homozygous for the sickle globin allele will show only the 1340 bp segment; normal DNA samples will show only the 1140 and 201 bp segments.

The α-thalassemia syndromes are basically four in number, and differ in severity according to the number of operative α-globin genes in the diploid genome of an individual (Orkin et al. 1979; Lauer et al. 1980). α-thalassemia defects are generally due to deletion, and with there being normally two functional α-globin genes per haploid genome, in tandem on chromosome 16, there are two classes of defect: α-thalassemia$_1$ involves deletion of part or all of both α-globin genes on a chromosome, while α-thalassemia$_2$ results from deletion of a single gene.

The silent carrier state and the α-thalassemia trait are the result of the absence of one or two α-globin genes, respectively, from a chromosome, and heterozygotes experience no phenotype, or possibly mild anemia in the thalassemia trait genotype (α-thal$_1$/+). Hemoglobin H disease is manifested by a moderate form of anemia, and develops in compound heterozygotes of α-thal$_1$ and α-thal$_2$. Homozygosity for α-thal$_1$, hydrops fetalis, results in fetal death. It is for this latter disorder that an early prenatal diagnosis of genotype in pregnancies at risk for thalassemia is particularly important.

Characterization of the thalassemia deletions in terms of the restriction enzyme map of the α-globin gene region allowed the definition of restriction enzyme cleavage patterns of DNA that are indicative in Southern blot hybridizations of each of the major forms of α-thalassemia in Asian and Mediterranean populations (Orkin et al. 1979). Such hybridization tests have been applied to the prenatal diagnosis of homozygous α-thalassemia (Dozy et al. 1979; Zeng and Huang 1985).

4 Tracking Disease Genes in Families with RFLPs

Some genetic diseases have been mapped to chromosome regions on the basis of the patterns of their segregation with DNA sequence microheterogeneities identified by the presence or absence of restriction enzyme cleavage sites (so-called *Restriction Fragment Length Polymorphisms; RFLPs*). There is often sufficiently tight genetic linkage between the locus of a heritable disorder and a polymorphic restriction enzyme cleavage site (at a distance of even several hundreds of thousands of base pairs) that pedigrees for a disease can be established according to the distribution in a family of the polymorphic restriction site, without direct reference to the gene itself. In such cases, a probe is any unique DNA sequence that lies in the polymorphic restriction fragment. Some laboratories have been establishing a bank of cloned human DNA segments that cover the human genome sufficiently well, and identifying restriction site polymorphisms that are sufficiently numerous and dispersed, that linkage relationships between genetic diseases and such polymorphisms can be exploited in a DNA hybridization system for the determination of genetic constitutions (White et al. 1983).

Examples of this technology are the recent identification of a human chromosome 4 probe for a polymorphic DNA segment that has a tight genetic linkage with the locus of Huntington's disease in a Venezuelan population (Gusella et al. 1983), and probes that identify RFLPs associated with the cystic fibrosis locus on chromosome 7 (Wainwright et al. 1985; White et al. 1985). One of the RFLP probes that can be used to monitor the inheritance of CF alleles in some families is the *met* cellular oncogene (Dean et al. 1985), which maps to within one centimorgan (ca. 10^6 bp) of the CF locus (White et al. 1985). There is also an indication that certain restriction fragment length polymorphisms associated with the Harvey *ras* oncogene segregate with DNAs of cancer patients and tumor tissue,

Table 2. Some major human genetic disorders for which there are DNA probes

A. Cystic fibrosis (Wainwright et al. 1985; White et al. 1985)
B. Diabetes
   1. Juvenile onset, insulin-dependent (Barbosa et al. 1980)
   2. Noninsulin-dependent
C. Hemoglobinopathies
   1. Sickle cell anemia, Hb C (Conner et al. 1983)
   2. β-Thalassemias (Orkin 1984)
   3. α-Thalassemia(s) (Orkin et al. 1979)
D. Hypercholesterolemia (Humphries et al. 1985)
E. Less frequent diseases, such as
   1. Antitrypsin deficiency (Kidd et al. 1983)
   2. Huntington's disease (Gusella et al. 1983)
   3. Duchenne's muscular dystrophy (Bakker et al. 1985)
   4. Hemophilia (Antonarakis et al. 1985)
   5. Phenylketonuria (Woo et al. 1983)

so that RFLP analysis may be applicable in some instances to the prediction of cancer susceptibility (Krontiris et al. 1985). Other examples are given in Table 2.

The RFLP approach to the diagnosis or prediction of genetic disorders has shortcomings, however. The significant disadvantages of RFLP tests are that (1) while common, linkages between disorders and restriction site polymorphisms are not always present nor informative, and (2) when present, the phase of such a linkage in a family must be established before a diagnosis can be made. This means that a disease gene will track with a polymorphism in a particular phase in one family but not another, and that DNA samples must be obtained from at least a few members of a family before a diagnosis can be made using proband DNA.

5 Detecting Point Mutations with Synthetic DNA Probes

If one has information on the DNA sequence of a gene and one or more mutant alleles, the most informative and simplest DNA diagnostic system is the use of oligonucleotides as hybridization probes.

The power of oligonucleotide probes is that they are sensitive to all DNA sequence alterations, from any single base-pair substitution to insertions, deletions, and translocations of any magnitude. Also, they probe a mutation directly, and there is not the requirement of family studies to evaluate genetic linkage of the mutation with another marker, as is necessary in RFLP analysis.

The first application of synthetic oligonucleotide hybridization to human genomic DNA for the detection of a point mutation was for the discrimination of sickle cell and normal β-globin genes (Conner et al. 1983). Nonakaidecanucleotide (19-mer) probes, labeled with [^{32}P], can be annealed to DNA using conditions under which only perfectly matched duplexes are stable, and hybrids containing even a single base mismatch can be dissociated (Wallace et al. 1981). Conner et al. (1983) used 19-mer probes representing the normal and sickle globin alleles, and differing by one nucleotide (Fig. 1), in hybridizations to human DNAs digested with BamHI and displayed in agarose gels. BamHI produces a ca. 2 kb segment of the β-globin gene that contains the site of the sickle mutation, and hybridization of normal and mutant probes to duplicate samples of DNA allows unambiguous determination of genotype in clinical samples (Fig. 2).

Since the work of Conner et al. (1983), synthetic oligonucleotides have been used as probes of amniocyte or chorionic villus DNA for the prenatal diagnosis of some β-thalassemias (Orkin et al. 1983; Pirastu et al. 1983) and α_1-antitrypsin deficiency (Kidd et al. 1983), for example, as well as sickle cell anemia.

However, just as a need for knowledge of the phase relationships of restriction fragment length polymorphisms with alleles of a gene limits the use of cloned sements of DNA as probes for genetic disorders, so the use of oligonucleotides as direct probes is limited by the extent to which genetic diseases are understood. Ideally, one should know the precise kind and location of a difference between a normal and a mutant allele of a gene in order to synthesize an appropriate oligonucleotide probe, as was done with the β-hemoglobinopathies. At least, one can hope to guess genotypic differences on the basis of comparisons of normal and mutant gene products, as was done in the development of a probe for α_1-antitrypsin deficiency by Kidd et al. (1983). They had the normal gene cDNA sequence, the mutant protein amino acid sequence, and a fortunate single amino acid change that must have arisen from a single base pair transition, and so were able to identify a single oligonucleotide probe sequence that would detect the mutation.

Nevertheless, for the future, DNA sequencing of extant and new genomic and cDNA clones, and advances in the microsequencing of proteins and nucleic acids will allow expansion of the repertoire of diagnostically significant oligonucleotide probes.

The exquisite feature of oligonucleotide hybridization analysis is that one can discriminate any particular single base-pair substitution in a genome comprised of about 3,000,000,000 bp. Further, it is not necessary to obtain and analyze DNA from members of a proband's family for an unambiguous diagnosis to be possible. In practice, given knowledge of the molecular differences between a normal and a mutant allele, one synthesizes an oligonucleotide that is sufficiently long that it represents a unique sequence of the human genome (and therefore will not give false signals with sequences that are related to the one of interest), and is sufficiently short that conditions of hybridization permit discrimination of the stabilities of a perfect hybrid from one having a single base-pair mismatch; a length that works is 19 nucleotides, and the useful range might be 17 to 22 nucleotides for human genome probes.

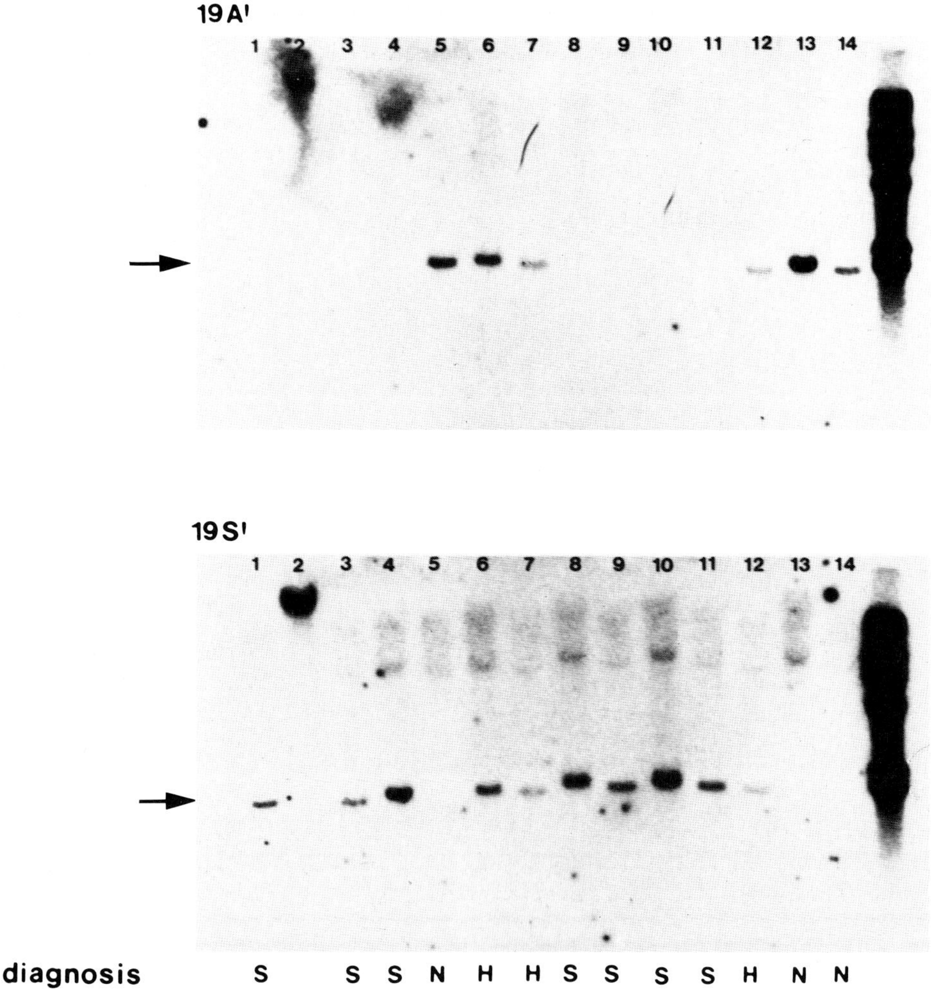

Fig. 2. Genotype determination by oligonucleotide hybridization. In this example, the test is for the sickle cell anemia allele of the beta-globin gene, which differs from the normal globin allele by a single base change in the codon for amino acid 6 (Fig. 1). The test is accomplished by hybridizing ^{32}P-labelled 19-mers representing the normal globin sequence that includes codon 6 (19A′), and the identical sequence but for a single base change at the sickle cell mutation site (19S′), to 2–3 micrograms each of genomic DNA samples that were digested with Bam HI and electrophoresed in duplicate agarose gels. DNA in the upper gel was challenged with the 19A′ oligonucleotide, and that in the lower gel with the 19S′ oligonucleotide. The autoradiographic exposure shown was overnight. The arrows indicate the position of a 1.8 kb Bam H segment of human DNA that contains the 5′ end of the beta-globin gene. Genotype is determined from whether one or the other or both probes form a stable hybrid with a DNA sample. For example, DNA in the sample lanes 4 bound only the 19S′ probe, so the individual is homozygous for the sickle mutation; DNA in lanes 5 bound only the 19A′ probe, so the individual is homozygous normal; and the sample in lanes 6 bound both probes stably so this individual is a carrier. The far right lane in the gels is ^{32}P-labelled Hind III segments of phage lambda DNA run as size markers

6 Molecular Cytogenetics Analysis of Chromosome Aberrations

Two applications of cytogenetics in diagnostics are in (1) the evaluation of fetal cell karyotypes in cultured amniocytes for such aneuploidies as trisomy 21, Down's syndrome, and (2) the characterization of chromosomal rearrangements in malignant tissues as a means of identifying particular cancers and following a course of treatment.

The field of karyotype analysis has made its greatest contribution to medicine in correlating at high cytological resolution particular chromosomal abnormalities with specific human malignancies (Pearson and Rowley 1985; see below). The various chromosome banding techniques have allowed precise identification of chromosome breakpoints involved in translocations and deletions that are associated at fairly high frequency with different leukemias, for example (Fig. 3). The knowledge from karyotype studies of which particular cancer is affecting a patient allows the oncologist to select the most appropriate course of treatment, and to make an informed prognosis.

The other very significant contribution of human cytogenetics has been in the evaluation of amniocyte karyotypes for aneuploidies, and for prenatal sex determination. Fetuses at risk for chromosome abnormalities such as may be associated with increasing maternal age, or may be found in particular families, are studied at the chromosome level using cells cultured from amniotic fluid. Similarly, pregnancies in families with a history of sex-linked disorders may be probed early for fetal sex by amniocyte analysis so that plans can be made for the eventuality of an affected male child.

Following the establishment of banding techniques for subchromosomal resolution, and also the introduction of mouse/human somatic cell hybrid panels for gene mapping (Kamarck et al. 1984), more recent advances in cytogenetics have involved development of in situ hybridization procedures for the chromosomal localization of particular DNA sequences, even single copy ones, and for the rapid identification of chromosome aberrations using, say, oncogenes as hybridization probes.

A very attractive new procedure for the nonradioisotopic detection of in situ hybrids involves a sensitive immunofluorescence detection system that should be applicable to the comparatively rapid detection of single copy gene hybridization (Pinkel et al. 1986). It has already been used for the identification of human chromosomes in human/rodent cell hybrids, for the localization of moderately repeated sequences, and for the determination of human cell gender using cloned Y chromosome DNA sequences as a probe (Pinkel et al. 1986; Fig. 4).

It has emerged in the field of oncology over the past few years that certain cancers, especially leukemias, are very frequently found to have associated with them particular chromosome abnormalities. The first such correlation, made some 25 years ago, was of chronic myelogenous leukemia with a translocation involving chromosomes 9 and 22, the so-called Philadelphia chromosome (Nowell and Hungerford 1960). Since that time, a number of other chromosomal rearrangements have been identified with particular malignancies, as shown in Fig. 3 (Pearson and Rowley 1985).

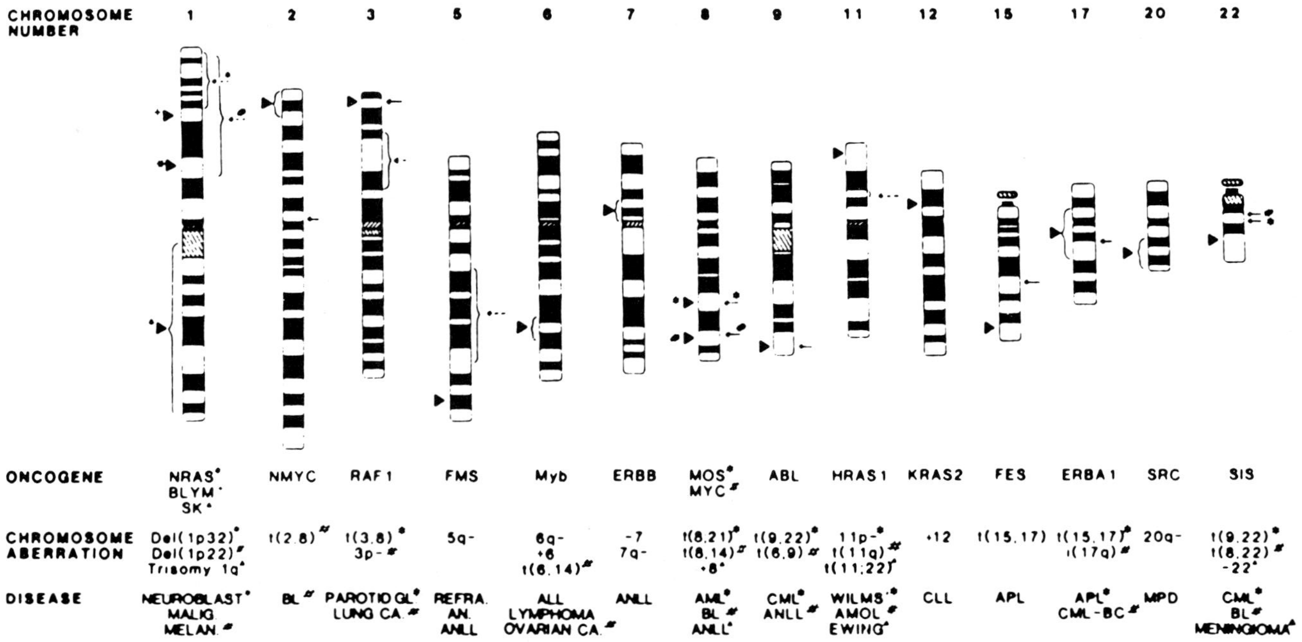

Fig. 3. Associations among oncogenes, chromosomal rearrangements and neoplasias. The cellular oncogenes localized to each chromosome are listed, and their positions are indicated by the heavy arrowheads to the left of each chromosome drawing. Involvements of some oncogenes with translocations frequently found in association with certain neoplasias are mentioned in the text. Chromosome breakpoints of translocations and deletions frequently associated with the diseases listed under each chromosome are indicated by solid or dashed arrows, respectively, to the right of each chromosome. For example, acute promyelocytic leukemia (APL) is often associated with a rearrangement involving chromosomes 15 and 17. (From Rowley 1984)

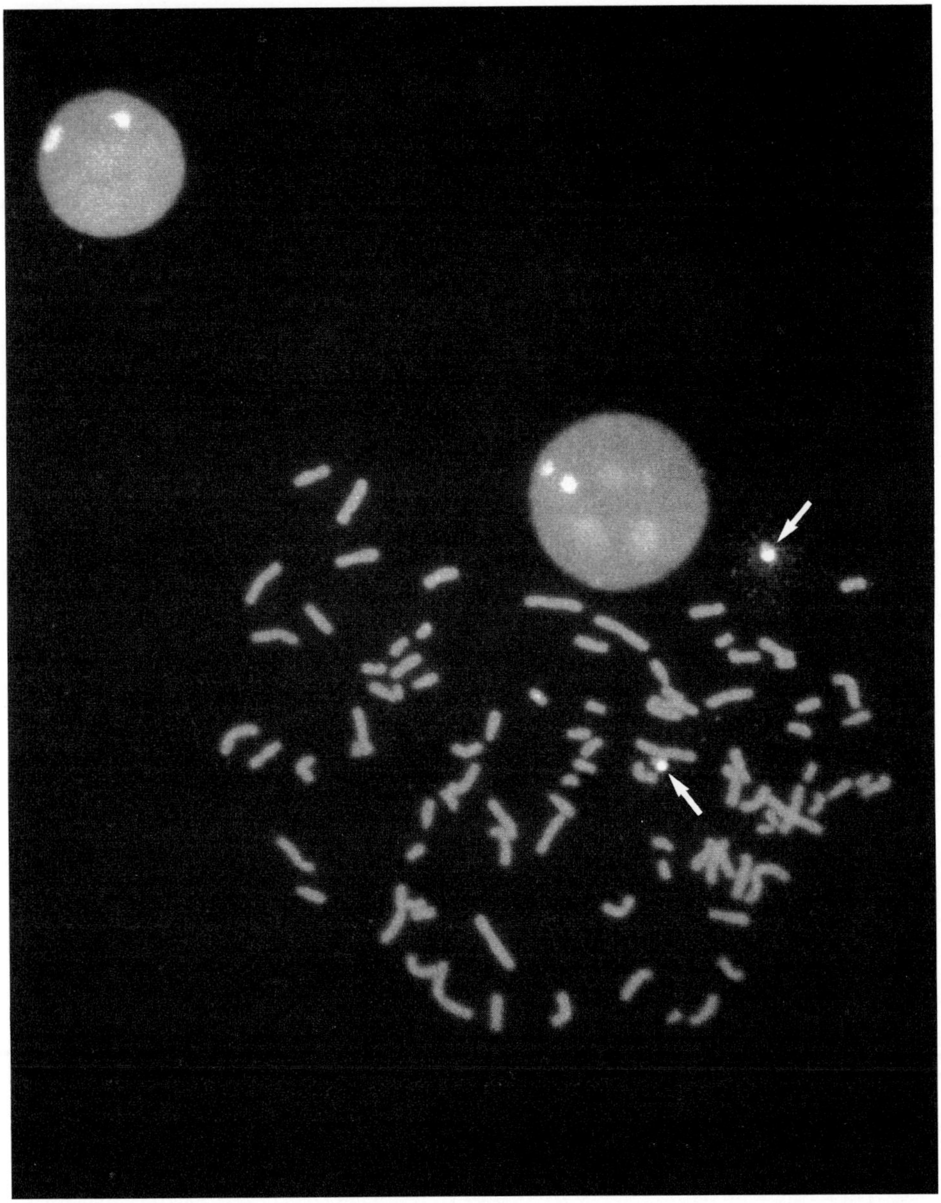

Fig. 4. In situ hybridization of a cloned chromosome-specific DNA segment to interphase nuclei and a metaphase chromosome spread. The cytological preparation was of a tetraploid lymphoblastoid cell line from a human male. The probe was a recombinant plasmic containing a 3.4 kb repetitive sequence of human Y chromosome origin. The plasmid was tagged with biotin, then hybridization was detected with fluorescein conjugated antibodies using a method comparable to that of Pinkel et al. (1986). The fluorescein was yellow green in the original color micrograph, and is white in this figure. It identifies two Y chromosomes in the tetraploid metaphase spread (arrows), and two condensed chromatin spots in each of the interphase nuclei. The rest of the chromatin fluoresced red due to counter-staining of the preparation by propidium iodide, and is gray in this figure. The original micrograph was kindly provided by John T. Hart

Further correlated with chromosome rearrangements and cancers is the proximity of some of the cellular oncogenes (Cooper 1982) to some translocation breakpoints. Burkitt's lymphoma is associated with translocations involving chromosome 8, most frequently t(8; 14); the breakpoints in chromosome 8 are near or within c-*myc*, and in other chromosomes in the variable regions of immunoglobulin genes (Leder et al. 1983). The translocations put *myc* in the context of one or another of the immunoglobulin genes, and alterations in affected cells in the expression of *myc*, the product of which is a nuclear DNA-binding protein, must participate in the transformation process.

Another well-characterized rearrangement that involves a cellular oncogene is the 9; 22 translocation of chronic myelogenous leukemia (Groffen et al. 1984). The breakpoint in chromosome 9 is near c-*abl*, and that in chromosome 22 is in a short region called *bcr* for breakpoint cluster region. Evidently, realignment of the c-*abl* gene, which codes for a plasma membrane tyrosine kinase activity, is involved in malignant transformation in CML.

It is the expectation of molecular cytogeneticists in biotechnology and medicine that a variety of panels of specific DNA hybridization probes will be developed so that rapid and accurate determinations of genotype can be made in the several prenatal, family counseling, and oncological applications indicated in this review. Certainly, considerable effort is being made in both commercial and academic research laboratories in the development of probes and hybridization and detection formats, and it is only a matter of time before simple and rapid dot hybridization and in situ hybridization procedures are standard tools of diagnosis and genetic counseling.

References

Antonarakis SE, Carpenter RJ, Hoyer LW, Toole JJ, Copeland KL, Carta CA, Caskey CT, Kazazian HH (1985) Prenatal diagnosis of haemophilia A by factor VIII gene analysis. Lancet i:1407–1409

Bakker E, Goor N, Wrogemann K, Kunkel LM, Fenton WA, Majoor-Krakauer D, Jahoda MGJ, van Ommen GJB, Hofker MH, Mandel JL, Davies KE, Willard HF, Sandkuyl L, v. Essen AJ, Sachs ES, Pearson PL (1985) Prenatal diagnosis and carrier detection of Duchenne muscular dystrophy with closely linked RFLPs. Lancet i:655–658

Barbosa J, Chern MM, Anderson VE, Noreen H, Johnson S, Reinsmoen N, McCarty R, King R, Greenberg L (1980) Linkage analysis between the major histocompatibility system and insulin dependent diabetes in families with patients in two consecutive generations. J Clin Invest 65:592–601

Boehm CD, Antonarakis SE, Phillips JA, Stetten G, Kazazian HH (1983) Prenatal diagnosis using DNA polymorphisms: Report on 95 pregnancies at risk for sickle cell disease or β-thalassemia. N Engl J Med 308:1054–1058

Botstein D, White RL, Skolnick M, Davis RW (1980) Construction of a genetic linkage map in man using restriction fragment length polymorphisms. Am J Hum Genet 32:314–331

Chang JC, Kan YW (1982) A sensitive test for prenatal diagnosis of sickle cell anemia: Direct analysis of amniocyte DNA with *Mst*II. Trans Assoc Am Phys 95:71–78

Conner BJ, Reyes AA, Morin C, Itakura K, Teplitz RL, Wallace RB (1983) Detection of sickle cell β^S-globin allele by hybridization with synthetic oligonucleotides. Proc Natl Acad Sci USA 80:278–282

Cooper GM (1982) Cellular transforming genes. Science 218:801–806

Dean M, Park M, Le Beau MM, Robins TS, Diaz MO, Rowley JD, Blair DG, Vande Woude GF (1985) The human *met* oncogene is related to the tyrosine kinase oncogenes. Nature 318:385–388

Dozy AM, Forman EN, Abuello DN, Barsel-Bowers G, Mahoney MJ, Forget BG, Kan YW (1979) Prenatal diagnosis of homozygous α-thalassemia. JAMA 241:1610–1612

Groffen J, Stephenson JR, Heisterkamp N, deKlein A, Bartram CR, Grosveld G (1984) Philadelphia chromosome breakpoints are clustered within a limited region, bcr, on chromosome 22. Cell 36:93–99

Gusella JF, Wexler NS, Conneally PM, Naylor SL, Anderson MA, Tanzi RE, Watkins PC, Ottina K, Wallace MR, Sakaguchi AY, Young AB, Shoulton I, Bonilla E, Martin JB (1983) A polymorphic DNA marker genetically linked to Huntington's disease. Nature 306:234–238

Humphries SE, Horsthemke B, Seed M, Holm M, Wynn V, Kewssling AM, Donald JA, Jowett N, Galton DJ, Williamson R (1985) A common DNA polymorphism of the low-density lipoprotein (LDL) receptor gene and its use in diagnosis. Lancet i:1003–1005

Kamarck ME, Barker PE, Miller RL, Ruddle FH (1984) Somatic cell hybrid mapping panels. Exp Cell Res 152:1–14

Kan YW, Lee KY, Furbetta M, Anguis A, Cao A (1980) Polymorphism of DNA sequence in the β-globin gene region. N Engl J Med 302:185–188

Kidd VJ, Wallace RB, Itakura K, Woo SLC (1983) α_1-antitrypsin deficiency detection by direct analysis of the mutation in the gene. Nature 304:230–234

Krontiris TG, DiMartino NA, Colb M, Parkinson DR (1985) Unique allelic restriction fragments of the human Ha-*ras* locus in leukocyte and tumor DNAs of cancer patients. Nature 313:369–375

Lauer J, Shen JC, Maniatis T (1980) The chromosomal arrangement of human α-like globin genes: Sequence homology and α-globin gene deletions. Cell 20:119–130

Lawn RM, Efstratiadis, O'Connell C, Maniatis T (1980) The nucleotide sequence of the human β-globin gene. Cell 21:647–651

Leder P, Battey J, Lenoir G, Moulding C, Murphy W, Potter H, Stewart T, Taub R (1983) Translocations among antibody genes in human cancer. Science 222:765–771

Nowell PC, Hungerford DA (1960) A minute chromosome in human granulocytic leukemia. Science 132:1497

Orkin SH (1984) Prenatal diagnosis of hemoglobin disorders by DNA analysis. Blood 63:249–253

Orkin SH, Old J, Lazarus H, Altay C, Gurgey A, Weatherall DJ, Nathan DG (1979) The molecular basis of α-thalassemias: Frequent occurrence of dysfunctional α loci among non-asians with Hb H disease. Cell 17:33–42

Orkin SH, Kazazian HH, Antonarakis SE, Goff SC, Boehm CD, Sexton JP, Waber PG, Giardina PJV (1982) Linkage of β-thalassaemia mutations and β globin gene polymorphisms with DNA polymorphisms in human β-globin gene cluster. Nature 296:627–631

Orkin SH, Markham AF, Kazazian HH (1983) Direct detection of the common Mediterranean β-thalassemia gene with synthetic DNA probes. J Clin Invest 71:775–779

Pearson M, Rowley JD (1985) The relation of oncogenesis and cytogenetics in leukemia and lymphoma. Annu Rev Med 36:471–483

Pinkel D, Straume T, Gray JW (1986) Cytogenetic analysis using quantitative, high-sensitivity, fluorescence hybridization. Proc Natl Acad Sci USA 83:2934–2938

Pirastu M, Kan YW, Cao A, Conner BJ, Teplitz RL, Wallace RB (1983) Prenatal diagnosis of β-thalassemia. Detection of a single nucleotide mutation in DNA. N Engl J Med 309:284–287

Rowley JD (1984) The biological implications of consistent chromosome rearrangements. Cancer Res 44:3159–3169

Schmidtke J, Cooper DN (1983) A list of cloned human DNA sequences. Hum Genet 65:19–26

Wainwright BJ, Scambler PJ, Schmidtke J, Watson EA, Law HY, Farrall M, Cooke HJ, Eiberg H, Williamson R (1985) Localization of cystic fibrosis to human chromosome 7cen-q22. Nature 318:384–385

Wallace RB, Johnson MJ, Hirose T, Kawashima EH, Itakura K (1981) The use of synthetic oligonucleotides as hybridization probes. II. Hybridization of oligonucleotides of mixed sequence to rabbit β-globin DNA. Nucl Acids Res 9:879–894

White RL, Barker D, Holm T, Berkowitz J, Leppert M, Cavenee W, Leach R, Drayna D (1983) Approaches to linkage analysis in the human. Banbury Report 14:235–250, Cold Spring Harbor Lab, New York

White R, Woodward S, Leppert M, O'Connell P, Hoff M, Herbst J, Lalouel JM, Dean M, Vande Woude G (1985) A closely linked genetic marker for cystic fibrosis. Nature 318:382–384

Wilson JT, Milner PF, Summer ME, Nallseth FS, Fadel HE, Reindollar RH, McDonough PG, Wilson LB (1982) Use of restriction endonucleases for mapping the allele for β^{S}-globin. Proc Natl Acad Sci USA 79:3628–3631

Woo SLC, Lidsky AS, Güttler F, Chandra T, Robson KJH (1983) Cloned human phenylalanine hydroxylase gene allows prenatal diagnosis and carrier detection of classical phenylketonuria. Nature 306:151–155

Zeng YT, Huang SZ (1985) α-globin gene organization and prenatal diagnosis of α-thalassemia in chinese. Lancet i:304–307

Closing Remarks

J. G. Gall[1]

In my closing remarks I would like to review some of the major contributions made by Wolfgang Beermann and his co-workers during the past 35 years and to point out how the studies reported at our Symposium relate to those contributions. I will apologize ahead of time for not mentioning all the speakers by name. Since my aim is to trace Beermann's influence, some of the talks illustrate my theme more directly than others.

Beermann's first major publication, which appeared in 1952, was entitled, „*Chromomerenkonstanz* ..." (its actual title was much longer, in keeping with German tradition, but I will use abbreviations for this and the other papers I cite). It was in part a response to a claim made by two workers in Turkey named Kosswig and Sengün, who maintained that the banding pattern of polytene chromosomes varied from tissue to tissue and, therefore, could not be used to map the location of genes on chromosomes. Against this "Istanbul Hypothesis" Beermann marshalled an impressive array of cytological observations, using the favorable, but hitherto little analyzed chromosomes of *Chironomus:*

1. By mapping the same chromosome regions in salivary gland, midgut, Malpighian tubule, and rectum he showed that the banding pattern was essentially invariant. Since then we have not worried *whether* the banding pattern reflects the underlying gene pattern, but *how* it does. For a time it was popular to suggest a one gene/one band model of chromosome organization; on several previous occasions Burke Judd has summarized the evidence for and against such a simple model. His discussion at this Symposium of the *white* locus of *Drosophila,* as well as Ted Wright's analysis of the *Dopa decarboxylase* locus, emphasized a much more complex picture. At a slightly coarser level of resolution the banding patterns on mitotic chromosomes were described for us by Dieter Schweizer.
2. On top of the invariant banding pattern Beermann recognized that a more loosely organized state, called a puff, characterized certain bands at specific times in specific tissues. I will come back to this most important observation in a moment.
3. Beermann followed the growth of the polytene chromosomes during larval development and concluded that their enormous increase in size was due fundamentally to an increase in the number of constituent subunits. He showed that these subunits could be visualized directly in the giant puffs known as Balbiani Rings (BRs). This view of chromosome structure contributed to the then novel

[1] Carnegie Institution, Baltimore, Maryland 21210, USA.

Results and Problems in Cell Differentiation 14
Structure and Function of Eukaryotic Chromosomes
Edited by W. Hennig

hypothesis of uninemy by emphasizing that a chromatid is the fundamental building block of chromosomes. It is appropriate that the architect of the uninemy hypothesis, J. Herbert Taylor, presented his most recent views on chromosome replication to us at this meeting.

Polytene chromosomes continue to be a source of fascination and new information. Dieter Ammermann summarized current knowledge about polytene chromosomes in the ciliated protozoa. E. R. Schmidt described repetitive sequences in the polytene chromosomes of *Chironomus;* and H. Saumweber presented a detailed computer analysis of the spatial arrangement of polytene chromosomes inside *Drosophila* salivary gland nuclei.

In 1960 Beermann published, *„Der Nukleolus ...",* a beautiful study of the nucleolar organizer in *Chironomus,* and perhaps the most elegant cytogenetic analysis of this organelle since Barbara McClintock's original description in 1934. Not only did this paper firmly establish that the nucleolar organizer is a vital genetic locus, but it provided hints that the nucleolus might be involved in ribosomal RNA metabolism. Only a few years later Edström and Beermann strengthened this hypothesis by showing that nucleolar RNA and cytoplasmic RNA (i.e., primarily ribosomal RNA) had similar base compositions.

Today the nucleolus still attracts interest as new techniques permit more and more detailed analysis. Ulrich Scheer discussed, among other topics, the use of the Miller spreading technique for visualizing rDNA transcription units by electron microscopy, and Michael Trendelenburg showed that these transcription units are visible even in the light microscope with the aid of video microscopy and on-line image processing. de Lucchini described molecular and cytological aspects of rDNA organization in the newt *Triturus.*

Among the many topics on which Beermann concentrated none has had more widespread impact or significance than the analysis of puffs and BRs of polytene chromosomes. A key paper appeared in 1961 under the title, *„Ein Balbiani-Ring ...".* Beermann's earlier studies had shown tissue and temporal specificity in the appearance and disappearance of puffs, leading quite naturally to the hypothesis that these changes represented gene activity. However, a rigorous demonstration demanded that a known genetic function be localized in a specific chromosome region, and that puffing or BR formation in this region be correlated with activation of the gene. Beermann did this using the SZ^+ locus, defined by the presence or absence of special secretory granules produced by four *„Sonderzellen"* in the salivary gland. By first mapping the SZ^+ locus cytologically and then showing that a BR appears at the locus in the *Sonderzellen* when they express the gene (and only in those cells) Beermann provided strong evidence for the idea that differentiation involves both temporal and spatial control of gene expression. Parenthetically, I have always found it difficult to explain this paper in my classes; gene mapping by classical cytogenetics is not easy and the fact that the SZ^+ locus could be mapped first without reference to the BR itself seems to be a difficult concept.

Today we argue very little about whether differential gene activity occurs, but instead we concentrate on how that activity is controlled. A good example was provided by Eric Davidson's insightful analysis of specific gene expression during development of the sea urchin embryo.

Following the initial cytogenetic analysis of puffs and BRs, Beermann and his co-workers concentrated on the biochemical significance of puffing. The messenger RNA hypothesis was receiving much attention in *E. coli,* and it was widely believed that gene activity in eukaryotic cells would also involve RNA. Using the recently introduced method of autoradiography with [^{3}H]-labeled compounds, Pelling (1964) showed that puffs and BRs, as well as nucleoli, were regions of active RNA synthesis. Although in these early studies the nature of the synthesized RNA could not be determined, it was correctly assumed that in some sense one was visualizing individual gene activity. Innumerable studies of puffing have appeared during the past 25 years and only a part of this vast literature was touched upon in our Symposium. Two unusual puffing situations were described. In his review of polytene chromosome structure Günther Korge mentioned puffs that are not associated with RNA synthesis, and Mary Lou Pardue gave a detailed analysis of heat shock puff 93D of *Drosophila melanogaster* at which a long noncoding repetitive sequence is transcribed.

The proteins secreted by the salivary gland received early attention in Beermann's laboratory. Ulrich Grossbach (1969) analyzed these proteins by gel electrophoresis at a time when the techniques for doing so were in their infancy. By correlating the appearance of specific BRs with protein bands in his gels, he provided further evidence for the idea that the BRs produce the messenger RNAs for the salivary gland secretions. In our Symposium Jan-Erik Edström presented detailed sequence data for several proteins controlled by BR RNAs in *Chironomus tentans.* Ulrich Grossbach also discussed proteins, not the secretory products of the gland, but proteins found in the chromosomes themselves; he used specific antibodies to localize HMG proteins and histone Hl variants in puffs and BRs.

We know very few cases where the immediate signal for gene activation is well characterized and easily controlled. One of the first of these was identified by Ulrich Clever (1961) in Beermann's laboratory. Clever showed that a specific band in chromosome I became puffed within minutes after *Chironomus* larvae were injected with the insect hormone ecdysone. The initial puffing was followed by a defined cascade of activation and regression of puffs during the ensuing 24–48 h. Although several other puffing sequences have been described since Clever's original observation, particularly the heat shock puffs of *Drosophila,* the molecular basis of the ecdysone effect has remained largely unexplored. A major step in clarifying the interaction of hormone and gene was described by Dave Hogness, whose group has identified and cloned the ecdysone-induced early puff genes of *Drosophila melanogaster.*

The organization of chromatids in the BRs occupied Beermann's attention in his earliest light microscopic studies as well as in later EM investigations. He concluded that the chromosome in the region of the BR is divided into smaller and smaller bundles of chromatids, and that individual chromatids form loops in the outer part of the BR. In thin sections of the BR one sees granules on these loops that represent the RNP transcription products of the locus. To a remarkable extent these BR loops resemble the paired loops on lampbrush chromosomes of oocytes, the major difference being in the number of chromatids involved, thousands in the BR, but only two in the lampbrush chromosome. We cannot thank Beermann for discovering the lampbrush chromosomes of oocytes; that was done by two other Germans, Flemming and Rückert in the latter part of the 19th century, as described for us by Mick Callan in his historical survey of lampbrush chromosome research. But those of us working on lampbrush chromosomes, including myself and Mick Callan, can thank him for stimulating our thinking and for drawing the analogy between polytene puffs and lampbrush loops. And we can thank the Beermann group for discovering the lampbrush loops of *Drosophila* spermatocytes. Starting about 25 years ago Günther Meyer and Oswald Hess (1961), and later Wolfgang Hennig (1967), gave us a remarkably complete analysis of the Y chromosome loops using a variety of approaches, including light and electron microscopy, cytochemistry, and cytogenetics. Some of the more recent molecular cloning and sequencing information for the Y chromosome was discussed at our Symposium by Hennig, and also by Peter Vogt and H. Bünemann.

In all his writing Beermann emphasized that puffs and BRs represent simply an unfolding of the basic chromosome structure and that they are regions of RNA, not DNA synthesis. They were to be contrasted with the DNA puffs of *Rhynchosciara* and *Sciara* which do, in fact, represent regions of localized DNA increase. We now know that the level of polyteny along *Drosophila* chromosomes is variable; satellite DNA, rDNA, and certain constricted regions in the euchromatin are underreplicated in the salivary gland, and at least two regions are overreplicated or amplified in the ovary. Allan Spradling summarized recent work on the amplification of chorion genes that takes place in the follicle cells of the ovary at the time when the egg shell is laid down.

So much for the puffs and BRs, which have occupied the better part of Beermann's scientific career. Beermann never worked on transposition as such, as far as I know. This currently active area was represented in our Symposium by Schwarz-Sommer's discussion of transposable elements in the development and evolution of plants, and by Heinz Tobler's account of a retrovirus-like transposable element from the nematode *Ascaris* and its probable relationship to chromosome diminution. Diminution, on the other hand, was studied in Beermann's laboratory quite extensively. In what I have long regarded as the most beautifully analyzed case of diminution, Sigrid Beermann (1977) described chromatin elimination during the cleavage divisions in eggs of the copepod *Cyclops*. What makes the *Cyclops* case so interesting is the clear evidence for loss of intercalary regions of the chromosomes, with the implication that the remaining pieces must be spliced together again into whole chromosomes. This case surely deserves further study at the molecular level.

Beermann's style was to work on his own projects and to have only a few associates at a time, each with his or her own particular area of study. One can argue about the relative merits of "little" science and "big" science for accomplishing certain goals. But certainly Beermann's approach is fully justified by the fact that over the years his small group provided the framework for our current view of polytene chromosome structure and function.

I am sure I speak for all participants of the Symposium when I say that we have enjoyed our 3 days in Tübingen helping Wolfgang Beermann celebrate his 65th birthday (belatedly), while learning more about the chromosomes he loves so well.

References

Beermann S (1977) The diminution of heterochromatic chromosomal segments in Cyclops (Crustacea, Copepoda). Chromosoma 60:297–344

Beermann W (1952) Chromomerenkonstanz und spezifische Modifikationen der Chromosomenstruktur in der Entwicklung und Organdifferenzierung von Chironomus tentans. Chromosoma 5:139–198

Beermann W (1960) Der Nukleolus als lebenswichtiger Bestandteil des Zellkernes. Chromosoma 11:263–296

Beermann W (1961) Ein Balbiani-Ring als Locus einer Speicheldrüsen-Mutation. Chromosoma 12:1–25

Clever U (1961) Genaktivitäten in den Riesenchromosomen von Chironomus tentans und ihre Beziehungen zur Entwicklung. I. Genaktivierungen durch Ecdyson. Chromosoma 12:607–675

Grossbach U (1969) Chromosomen-Aktivität und biochemische Zelldifferenzierung in den Speicheldrüsen von Camptochironomus. Chromosoma 28:136–187

Hennig W (1967) Untersuchungen zur Struktur und Funktion des Lampenbürsten-Y-Chromosoms in der Spermatogenese von Drosophila. Chromosoma 22:294–357

Meyer GF, Hess O, Beermann W (1961) Phasenspezifische Funktionsstrukturen in Spermatocytenkernen von Drosophila melanogaster und ihre Abhängigkeit vom Y-Chromosom. Chromosoma 12:676–716

Pelling C (1964) Ribonukleinsäure-Synthese der Riesenchromosomen. Autoradiographische Untersuchungen an Chironomus tentans. Chromosoma 15:71–122

Subject Index